In Situ–Produced Cosmogenic Nuclides and Quantification of Geological Processes

Edited by

Lionel L. Siame
CEREGE
(Centre Européen de Recherche et d'Enseignement des Géosciences de l'Environnement)
CEREGE: UMR 6635–CNRS, IRD, Université Paul Cézanne
Plateau de l'Arbois, BP 80
13545 Aix-en-Provence cedex 04
France

Didier L. Bourlès
CEREGE
(Centre Européen de Recherche et d'Enseignement des Géosciences de l'Environnement)
CEREGE: UMR 6635–CNRS, IRD, Université Paul Cézanne
Plateau de l'Arbois, BP 80
13545 Aix-en-Provence cedex 04
France

Erik T. Brown
Large Lakes Observatory, RLB-207
University of Minnesota
2205 East 5th Street
Duluth, Minnesota 55812
USA

THE
GEOLOGICAL
SOCIETY
OF AMERICA

Special Paper 415

3300 Penrose Place, P.O. Box 9140 ▪ Boulder, Colorado 80301-9140 USA

2006

Copyright © 2006, The Geological Society of America, Inc. (GSA). All rights reserved. GSA grants permission to individual scientists to make unlimited photocopies of one or more items from this volume for noncommercial purposes advancing science or education, including classroom use. For permission to make photocopies of any item in this volume for other noncommercial, nonprofit purposes, contact the Geological Society of America. Written permission is required from GSA for all other forms of capture or reproduction of any item in the volume including, but not limited to, all types of electronic or digital scanning or other digital or manual transformation of articles or any portion thereof, such as abstracts, into computer-readable and/or transmittable form for personal or corporate use, either noncommercial or commercial, for-profit or otherwise. Send permission requests to GSA Copyright Permissions, 3300 Penrose Place, P.O. Box 9140, Boulder, Colorado 80301-9140, USA.

Copyright is not claimed on any material prepared wholly by government employees within the scope of their employment.

Published by The Geological Society of America, Inc.
3300 Penrose Place, P.O. Box 9140, Boulder, Colorado 80301-9140, USA
www.geosociety.org

Printed in U.S.A.

GSA Books Science Editors: Marion E. Bickford and Abhijit Basu

Library of Congress Cataloging-in-Publication Data

In situ-produced cosmogenic nuclides and quantification of geological processes / edited by Lionel Siame, Didier L. Bourlès, Erik T. Brown.
    p. cm. — (Special paper ; 415)
  Includes bibliographical references.
  ISBN-13 978-0-8137-2415-7 (pbk.)
  ISBN-10 0-8137-2415-5 (pbk.)
    1. Radioisotopes in geology. 2. Radioisotopes in glaciology. 3. Radioisotopes in geology--Asia.
  I. Siame, Lionel, 1970-. II. Bourlès, Didier L., 1955- . III. Brown, Erik T., 1963-. IV. Special papers (Geological Society of America) ; 415.
QE501.4.N9.I5 2006
551.028/4 22

2006049567

Cover, top right: The Kromer valley (Vorarlberg, Austria). Photo by H. Kerschner. **Middle right:** A tectonic escarpment along the Gurvan-Bulag Fault (Gobi-Altay, Mongolia). Photo by J.-F. Ritz. **Bottom right:** Canyon incision by the Bitut River (Gobi-Altay, Mongolia). Photo by R. Braucher. Earth image courtesy of the National Aeronautics and Space Administration (NASA; http://visibleearth.nasa.gov/view_detail.php?id=2429). **Background:** In visible light, the bulk of the Milky Way galaxy's stars are eclipsed behind thick clouds of galactic dust and gas. But to the infrared eyes of NASA's Spitzer Space Telescope, distant stars and dust clouds shine with unparalleled clarity and color. Red clouds indicate the presence of large organic molecules mixed with dust. Black patches are dense obscuring dust clouds impenetrable even to Spitzer. Bright arcs of white are massive stellar incubators. Courtesy of NASA (http://www.jpl.nasa.gov/multimedia/slideshows/spitzer-200605/).

10 9 8 7 6 5 4 3 2 1

Contents

Preface v

Recent Papers vii

I. Extending Geographical and Temporal Applicability of In Situ–Produced Cosmogenic Nuclides

1. A review of burial dating methods using ^{26}Al and ^{10}Be 1
   D.E. Granger

2. Extending ^{10}Be applications to carbonate-rich and mafic environments 17
   R. Braucher, P.-H. Blard, L. Benedetti, and D.L. Bourlès

II. Glacial Geology

3. Applications of cosmogenic nuclides to Laurentide Ice Sheet history and dynamics 29
   J.P. Briner, J.C. Gosse, and P.R. Bierman

4. The timing of glacier advances in the northern European Alps based on surface exposure dating with cosmogenic ^{10}Be, ^{26}Al, ^{36}Cl, and ^{21}Ne 43
   S. Ivy-Ochs, H. Kerschner, A. Reuther, M. Maisch, R. Sailer, J. Schaefer, P.W. Kubik,
   H.-A. Synal, and C. Schlüchter

III. Applying Cosmogenic Nuclides to Active Tectonics in Asia

5. Applications of morphochronology to the active tectonics of Tibet 61
   F.J. Ryerson, P. Tapponnier, R.C. Finkel, A.-S. Mériaux, J. Van der Woerd, C. Lasserre,
   M.-L. Chevalier, Xi-wei Xu, Hai-bing Li, and G.C.P. King

6. Using in situ–produced ^{10}Be to quantify active tectonics in the Gurvan Bogd mountain range (Gobi-Altay, Mongolia) 87
   J.-F. Ritz, R. Vassallo, R. Braucher, E.T. Brown, S. Carretier, and D.L. Bourlès

IV. Landscape Evolution

7. Eroding the land: Steady-state and stochastic rates and processes through a cosmogenic lens 111
   A.M. Heimsath

8. Exposure dating (^{10}Be, ^{26}Al) of natural terrain landslides in Hong Kong, China 131
   R.J. Sewell, T.T. Barrows, S.D.G. Campbell, and L.K. Fifield

Preface

As accelerator mass spectrometric techniques matured in the 1980s, measurement of the exceedingly low levels of cosmogenic nuclides present in natural materials became relatively routine. These new analytical capabilities led to expansion of the application of cosmogenic nuclides in earth sciences, as manifested by the rapid increase in peer-reviewed publications (Fig. 1). Since the mid-1990s, most of this growth came from studies utilizing in situ–produced cosmogenic nuclides, which accumulate within surface rocks and soils during exposure to cosmic rays. In situ–produced cosmogenic nuclides can provide chronologies of environmental change over the past few thousand to several millions of years, and may be used to quantify a wide range of weathering and sediment transport processes. These nuclides are thus now used across a broad spectrum of earth science disciplines, including paleoclimatology, geomorphology, and active tectonics. As chronometers, they have been successfully applied to determine: (*i*) age, extension, and dynamics of glaciers and ice-shields; (*ii*) the timing of riverine responses to climate change or tectonic movement; (*iii*) the timing of landslides; (*iv*) ages and process rates of landscapes; and (*v*) ages and rates of earthquake fault displacements. As tools for quantitative study of surface processes, they may be used to evaluate collapse, denudation, burial, bioturbation, and soil creep. They may also provide a qualitative basis for distinguishing allochthonous from autochthonous materials. In addition, new methods are being developed to measure cosmogenic nuclides in a broader suite of mineral phases (e.g., ^{10}Be in sanidine and carbonates, ^{10}Be and ^{36}Cl in clinopyroxenes). Study of other lithologies will permit application of wider range of surface environments, expanding both spatially and temporally their applications. A recent review article on cosmogenic nuclides (Gosse and Phillips, 2001) included a compilation of studies that employed cosmogenic nuclides; to update that reference list we propose a compilation of the most prominent papers published since their review article.

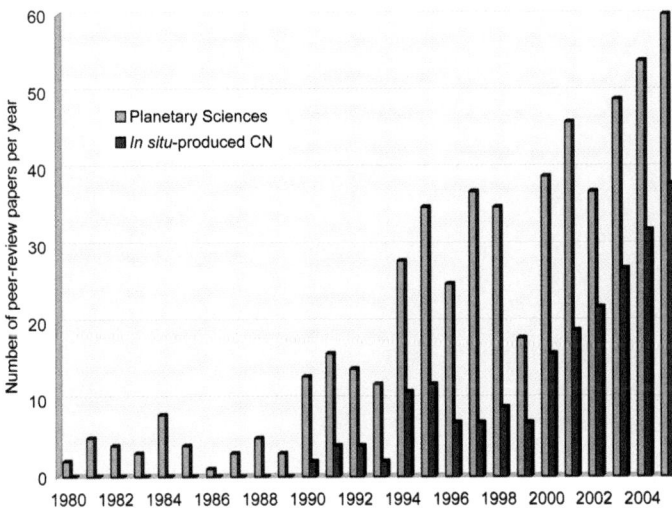

Figure 1. Growth of the total number of published peer-review papers utilizing cosmogenic nuclides (CN) in planetary sciences compared to the growth in the subset of publications that employ in situ–produced cosmogenic nuclides (source: ISI Web of Knowledge^{SM}).

Recognizing the growing interest in this field, a session titled "Application of Cosmogenic Nuclides to the Study of Earth Surface Processes: The Practice and the Potential" was held during the 32nd International Geological Congress (Florence, 21–28 August 2004). This symposium was not only aimed at researchers already applying cosmogenic nuclides but also at scientists whose research might benefit from cosmogenic nuclide analyses, but who have little knowledge of the technique. The creation of a Geological Society of America publication built around the philosophy of the session in Florence was then suggested by Abhijit Basu (Indiana University; GSA books science editor). The present volume represents the fruits of that endeavor.

This book is organized around sections that focus on specific aspects of the utilization of cosmogenic nuclides in earth sciences. The first section presents development of new methods for application of in situ produced cosmogenic nuclides. The article by D. Granger reviews the development of burial dating methods over the past 50 years and presents ways in which multiple cosmogenic nuclides with differing decay rates can be used to quantify the time since burial of rocks in a range of geological settings. The paper by R. Braucher et al. presents new perspectives to extend the time span and erosion rate range quantifiable in carbonate-rich environments, and provide new insights for the possibility of deciphering complex exposure histories by differential radioactive decay over several m.y. in mafic environments. The second section is dedicated to glacial geology. J. Briner et al. and S. Ivy-Ochs et al. review the application of in situ cosmogenic nuclides to study of the Laurentide Ice Sheet during the last glacial cycle and of the timing of glacial advances in the northern Alps, respectively. The third section deals with active tectonics, focusing on applications of in situ–produced cosmogenic nuclides to constrain slip rates of active faults in Asia. F. Ryerson et al. present a review of their studies in Tibet and discuss the apparent disparity between geomorphic and geodetic slip-rates. J.-F. Ritz et al. present an updated synthesis of morphotectonic studies that quantify active tectonics along the Gurvan Bogd mountain range in the Mongolian Gobi-Altay. The final section is dedicated to landscape development. A. Heimsath's paper focuses on the use of cosmogenic ^{10}Be and ^{26}Al, extracted from quartz in bedrock, saprolite, and detrital material to quantify sediment production or erosion rates and processes. The paper by R. Sewell et al. presents the application of exposure dating using cosmogenic nuclides to natural terrain landslides in Hong Kong.

We would like to acknowledge all the people that greatly helped us in editing the present book. At the Geological Society of America, we thank Sara Colvard, Abhijit Basu, and the GSA publications staff for their constant help during the editing process. We also would like to thank the individual reviewers whose constructive and helpful comments greatly improved this work.

REFERENCE CITED

Gosse, J.C., and Phillips, F.M., 2001, Terrestrial in situ cosmogenic nuclides: theory and application: Quaternary Science Review, v. 20, p. 1475–1560, doi: 10.1016/S0277-3791(00)00171-2.

Lionel L. Siame
Didier L. Bourlès
Erik T. Brown

RECENT PAPERS

A review article on cosmogenic nuclides by J.C. Gosse and F.M. Phillips in 2001 included a compilation of studies that employed cosmogenic nuclides. Here, we update that list with a proposed compilation of the most prominent papers published since that 2001 review article, divided into sections that correspond to the sections of this volume, plus a section of recent review papers.

Section 1. Extending Geographical and Temporal Applicability of In Situ–Produced Cosmogenic Nuclides

Altmaier, M., Klas, W., and Herpers, U., 2001, Surface exposure dating by in–situ produced cosmogenic nuclides: chemical mineral separation of purified quartz: Radiochimica Acta, v. 89, p. 779–782.

Blard, P.H., Lave, J., Pik, R., Quidelleur, X., Bourles, D., and Kieffer, G., 2005, Fossil cosmogenic He-3 record from K-Ar dated basaltic flows of Mount Etna volcano (Sicily, 38 degrees N): Evaluation of a new paleoaltimeter: Earth and Planetary Science Letters, v. 236, p. 613–631, doi: 10.1016/j.epsl.2005.05.028.

Braucher, R., Benedetti, L., Bourlès, D.L., Brown, E.T., and Chardon, D., 2005, Use of in situ–produced Be-10 in carbonate-rich environments: A first attempt: Geochimica et Cosmochimica Acta, v. 69, p. 1473–1478, doi: 10.1016/j.gca.2004.09.010.

Chmiel, G., Fritz, S.J., and Elmore, D., 2003, Control of Cl-36 production in carbonaceous shales by phosphate minerals: Geochimica et Cosmochimica Acta, v. 67, p. 2377–2395, doi: 10.1016/S0016-7037(03)00022-X.

Desilets, D., and Zreda, M., 2001, On scaling cosmogenic nuclide production rates for altitude and latitude using cosmic-ray measurements: Earth and Planetary Science Letters, v. 193, p. 213–225, doi: 10.1016/S0012-821X(01)00477-0.

Desilets, D., Zreda, M., and Lifton, N.A., 2001, Comment on 'Scaling factors for production rates of in situ produced cosmogenic nuclides: a critical reevaluation' by Tibor J. Dunai: Earth and Planetary Science Letters, v. 188, p. 283–287, doi: 10.1016/S0012-821X(01)00302-8.

Dunai, T.J., 2000, Scaling factors for production rates of in situ produced cosmogenic nuclides: a critical reevaluation (v. 176, p. 157–169, 2000) [Erratum]: Earth and Planetary Science Letters, v. 178, p. 425–425, doi: 10.1016/S0012-821X(00)00092-3.

Dunai, T.J., 2000, Scaling factors for production rates of in situ produced cosmogenic nuclides: a critical reevaluation: Earth and Planetary Science Letters, v. 176, p. 157–169, doi: 10.1016/S0012-821X(99)00310-6.

Dunai, T.J., 2001, Influence of secular variation of the geomagnetic field on production rates of in situ produced cosmogenic nuclides: Earth and Planetary Science Letters, v. 193, p. 197–212, doi: 10.1016/S0012-821X(01)00503-9.

Dunai, T.J., 2001, Reply to comment on 'Scaling factors for production rates of in situ produced cosmogenic nuclides: a critical reevaluation' by Darin Desilets, Marek Zreda and Nathaniel Lifton: Earth and Planetary Science Letters, v. 188, p. 289–298, doi: 10.1016/S0012-821X(01)00303-X.

Gordon, S.I., and Dorn, R.I., 2005, In situ weathering rind erosion: Geomorphology, v. 67, p. 97–113, doi: 10.1016/j.geomorph.2004.06.011.

Graham, I.J., Barry, B.J., Ditchburn, R.G., and Whitehead, N.E., 2000, Validation of cosmogenic nuclide production rate scaling factors through direct measurement: Nuclear Instruments and Methods in Physics Research, Section B—Beam Interactions with Materials and Atoms, v. 172, p. 802–805.

Heimsath, A.M., and Farid, H., 2002, Hillslope topography from unconstrained photographs: Mathematical Geology, v. 34, p. 929–952, doi: 10.1023/A:1021364623017.

Hetzel, R., Niedermann, S., Ivy-Ochs, S., Kubik, P.W., Tao, M.X., and Gao, B., 2002, Ne-21 versus Be-10 and Al-26 exposure ages of fluvial terraces: the influence of crustal Ne in quartz: Earth and Planetary Science Letters, v. 201, p. 575–591, doi: 10.1016/S0012-821X(02)00748-3.

Kim, K.J., and Englert, P.A.J., 2004, Profiles of in situ Be-10 and Al-26 at great depths at the Macraes Flat, East Otago, New Zealand: Earth and Planetary Science Letters, v. 223, p. 113–126, doi: 10.1016/j.epsl.2004.04.006.

Kim, K.J., and Imamura, M., 2004, Exposure dating of underwater rocks: potential application to studies of land bridges during the Ice Ages: Nuclear Instruments and Methods in Physics Research, Section B—Beam Interactions with Materials and Atoms, v. 223–24, p. 608–612.

Kober, F., Ivy-Ochs, S., Leya, I., Baur, H., Magna, T., Wieler, R., and Kubik, P.W., 2005, In situ cosmogenic Be-10 and Ne-21 in sanidine and in situ cosmogenic He-3 in Fe-Ti-oxide minerals: Earth and Planetary Science Letters, v. 236, p. 404–418, doi: 10.1016/j.epsl.2005.05.020.

Liu, T.Z., 2003, Blind testing of rock varnish microstratigraphy as a chronometric indicator: results on late Quaternary lava flows in the Mojave Desert, California: Geomorphology, v. 53, p. 209–234, doi: 10.1016/S0169-555X(02)00331-8.

Masarik, J., Frank, M., Schafer, J.M., and Wieler, R., 2001, Correction of in situ cosmogenic nuclide production rates for geomagnetic field intensity variations during the past 800,000 years: Geochimica et Cosmochimica Acta, v. 65, p. 2995–3003, doi: 10.1016/S0016-7037(01)00652-4.

Masarik, J., Kollar, D., and Vanya, S., 2000, Numerical simulation of in situ production of cosmogenic nuclides: Effects of irradiation geometry: Nuclear Instruments and Methods in Physics Research, Section B—Beam Interactions with Materials and Atoms, v. 172, p. 786–789.

Muzikar, P., 2005, Geomagnetic field variations and the accumulation of in-situ cosmogenic nuclides in an eroding landform: Geochimica et Cosmochimica Acta, v. 69, p. 4127–4131, doi: 10.1016/j.gca.2004.12.027.

Phillips, W.M., 2000, Estimating cumulative soil accumulation rates with in situ-produced cosmogenic nuclide depth profiles: Nuclear Instruments and Methods in Physics Research, Section B—Beam Interactions with Materials and Atoms, v. 172, p. 817–821.

Pigati, J.S., and Lifton, N.A., 2004, Geomagnetic effects on time-integrated cosmogenic nuclide production with emphasis on in situ C-14 and Be-10: Earth and Planetary Science Letters, v. 226, p. 193–205, doi: 10.1016/j.epsl.2004.07.031.

Reedy, R.C., 2000, Predicting the production rates of cosmogenic nuclides: Nuclear Instruments and Methods in Physics Research, Section B—Beam Interactions with Materials and Atoms, v. 172, p. 782–785.

Schaller, M., von Blanckenburg, F., Veit, H., and Kubik, P.W., 2003, Influence of periglacial cover beds on in situ-produced cosmogenic Be-10 in soil sections: Geomorphology, v. 49, p. 255–267, doi: 10.1016/S0169-555X(02)00189-7.

Siame, L., Belllier, O., Braucher, R., Sébrier, M., Cushing, M., Bourlès, D., Hamelin, B., Baroux, E., de Voogd, B., Raisbeck, G., and Yiou, F., 2004, Local erosion rates versus active tectonics: cosmic ray exposure modelling in Provence (south-east France): Earth and Planetary Science Letters, v. 220, p. 345–364, doi: 10.1016/S0012-821X(04)00061-5.

Snowball, I., and Sandgren, P., 2002, Geomagnetic field variations in northern Sweden during the Holocene quantified from varved lake sediments and their implications for cosmogenic nuclide production rates: The Holocene, v. 12, p. 517–530, doi: 10.1191/0959683602hl562rp.

Stock, G.M., Granger, D.E., Sasowsky, I.D., Anderson, R.S., and Finkel, R.C., 2005, Comparison of U-Th, paleomagnetism, and cosmogenic burial methods for dating caves: Implications for landscape evolution studies: Earth and Planetary Science Letters, v. 236, p. 388–403, doi: 10.1016/j.epsl.2005.04.024.

Yokoyama, Y., Caffee, M.W., Southon, J.R., and Nishiizumi, K., 2004, Measurements of in situ produced C-14 in terrestrial rocks: Nuclear Instruments and Methods in Physics Research, Section B—Beam Interactions with Materials and Atoms, v. 223–24, p. 253–258.

Section 2. Glacial Geology

Balco, G., Stone, J.O.H., Jennings, C, 2005. Dating Plio-Pleistocene glacial sediments using the cosmic-ray-produced radionuclides Be-10 and Al-26: American Journal of Science, v. 305, no. 1, p. 41.

Balco, G., Stone, J.O.H., and Mason, J.A., 2005, Numerical ages for Plio-Pleistocene glacial sediment sequences by Al-26/Be-10 dating of quartz in buried paleosols: Earth and Planetary Science Letters, v. 232, p. 179–191, doi: 10.1016/j.epsl.2004.12.013.

Beer, J., Muscheler, R., Wagner, G., Laj, C., Kissel, C., Kubik, P.W., and Synal, H.A., 2002, Cosmogenic nuclides during Isotope Stages 2 and 3: Quaternary Science Reviews, v. 21, p. 1129–1139, doi: 10.1016/S0277-3791(01)00135-4.

Brigham-Grette, J., Gualtieri, L.M., Glushkova, O.Y., Hamilton, T.D., Mostoller, D., and Kotov, A., 2003, Chlorine-36 and C-14 chronology support a limited last glacial maximum across central Chukotka, northeastern Siberia, and no Beringian ice sheet: Quaternary Research, v. 59, p. 386–398, doi: 10.1016/S0033-5894(03)00058-9.

Briner, J.P., Kaufman, D.S., Werner, A., Caffee, M., Levy, L., Manley, W.F., Kaplan, M.R., and Finkel, R.C., 2002, Glacier readvance during the late glacial (Younger Dryas?) in the Ahklun Mountains, southwestern Alaska: Geology, v. 30, p. 679–682, doi: 10.1130/0091-7613(2002)030<0679:GRDTLG>2.0.CO;2.

Briner, J.P., Miller, G.H., Davis, P.T., Bierman, P.R., and Caffee, M., 2003, Last Glacial Maximum ice sheet dynamics in Arctic Canada inferred from young erratics perched on ancient tors: Quaternary Science Reviews,. v. 22, p. 437–444, doi: 10.1016/S0277-3791(03)00003-9.

Brown, E.T., Bendick, R., Bourlès, D.L., Gaur, V., Molnar, P., Raisbeck, G.M., and Yiou, F., 2003, Early Holocene climate recorded in geomorphological features in Western Tibet: Paleogeography Paleoclimatology Paleoecology, v. 199, p. 141–151, doi: 10.1016/S0031-0182(03)00501-7.

Colgan, P.M., Bierman, P.R., Mickelson, D.M., and Caffee, M., 2002, Variation in glacial erosion near the southern margin of the Laurentide Ice Sheet, south-central Wisconsin, USA: Implications for cosmogenic dating of glacial terrains: Geological Society of America Bulletin, v. 114, p. 1581–1591, doi: 10.1130/0016-7606(2002)114<1581:VIGENT>2.0.CO;2.

Fabel, D., Harbor, J., Dahms, D., James, A., Elmore, D., Horn, L., Daley, K., and Steele, C., 2004, Spatial patterns of glacial erosion at a valley scale derived from terrestrial cosmogenic Be-10 and Al-26 concentrations in rock: Annals of the Association of American Geographers. Association of American Geographers, v. 94, p. 241–255, doi: 10.1111/j.1467-8306.2004.09402001.x.

Fabel, D., Stroeven, A.P., Harbor, J., Kleman, J., Elmore, D., and Fink, D., 2002, Landscape preservation under Fennoscandian ice sheets determined from in situ produced Be-10 and Al-26: Earth and Planetary Science Letters, v. 201, p. 397–406, doi: 10.1016/S0012-821X(02)00714-8.

Finkel, R.C., Owen, L.A., Barnard, P.L., and Caffee, M.W., 2003, Beryllium-10 dating of Mount Everest moraines indicates a strong monsoon influence and glacial synchroneity throughout the Himalaya: Geology, v. 31, p. 561–564, doi: 10.1130/0091-7613(2003)031<0561:BDOMEM>2.0.CO;2.

Hughen, K.A., Southon, J.R., Lehman, S.J., and Overpeck, J.T., 2000, Synchronous radiocarbon and climate shifts during the last deglaciation: Science, v. 290, p. 1951–1954, doi: 10.1126/science.290.5498.1951.

Ivins, E.R., and James, T.S., 2005, Antarctic glacial isostatic adjustment: a new assessment: Antarctic Science, v. 17, p. 541–553, doi: 10.1017/S0954102005002968.

Kaplan, M.R., and Miller, G.H., 2003, Early Holocene delevelling and deglaciation of the Cumberland Sound region, Baffin Island, Arctic Canada: Geological Society of America Bulletin, v. 115, p. 445–462, doi: 10.1130/0016-7606(2003)115<0445:EHDADO>2.0.CO;2.

Kaplan, M.R., Miller, G.H., and Steig, E.J., 2001, Low-gradient outlet glaciers lice streams drained the Laurentide ice sheet: Geology, v. 29, p. 343–346, doi: 10.1130/0091-7613(2001)029<0343:LGOGIS>2.0.CO;2.

Li, Y.K., Harbor, J., Stroeven, A.P., Fabel, D., Kleman, J., Fink, D., Caffee, M., and Elmore, D., 2005, Ice sheet erosion patterns in valley systems in northern Sweden investigated using cosmogenic nuclides: Earth Surface Processes and Landforms, v. 30, p. 1039–1049, doi: 10.1002/esp.1261.

McCulloch, R.D., Fogwill, C.J., Sugden, D.E., Bentley, M.J., and Kubik, P.W., 2005, Chronology of the last glaciation in central Strait of Magellan and Bahia Inutil, southernmost South America: Geografiska Annaler: Series A, Physical Geography, v. 87A, p. 289–312.

Oberholzer, P., Baroni, C., Schaefer, J.M., Orombelli, G., Ochs, S.I., Kubik, P.W., Baur, H., and Wieler, R., 2003, Limited Pliocene/Pleistocene glaciation in Deep Freeze Range, northern Victoria Land, Antarctica, derived from in situ cosmogenic nuclides: Antarctic Science, v. 15, p. 493–502, doi: 10.1017/S0954102003001603.

Pratt, B., Burbank, D.W., Heimsath, A., and Ojha, T., 2002, Impulsive alluviation during early Holocene strengthened monsoons, central Nepal Himalaya: Geology, v. 30, p. 911–914, doi: 10.1130/0091-7613(2002)030<0911:IADEHS>2.0.CO;2.

Schafer, J.M., Tschudi, S., Zhao, Z.Z., Wu, X.H., Ivy-Ochs, S., Wieler, R., Baur, H., Kubik, P.W., and Schluchter, C., 2002, The limited influence of glaciations in Tibet on global climate over the past 170 000 yr: Earth and Planetary Science Letters, v. 194, p. 287–297, doi: 10.1016/S0012-821X(01)00573-8.

Smith, J.A., Seltzer, G.O., Farber, D.L., Rodbell, D.T., and Finkel, R.C., 2005, Early local last glacial maximum in the tropical Andes: Science, v. 308, p. 678–681, doi: 10.1126/science.1107075.

Stroeven, A.P., Fabel, D., Hattestrand, C., and Harbor, J., 2002, A relict landscape in the centre of Fennoscandian glaciation: cosmogenic radionuclide evidence of tors preserved through multiple glacial cycles: Geomorphology, v. 44, p. 145–154, doi: 10.1016/S0169-555X(01)00150-7.

Tschudi, S., Schafer, J.M., Borns, H.W., Ivy-Ochs, S., Kubik, P.W., and Schluchter, C., 2003, Surface exposure dating of Sirius Formation at Allan Hills nunatak, Antarctica: New evidence for long-term ice-sheet stability: Eclogae Geologica Helvetica v. 96, p. 109–114.

Tschudi, S., Schafer, J.M., Zhao, Z.Z., Wu, X.H., Ivy-Ochs, S., Kubik, P.W., and Schluchter, C., 2003, Glacial advances in Tibet during the Younger Dryas? Evidence from cosmogenic Be-10, Al-26, and Ne-21: Journal of Asian Earth Science, v. 22, p. 301–306, doi: 10.1016/S1367-9120(03)00035-X.

Zech, R., Abramowski, U., Glaser, B., Sosin, P., Kubik, P.W., and Zech, W., 2005, Late Quaternary glacial and climate history of the Pamir Mountains derived from cosmogenic Be-10 exposure ages: Quaternary Research, v. 64, p. 212–220, doi: 10.1016/j.yqres.2005.06.002.

Zech, R., Glaser, B., Sosin, P., Kubik, P.W., and Zech, W., 2005, Evidence for long-lasting landform surface instability on hummocky moraines in the Pamir Mountains (Tajikistan) from Be-10 surface exposure dating: Earth and Planetary Science Letters, v. 237, p. 453–461, doi: 10.1016/j.epsl.2005.06.031.

Zech, W., Glaser, B., Abramowski, U., Dittmar, C., and Kubik, P.W., 2003, Reconstruction of the Late Quaternary Glaciation of the Macha Khola valley (Gorkha Himal, Nepal) using relative and absolute (C-14, Be-10, dendrochronology) dating techniques: Quaternary Science Reviews, v. 22, p. 2253–2265, doi: 10.1016/S0277-3791(03)00008-8.

Section 3. Applying Cosmogenic Nuclides to Active Tectonics in Asia

Brown, E.T., Bendick, R., Bourlès, D.L., Gaur, V., Molnar, P., Raisbeck, G.M., and Yiou, F., 2002, Slip rates of the Karakorum fault, Ladakh, India, determined using cosmic ray exposure dating of debris flows and moraines: Journal of Geophysical Research, Solid Earth, v. 107, no. B9, 2192, doi: 10.1029/2000JB000100.

Chevalier, M.L., Ryerson, F.J., Tapponnier, P., Finkel, R.C., Van der Woerd, J., Li, H.B., and Liu, Q., 2005, Slip-rate measurements on the Karakorum Fault may imply secular variations in fault motion: Science, v. 307, p. 411–414, doi: 10.1126/science.1105466.

Daeron, M., Benedetti, L., Tapponnier, P., Sursock, A., and Finkel, R.C., 2004, Constraints on the post-25-ka slip rate of the Yammouneh fault (Lebanon) using in situ cosmogenic Cl-36 dating of offset limestone-clast fans: Earth and Planetary Science Letters, v. 227, p. 105–119, doi: 10.1016/j.epsl.2004.07.014.

Hetzel, R., Niedermann, S., Tao, M., Kubik, P.W., Ivy-Ochs, S., Gao, B., and Strecker, M.R., 2002, Low slip rates and long-term preservation of geomorphic features in Central Asia: Nature, v. 417, p. 428–432, doi: 10.1038/417428a.

Hubert-Ferrari, A., Suppe, J., Van der Woerd, J., Wang, X., and Lu, H.F., 2005, Irregular earthquake cycle along the southern Tianshan front, Aksu area, China: Journal of Geophysical Research, Solid Earth, v. 110, B06402, doi: 10.1029/2003JB002603.

Lasserre, C., Gaudemer, Y., Tapponnier, P., Mériaux, A.S., Van der Woerd, J., Yuan, D.Y., Ryerson, F.J., Finkel, R.C., and Caffee, M.W., 2002, Fast late Pleistocene slip rate on the Leng Long Ling segment of the Haiyuan fault, Qinghai, China: Journal of Geophysical Research, Solid Earth, v. 107, no. B11, 2276, doi: 10.1029/2000JB000060.

Mériaux, A.S., Ryerson, F.J., Tapponnier, P., Van der Woerd, J., Finkel, R.C., Xu, X.W., Xu, Z.Q., and Caffee, M.W., 2004, Rapid slip along the central Altyn Tagh Fault: Morphochronologic evidence from Cherchen He and Sulamu Tagh: Journal of Geophysical Research, Solid Earth, v. 109, B06401, doi: 10.1029/2003JB002558.

Mériaux, A.S., Tapponnier, P., Ryerson, F.J., Xu, X.W., King, G., Van der Woerd, J., Finkel, R.C., Li, H.B., Caffee, M.W., Xu, Z.Q., and Chen, W.B., 2005, The Aksay segment of the northern Altyn Tagh fault: Tectonic geomorphology, landscape evolution, and Holocene slip rate: Journal of Geophysical Research, Solid Earth, v. 110, B04404, doi: 10.1029/2004JB003210.

Palumbo, L., Benedetti, L., Bourles, D., Cinque, A., and Finkel, R., 2004, Slip history of the Magnola fault (Apennines, Central Italy) from Cl-36 surface exposure dating: evidence for strong earthquakes over the Holocene: Earth and Planetary Science Letters, v. 225, p. 163–176, doi: 10.1016/j.epsl.2004.06.012.

Phillips, F.M., Ayarbe, J.P., Harrison, J.B.J., and Elmore, D., 2003, Dating rupture events on alluvial fault scarps using cosmogenic nuclides and scarp morphology: Earth and Planetary Science Letters, v. 215, p. 203–218, doi: 10.1016/S0012-821X(03)00419-9.

Siame, L.L., Bellier, O., Sébrier, M., Bourlès, D.L., Leturmy, P., Perez, M., and Araujo, M., 2002, Seismic hazard reappraisal from combined structural geology, geomorphology and cosmic ray exposure dating analyses: the Eastern Precordillera thrust system (NW Argentina): Geophysical Journal International, v. 150, p. 241–260, doi: 10.1046/j.1365-246X.2002.01701.x.

Siame, L.L., Bellier, O., Sébrier, M., and Araujo, M., 2005, Deformation partitioning in flat subduction setting: the case of the Andean foreland of western Argentina (28°S–33°S): Tectonics, v. 24, TC5003, p. 1–24, doi: 10.1029/2005TC001787.

Siame, L.L., Braucher, R., Bourlès, D.L., Bellier, O., and Sébrier, M., 2001, Cosmic ray exposure dating of geomorphic surface features using in situ-production Be-10: tectonic and climatic implications: Bulletin de la Société Géologique de France, v. 172, p. 223–236, doi: 10.2113/172.2.223.

Wobus, C., Heimsath, A., Whipple, K., and Hodges, K., 2005, Active out-of-sequence thrust faulting in the central Nepalese Himalaya: Nature, v. 434, p. 1008–1011, doi: 10.1038/nature03499.

Xu, X.W., Tapponnier, P., Van Der Woerd, J., Ryerson, F.J., Wang, F., Zheng, R.Z., Chen, W.B., Ma, W.T., Yu, G.H., Chen, G.H., and Meriaux, A.S., 2005, Late quaternary sinistral slip rate along the Altyn Tagh Fault and its structural transformation model. Science in China Series D: Earth Science, v. 48, p. 384–397.

Section 4. Landscape Evolution

Anderson, S.P., Dietrich, W.E., and Brimhall, G.H., 2002, Weathering profiles, mass-balance analysis, and rates of solute loss: Linkages between weathering and erosion in a small, steep catchment: Geological Society of America Bulletin, v. 114, p. 1143–1158, doi: 10.1130/0016-7606(2002)114<1143:WPMBAA>2.0.CO;2.

Ballantyne, C.K., and Stone, J.O., 2004, The Beinn Alligin rock avalanche, NW Scotland: cosmogenic Be-10 dating, interpretation and significance: The Holocene, v. 14, p. 448–453, doi: 10.1191/0959683604hl720rr.

Belton, D.X., Brown, R.W., Kohn, B.P., Fink, D., and Farley, K.A., 2004, Quantitative resolution of the debate over antiquity of the central Australian landscape: implications for the tectonic and geomorphic stability of cratonic interiors: Earth and Planetary Science Letters, v. 219, p. 21–34, doi: 10.1016/S0012-821X(03)00705-2.

Bierman, P.R., and Caffee, M., 2001, Slow rates of rock surface erosion and sediment production across the Namib Desert and escarpment, southern Africa: American Journal of Science, v. 301, p. 326–358.

Bierman, P.R., and Caffee, M., 2002, Cosmogenic exposure and erosion history of Australian bedrock landforms: Geological Society of America Bulletin, v. 114, p. 787–803, doi: 10.1130/0016-7606(2002)114<0787:CEAEHO>2.0.CO;2.

Bierman, P.R., Reuter, J.M., Pavich, K., Gellis, A.C., Caffee, M.W., and Larsen, J., 2005, Using cosmogenic nuclides to contrast rates of erosion and sediment yield in a semi-arid, arroyo-dominated landscape, Rio Puerco Basin, New Mexico: Earth Surface Processes and Landforms, v. 30, p. 935–953, doi: 10.1002/esp.1255.

Bigot-Cormier, F., Braucher, R., Bourlès, D., Guglielmi, Y., Dubar, M., and Stephan, J.F., 2005, Chronological constraints on processes leading to large active landslides: Earth and Planetary Science Letters, v. 235, p. 141–150, doi: 10.1016/j.epsl.2005.03.012.

Braucher, R., Bourlès, D.L., Brown, E.T., Colin, F., Muller, J.P., Braun, J.J., Delaune, M., Minko, A.E., Lescouet, C., Raisbeck, G.M., and Yiou, F., 2000, Application of in situ-produced cosmogenic Be-10 and Al-26 to the study of lateritic soil development in tropical forest: theory and examples from Cameroon and Gabon: Chemical Geology, v. 170, p. 95–111, doi: 10.1016/S0009-2541(99)00243-0.

Braucher, R., Lima, C.V., Bourles, D.L., Gaspar, J.C., and Assad, M.L.L., 2004, Stone-line formation processes documented by in situ-produced Be-10 distribution, Jardim River basin, DF, Brazil: Earth and Planetary Science Letters, v. 222, p. 645–651, doi: 10.1016/j.epsl.2004.02.033.

Braucher, R., Siame, L., Bourlès, D., and Colin, F., 2000, Use of in-situ produced cosmogenic Be-10 to study the lateritic soil evolution: Bulletin de la Société Géologique de France, v. 171, p. 511–520, doi: 10.2113/171.5.511.

Brocard, G.Y., van der Beek, P.A., Bourlès, D.L., Siame, L.L., and Mugnier, J.L., 2003, Long-term fluvial incision rates and postglacial river relaxation time in the French Western Alps from Be-10 dating of alluvial terraces with assessment of inheritance, soil development and wind ablation effects: Earth and Planetary Science Letters, v. 209, p. 197–214, doi: 10.1016/S0012-821X(03)00031-1.

Burbank, D.W., 2002, Rates of erosion and their implications for exhumation: Mineralogical Magazine, v. 66, p. 25–52, doi: 10.1180/0026461026610014.

Clapp, E.M., Bierman, P.R., and Caffee, M., 2002, Using Be-10 and Al-26 to determine sediment generation rates and identify sediment source areas in an arid region drainage basin: Geomorphology, v. 45, p. 89–104, doi: 10.1016/S0169-555X(01)00191-X.

Dorn, R.I., 2003, Boulder weathering and erosion associated with a wildfire, Sierra Ancha Mountains, Arizona: Geomorphology, v. 55, p. 155–171, doi: 10.1016/S0169-555X(03)00138-7.

Ferrier, K.L., Kirchner, J.W., and Finkel, R.C., 2005, Erosion rates over millennial and decadal timescales at Caspar Creek and Redwood Creek: Northern California Coast Ranges: Earth Surface Processes and Landforms, v. 30, p. 1025–1038, doi: 10.1002/esp.1260.

Fogwill, C.J., Bentley, M.J., Sugden, D.E., Kerr, A.R., and Kubik, P.W., 2004, Cosmogenic nuclides Be-10 and Al-26 imply limited Antarctic Ice Sheet thickening and low erosion in the Shackleton Range for > 1 m.y: Geology, v. 32, p. 265–268, doi: 10.1130/G19795.1.

Friend, D.A., Phillips, F.M., Campbell, S.W., Liu, T.H., and Sharma, P., 2000, Evolution of desert colluvial boulder slopes: Geomorphology, v. 36, p. 19–45, doi: 10.1016/S0169-555X(00)00045-3.

Gellis, A.C., Pavich, M.J., Bierman, P.R., Clapp, E.M., Ellevein, A., and Aby, S., 2004, Modern sediment yield compared to geologic rates of sediment production in a semi-arid basin, New Mexico: Assessing the human impact: Earth Surface Processes and Landforms, v. 29, p. 1359–1372, doi: 10.1002/esp.1098.

Heimsath, A.M., Chappell, J., Dietrich, W.E., Nishiizumi, K., and Finkel, R.C., 2000, Soil production on a retreating escarpment in southeastern Australia: Geology, v. 28, p. 787–790, doi: 10.1130/0091-7613(2000)028<0787: SPOARE>2.3.CO;2.

Heimsath, A.M., Chappell, J., Dietrich, W.E., Nishiizumi, K., and Finkel, R.C., 2001, Late Quaternary erosion in southeastern Australia: a field example using cosmogenic nuclides: Quaternary International, v. 83–5, p. 169–185, doi: 10.1016/S1040-6182(01)00038-6.

Heimsath, A.M., Chappell, J., Spooner, N.A., and Questiaux, D.G., 2002, Creeping soil: Geology, v. 30, p. 111–114, doi: 10.1130/0091-7613(2002)030<0111:CS>2.0.CO;2.

Heimsath, A.M., Dietrich, W.E., Nishiizumi, K., and Finkel, R.C., 2001, Stochastic processes of soil production and transport: Erosion rates, topographic variation and cosmogenic nuclides in the Oregon Coast Range: Earth Surface Processes and Landforms, v. 26, p. 531–552, doi: 10.1002/esp.209.

Heimsath, A.M., Furbish, D.J., and Dietrich, W.E., 2005, The illusion of diffusion: Field evidence for depth-dependent sediment transport: Geology, v. 33, p. 949–952, doi: 10.1130/G21868.1.

Hewawasam, T., von Blanckenburg, F., Schaller, M., and Kubik, P., 2003, Increase of human over natural erosion rates in tropical highlands constrained by cosmogenic nuclides: Geology, v. 31, p. 597–600, doi: 10.1130/0091-7613(2003)031<0597:IOHONE>2.0.CO;2.

Kim, K.J., and Englert, P.A.J., 2004, In situ cosmogenic nuclide production of Be-10 and Al-26 in marine terraces, Fiordland, New Zealand: Nuclear Instruments and Methods of Physics Research, Section B—Beam Interactions with Materials and Atoms, v. 223–24, p. 639–644.

Kim, K.J., and Sutherland, R., 2004, Uplift rate and landscape development in southwest Fiordland, New Zealand, determined using Be-10 and Al-26 exposure dating of marine terraces: Geochimica et Cosmochimica Acta, v. 68, p. 2313–2319, doi: 10.1016/j.gca.2003.11.005.

Kobor, J.S., and Roering, J.J., 2004, Systematic variation of bedrock channel gradients in the central Oregon Coast Range: implications for rock uplift and shallow landsliding: Geomorphology, v. 62, p. 239–256, doi: 10.1016/j.geomorph.2004.02.013.

Lal, D., 2000, Cosmogenic nuclide production rate systematics in terrestrial materials: Present knowledge, needs and future actions for improvement: Nuclear Instruments and Methods of Physics Research, Section B—Beam Interactions with Materials and Atoms, v. 172, p. 772–781.

Ma, P., Aggrey, K., Tonzola, C., Schnabel, C., De Nicola, P., Herzog, G.F., Wasson, J.T., Glass, B.P., Brown, L., Tera, F., Middleton, R., and Klein, J., 2004, Beryllium-10 in Australasian tektites: constraints on the location of the source crater: Geochimica et Cosmochimica Acta, v. 68, p. 3883–3896, doi: 10.1016/j.gca.2004.03.026.

Matmon, A., Bierman, P.R., Larsen, J., Southworth, S., Pavich, M., and Caffee, M., 2003, Temporally and spatially uniform rates of erosion in the southern Appalachian Great Smoky Mountains: Geology, v. 31, p. 155–158.

Matmon, A., Bierman, P.R., Larsen, J., Southworth, S., Pavich, M., Finkel, R., and Caffee, M., 2003, Erosion of an ancient mountain range, the Great Smoky Mountains, North Carolina and Tennessee: American Journal of Science, v. 303, p. 817–855.

Matmon, A., Crouvi, O., Enzel, Y., Bierman, P., Larsen, J., Porat, N., Amit, R., and Caffee, M., 2003, Complex exposure histories of chert clasts in the late Pleistocene shorelines of Lake Lisan, southern Israel: Earth Surface Processes and Landforms, v. 28, p. 493–506, doi: 10.1002/esp.454.

Morel, P., von Blanckenburg, F., Schaller, M., Kubik, P.W., and Hinderer, M., 2003, Lithology, landscape dissection and glaciation controls on catchment erosion as determined by cosmogenic nuclides in river sediment (the Wutach Gorge, Black Forest). Terra Nova, v. 15, p. 398–404.

Nichols, K.K., Bierman, P.R., Eppes, M.C., Caffee, M., Finkel, R., and Larsen, J., 2005, Late Quaternary history of the Chemehuevi Mountain piedmont, Mojave Desert, deciphered using Be-10 and Al-26: American Journal of Science, v. 305, p. 345–368.

Nichols, K.K., Bierman, P.R., Hooke, R.L., Clapp, E.M., and Caffee, M., 2002, Quantifying sediment transport on desert piedmonts using Be-10 and Al-26: Geomorphology, v. 45, p. 105–125, doi: 10.1016/S0169-555X(01)00192-1.

Niemi, N.A., Oskin, M., Burbank, D.W., Heimsath, A.M., and Gabet, E.J., 2005, Effects of bedrock landslides on cosmogenically determined erosion rates: Earth and Planetary Science Letters, v. 237, p. 480–498, doi: 10.1016/j.epsl.2005.07.009.

Nishiizumi, K., Caffee, M.W., Finkel, R.C., Brimhall, G., and Mote, T., 2005, Remnants of a fossil alluvial fan landscape of Miocene age in the Atacama Desert of northern Chile using cosmogenic nuclide exposure age dating: Earth and Planetary Science Letters, v. 237, p. 499–507, doi: 10.1016/j.epsl.2005.05.032.

Parker, G., and Perg, L.A., 2005, Probabilistic formulation of conservation of cosmogenic nuclides: effect of surface elevation fluctuations on approach to steady state: Earth Surface Processes and Landforms, v. 30, p. 1127–1144, doi: 10.1002/esp.1266.

Phillips, J., 2003, Alluvial storage and the long-term stability of sediment yields: Basin Research, v. 15, p. 153–163, doi: 10.1046/j.1365-2117.2003.00204.x.

Phillips, J.D., Marion, D.A., Luckow, K., and Adams, K.R., 2005, Nonequilibrium regolith thickness in the Ouachita Mountains: The Journal of Geology, v. 113, p. 325–340, doi: 10.1086/428808.

Pratt-Sitaula, B., Burbank, D.W., Heimsath, A., and Ojha, T., 2004, Landscape disequilibrium on 1000–10,000 year scales Marsyandi River, Nepal, central Himalaya: Geomorphology, v. 58, p. 223–241, doi: 10.1016/j.geomorph.2003.07.002.

Riebe, C.S., Kirchner, J.W., and Finkel, R.C., 2003, Long-term rates of chemical weathering and physical erosion from cosmogenic nuclides and geochemical mass balance: Geochimica et Cosmochimica Acta, v. 67, p. 4411–4427, doi: 10.1016/S0016-7037(03)00382-X.

Riebe, C.S., Kirchner, J.W., and Finkel, R.C., 2004, Erosional and climatic effects on long-term chemical weathering rates in granitic landscapes spanning diverse climate regimes: Earth and Planetary Science Letters, v. 224, p. 547–562, doi: 10.1016/j.epsl.2004.05.019.

Riebe, C.S., Kirchner, J.W., and Finkel, R.C., 2004, Sharp decrease in long-term chemical weathering rates along an altitudinal transect: Earth and Planetary Science Letters, v. 218, p. 421–434, doi: 10.1016/S0012-821X(03)00673-3.

Riebe, C.S., Kirchner, J.W., and Granger, D.E., 2001, Quantifying quartz enrichment and its consequences for cosmogenic measurements of erosion rates from alluvial sediment and regolith: Geomorphology, v. 40, p. 15–19, doi: 10.1016/S0169-555X(01)00031-9.

Riebe, C.S., Kirchner, J.W., Granger, D.E., and Finkel, R.C., 2000, Erosional equilibrium and disequilibrium in the Sierra Nevada, inferred from cosmogenic Al-26 and Be-10 in alluvial sediment: Geology, v. 28, p. 803–806, doi: 10.1130/0091-7613(2000)028<0803:EEADIT>2.3.CO;2.

Riebe, C.S., Kirchner, J.W., Granger, D.E., and Finkel, R.C., 2001, Minimal climatic control on erosion rates in the Sierra Nevada, California: Geology, v. 29, p. 447–450, doi: 10.1130/0091-7613(2001)029<0447:MCCOER>2.0.CO;2.

Riebe, C.S., Kirchner, J.W., Granger, D.E., and Finkel, R.C., 2001, Strong tectonic and weak climatic control of long-term chemical weathering rates: Geology, v. 29, p. 511–514, doi: 10.1130/0091-7613(2001)029<0511:STAWCC>2.0.CO;2.

Safran, E.B., Bierman, P.R., Aalto, R., Dunne, T., Whipple, K.X., and Caffee, M., 2005, Erosion rates driven by channel network incision in the Bolivian Andes: Earth Surface Processes and Landforms, v. 30, p. 1007–1024, doi: 10.1002/esp.1259.

Schaller, M., Hovius, N., Willett, S.D., Ivy-Ochs, S., Synal, H.A., and Chen, M.C., 2005, Fluvial bedrock incision in the active mountain belt of Taiwan from in situ-produced cosmogenic nuclides: Earth Surface Processes and Landforms, v. 30, p. 955–971, doi: 10.1002/esp.1256.

Schaller, M., von Blanckenburg, F., Hovius, N., and Kubik, P.W., 2001, Large-scale erosion rates from in situ-produced cosmogenic nuclides in European river sediments: Earth and Planetary Science Letters, v. 188, p. 441–458, doi: 10.1016/S0012-821X(01)00320-X.

Schaller, M., von Blanckenburg, F., Hovius, N., Veldkamp, A., van den Berg, M.W., and Kubik, P.W., 2004, Paleoerosion rates from cosmogenic Be-10 in a 1.3 Ma terrace sequence: Response of the River Meuse to changes in climate and rock uplift: The Journal of Geology, v. 112, p. 127–144, doi: 10.1086/381654.

Schaller, M., von Blanckenburg, F., Veldkamp, A., Tebbens, L.A., Hovius, N., and Kubik, P.W., 2002, A 30 000 yr record of erosion rates from cosmogenic Be-10 in Middle European river terraces: Earth and Planetary Science Letters, v. 204, p. 307–320, doi: 10.1016/S0012-821X(02)00951-2.

Schildgen, T., Dethier, D.P., Bierman, P., and Caffee, M., 2002, Al-26 and Be-10 dating of late Pleistocene and Holocene fill terraces: A record of fluvial deposition and incision: Colorado Front Range: Earth Surface Processes and Landforms, v. 27, p. 773–787, doi: 10.1002/esp.352.

Schroeder, P.A., Melear, N.D., Bierman, P., Kashgarian, M., and Caffee, M.W., 2001, Apparent gibbsite growth ages for regolith in the Georgia Piedmont: Geochimica et Cosmochimica Acta, v. 65, p. 381–386, doi: 10.1016/S0016-7037(00)00541-X.

Sharp, W.D., Ludwig, K.R., Chadwick, O.A., Amundson, R., and Glaser, L.L., 2003, Dating fluvial terraces by Th-230/U on pedogenic carbonate, Wind River Basin, Wyoming: Quaternary Research, v. 59, p. 139–150, doi: 10.1016/S0033-5894(03)00003-6.

Stock, G.M., Anderson, R.S., and Finkel, R.C., 2005, Rates of erosion and topographic evolution of the Sierra Nevada, California, inferred from cosmogenic Al-26 and Be-10 concentrations: Earth Surface Processes and Landforms, v. 30, p. 985–1006, doi: 10.1002/esp.1258.

Van der Wateren, F.M., Dunai, T.J., Van Balen, R.T., Klas, W., Verbers, A.L.L.M., Passchier, S., and Herpers, U., 1999, Contrasting Neogene denudation histories of different structural regions in the Transantarctic Mountains rift flank constrained by cosmogenic isotope measurements: Global Planetary Change, v. 23, p. 145–172, doi: 10.1016/S0921-8181(99)00055-7.

Vance, D., Bickle, M., Ivy-Ochs, S., and Kubik, P.W., 2003, Erosion and exhumation in the Himalaya from cosmogenic isotope inventories of river sediments: Earth and Planetary Science Letters, v. 206, p. 273–288, doi: 10.1016/S0012-821X(02)01102-0.

von Blanckenburg, F., Hewawasam, T., and Kubik, P.W., 2004, Cosmogenic nuclide evidence for low weathering and denudation in the wet, tropical highlands of Sri Lanka: Journal of Geophysical Research, Earth Surfaces, v. 109, p. F03008, doi: 10.1029/2003JF000049, doi: 10.1029/2003JF000049.

Ward, I.A.K., Nanson, G.C., Head, L.M., Fullagar, R.L.K., Price, D.M., and Fink, D., 2005, Late quaternary landscape evolution in the Keep River region, northwestern Australia: Quaternary Science Review, v. 24, p. 1906–1922, doi: 10.1016/j.quascirev.2004.11.004.

Wilkinson, M.T., Chappell, J., Humphreys, G.S., Fifield, K., Smith, B., and Hesse, P., 2005, Soil production in heath and forest, Blue Mountains, Australia: influence of lithology and palaeoclimate: Earth Surface Processes and Landforms, v. 30, p. 923–934, doi: 10.1002/esp.1254.

Wilkinson, M.T., and Humphreys, G.S., 2005, Exploring pedogenesis via nuclide-based soil production rates and OSL-based bioturbation rates: Australian Journal of Soil Research, v. 43, p. 767–779, doi: 10.1071/SR04158.

Wolkowinsky, A.J., and Granger, D.E., 2004, Early Pleistocene incision of the San Juan River, Utah, dated with Al-26 and Be-10: Geology, v. 32, p. 749–752, doi: 10.1130/G20541.1.

Review Papers

Bednarik, R.G., 2002, The dating of rock art: A critique: Journal of Archaeological Science, v. 29, p. 1213–1233, doi: 10.1006/jasc.2001.0711.

Bierman, P.R., 2004, Rock to sediment - Slope to sea with Be-10 - Rates of landscape change: Annual Review of Earth and Planetary Sciences, v. 32, p. 215–255, doi: 10.1146/annurev.earth.32.101802.120539.

Brown, E.T., Colin, F., and Bourlès, D.L., 2003, Quantitative evaluation of soil processes using in situ-produced cosmogenic nuclides: Comptes Rendus Geosciences, v. 335, p. 1161–1171, doi: 10.1016/j.crte.2003.10.004.

Cockburn, H.A.P., and Summerfield, M.A., 2004, Geomorphological applications of cosmogenic isotope analysis: Progress in Physical Geography, v. 28, p. 1–42, doi: 10.1191/0309133304pp395oa.

Englert, P.A.J., 2001, Discovery of cosmogenic nuclides: Early history and science applications: Journal of the Korean Physical Society, v. 39, p. 747–754.

Gosse, J.C., and Phillips, F.M., 2001, Terrestrial in situ cosmogenic nuclides: theory and application: Quaternary Science Review, v. 20, p. 1475–1560, doi: 10.1016/S0277-3791(00)00171-2.

Granger, D.E., and Muzikar, P.F., 2001, Dating sediment burial with in situ-produced cosmogenic nuclides: theory, techniques, and limitations: Earth and Planetary Science Letters, v. 188, p. 269–281, doi: 10.1016/S0012-821X(01)00309-0.

Kim, K.J., 2001, In situ cosmogenic isotopes in geological applications: Journal of the Korean Physical Society, v. 39, p. 783–789.

Lal, D., 2004, Assessing past climate changes from proxy records: an iterative process between discovery and observations: Quaternary International, v. 117, p. 5–16, doi: 10.1016/S1040-6182(03)00111-3.

Lal, D., and Chen, J., 2005, Cosmic ray labeling of erosion surfaces II: Special cases of exposure histories of boulders, soils and beach terraces: Earth and Planetary Science Letters, v. 236, p. 797–813, doi: 10.1016/j.epsl.2005.05.025.

Masarik, J., and Wieler, R., 2003, Production rates of cosmogenic nuclides in boulders: Earth and Planetary Science Letters, v. 216, p. 201–208, doi: 10.1016/S0012-821X(03)00476-X.

Muzikar, P., Elmore, D., and Granger, D.E., 2003, Accelerator mass spectrometry in geologic research: Geological Society of America Bulletin, v. 115, p. 643–654, doi: 10.1130/0016-7606(2003)115<0643:AMSIGR>2.0.CO;2.

Siame, L.L., Braucher, R., and Bourlès, D.L., 2000, Applications of in situ-produced cosmogenic nuclides to quantitative geomorphology: Bulletin de la Société Géologique de France, v. 171, p. 383–396, doi: 10.2113/171.4.383.

von Blanckenburg, F., 2005, The control mechanisms of erosion and weathering at basin scale from cosmogenic nuclides in river sediment: Earth and Planetary Science Letters, v. 237, p. 462–479, doi: 10.1016/j.epsl.2005.06.030.

A review of burial dating methods using ^{26}Al and ^{10}Be

Darryl E. Granger[†]
Department of Earth and Atmospheric Sciences, Purdue University, West Lafayette, Indiana 47907, USA

ABSTRACT

Multiple cosmogenic nuclides with different decay rates can be used to date exposure and burial of rocks over the timescales of radioactive decay. This paper reviews the development of such dating methods over the past ~50 years, beginning with a historical perspective on early meteorite studies, and later focusing on recent examples in the terrestrial field using the ^{26}Al-^{10}Be pair in quartz.

Two classes of terrestrial applications are discussed in detail. The first involves the use of ^{26}Al and ^{10}Be in rock or sediment that has experienced a complex history of repeated exposure and burial. In these cases, the cosmogenic nuclides can only provide a minimum near-surface age. Examples include sediment from beneath desert sand dunes, and rocks from beneath cold-based glaciers. The second class of application uses ^{26}Al and ^{10}Be to date discrete burial events, in cases where sediment has experienced a simple history of exposure followed by rapid burial. Examples include cave sediments, alluvial deposits, and sediment buried beneath glacial till. Finally, the half-lives of ^{26}Al and ^{10}Be are discussed, with special attention given to discrepant estimates of the ^{10}Be half-life. It is shown that geologic data are consistent with either half-life estimate of 1.51 m.y. or 1.34 m.y., but more closely conform to the shorter half-life.

Keywords: cosmogenic nuclides, burial, sediments, half-life, Quaternary.

INTRODUCTION

In any given rock, the concentration of cosmogenic nuclides is governed by the balance between production and radioactive decay. It is the goal of practically all cosmogenic nuclide analyses to reconstruct the history of production rates over time. Production rates may vary through time as a function of erosion rate, uplift rate, or burial and shielding. Whether in geologic or meteoritic applications, we seek samples that allow simplifying assumptions to be made regarding the cosmic-ray exposure history, such as constant exposure or steady-state erosion. From the very beginning of cosmogenic nuclide work, it was recognized that multiple cosmogenic nuclides, with different production rates and decay constants, may be measured in the same sample to decipher increasingly complex exposure histories. Although deciphering complex exposure and burial histories from multiple cosmogenic nuclides is relatively new in terrestrial studies, the approach comes from a long heritage of meteorite work. In order to place modern advances within the context of prior work, this paper begins with a brief historical review of methods development in the meteoritic field, followed by a summary of more recent terrestrial applications.

History and Development of Multiple-Nuclide Methods

Cosmogenic nuclide geochronology can trace its roots to Bauer (1947), who suggested a cosmic-ray origin for excess helium in iron meteorites previously detected by Arrol (1942) and others. It was immediately apparent that the buildup of cosmogenic helium could provide a chronometer for the exposure of

[†]E-mail: dgranger@purdue.edu

meteorites to cosmic rays in space. Over the following decade, many attempts were made to calculate the cosmic-ray exposure age of meteorites from the buildup of ^{3}He, but they were hampered by poor knowledge of ^{3}He production rates and by variation in production rates within the interior of the meteorite parent body. It was an underconstrained problem, with two unknowns (production rate and exposure time) but only one equation (for ^{3}He buildup).

Resolution to the production rate problem came in 1957, when two groups realized that a pair of cosmogenic nuclides could be used simultaneously to solve for a sample's production rate and its exposure time. Both Begemann et al. (1957) and Fireman and Schwarzer (1957) measured stable ^{3}He, which accumulates monotonically over time, and then normalized the measured concentration against that of tritium, a radioactive nuclide that accumulates at a rate similar to that of ^{3}He, but decays with a half-life of ~12 yr. Because tritium is short-lived, it achieves secular equilibrium within a century of exposure, and thus its concentration depends only on production rate. The concentration of ^{3}He, on the other hand, depends on both production rate and exposure time. In an elegant twist, it was shown that the ratio of these two cosmogenic nuclides indicates the exposure age of the meteorite, regardless of the absolute production rates. All that is required to solve for the cosmic-ray exposure age is knowledge of the two nuclides' relative production rates. The use of multiple-nuclide methods is nearly as old as the knowledge of cosmogenic nuclides themselves.

It was quickly realized by Whipple and Fireman (1959) that cosmogenic nuclide concentrations can be interpreted as either minimum exposure ages or maximum erosion rates of the meteorites in space (e.g., by micrometeorite impacts). Just as in terrestrial applications today, the concentration of cosmogenic nuclides at the surface can be due to long-term exposure at a constant production rate, or to a gradually increasing production rate as the present-day surface was exhumed by erosion. For samples with a constant erosion rate and an exponential production rate profile, the steady-state concentration (N) at the surface is given by equation 1:

$$N = P/[(1/\tau) + \rho\varepsilon/\Lambda], \qquad (1)$$

where P is the production rate at the surface, τ is the radioactive mean-life, ρ is the density of the rock, Λ is the penetration length, and ε is the erosion rate. This same equation is used today for terrestrial applications.

The first use of the radioactive decay of cosmogenic nuclides to date a shielding event appeared in Honda et al. (1961). In this paper, multiple cosmogenic nuclides ^{10}Be ($t_{1/2}$ = 1.5 m.y. or 1.3 m.y.; see later section on half-lives), ^{26}Al ($t_{1/2}$ = 0.7 m.y.), ^{36}Cl ($t_{1/2}$ = 0.3 m.y.), ^{40}K ($t_{1/2}$ = 1.25 b.y.), and ^{53}Mn ($t_{1/2}$ = 3.7 m.y.) were measured in four different iron meteorites. Although each meteorite had different concentrations of these radionuclides, they were present in a constant ratio, as expected, with the notable exception in some samples of ^{36}Cl, the most short-lived nuclide.

In a note added in proof, the authors wrote that the low ^{36}Cl concentration in the Williamstown meteorite "is difficult to interpret; perhaps Williamstown has a long terrestrial age." This brief note was expanded upon by Arnold (1961), who pointed out that the concentrations of short-lived radionuclides decrease exponentially after a meteorite fall, because production rates decrease a thousandfold once the meteorite is blanketed by Earth's atmosphere. By measuring the ratio of a short-lived to a long-lived radionuclide, the terrestrial age of the meteorite can be easily calculated from the deficit in the activity of the short-lived species, assuming that prior to landing all cosmogenic nuclides were at secular equilibrium. Arnold (1961) showed that the deficit of ^{36}Cl and the absence of shorter-lived nuclides such as ^{39}Ar ($t_{1/2}$ = 269 yr) indicated a terrestrial age of 600,000 yr for the Williamstown find.

The arguments above can be expressed in very simple equations governed by production and decay. The concentration of a cosmogenic nuclide with a single exposure history is given by

$$N = P\,\tau\,(1 - e^{-t/\tau}). \qquad (2)$$

Equation 2 can be simplified for certain situations. For example, for a long exposure history ($t \gg \tau$), as is common in meteorites, the concentration reaches secular equilibrium and

$$N = P\,\tau. \qquad (3)$$

For a stable cosmogenic nuclide ($\tau = \infty$), the concentration rises steadily over time (t) according to

$$N = P\,t. \qquad (4)$$

Thus, following Begemann et al. (1957) and Fireman and Schwarzer (1957), by measuring two nuclides, one stable (such as ^{3}He) and the other at secular equilibrium (such as tritium), their ratio can be used to calculate the cosmic-ray exposure age, t:

$$t = (N_a/N_b)\,\tau_b\,(P_b/P_a), \qquad (5)$$

where the subscripts a and b denote the two nuclides.

These equations can be modified for more complex exposure histories. For example, in a sample with a two-stage exposure history, such as a meteorite that was first exposed in space and later on Earth, the cosmogenic nuclide concentration can be expressed by

$$N = P_1\,\tau\,(1 - e^{-t_1/\tau})\,e^{-t_2/\tau} + P_2\,\tau\,(1 - e^{-t_2/\tau}), \qquad (6)$$

where the subscripts 1 and 2 represent two different times with different production rates.

For the simplifying case in which t_1 is long, and P_2 is small, for example in a meteorite that landed at an unknown time,

$$N = P_1\,\tau\,e^{-t_2/\tau}. \qquad (7)$$

Following Arnold et al. (1961), it can be seen that the ratio of two cosmogenic radionuclides is directly related to the terrestrial age t_2:

$$(N_a/N_b) = (P_a/P_b)(\tau_a/\tau_b) e^{-(t_2/\tau_a - t_2/\tau_b)}$$

$$t_2/\tau_b - t_2/\tau_a = ln(A_a P_b / A_b P_a)$$

$$t_2 = ln(A_a P_b / A_b P_a) / (1/\tau_b - 1/\tau_a), \quad (8)$$

where A_i represents the activity of the radionuclide i, or the number of radioactive decays per unit time.

Continuing studies of cosmogenic nuclides in meteorites and lunar rocks have revealed increasingly complex exposure histories over the timescales of radioactive decay. Nishiizumi et al. (1979) were some of the first to explore multiple exposure histories of meteorites using ^{53}Mn, ^{10}Be, and ^{26}Al. Later, Nishiizumi et al. (1991a, 1996) employed ^{14}C, ^{41}Ca, ^{36}Cl, ^{26}Al, ^{10}Be, and ^{53}Mn to reconstruct the exposure history of lunar meteorites in remarkable detail. They used these five radionuclides plus cosmogenic noble gases to solve for five unknowns in several lunar meteorites' irradiation histories: preejection exposure depth on the moon, depth in the meteoritic body in space, exposure time on the moon (t_m), exposure time in space (t_s), and terrestrial age since the time of fall (t_a). Each cosmogenic nuclide's activity is determined by equation 9:

$$A = P_m(1 - e^{-t_m/\tau}) e^{-(t_s + t_a)/\tau} + P_s(1 - e^{-t_s/\tau}) e^{-t_a/\tau}. \quad (9)$$

By solving equation 9 simultaneously for all radionuclides, and minimizing the misfit between measurements and the model, all five of the unknowns can be constrained. As with any inverse problem, it is important to realize that the number of measurements must meet or exceed the number of unknowns, and that the unknowns can only be resolved over timescales during which the cosmogenic nuclides behave differently. Nevertheless, the use of multiple cosmogenic nuclides with different radioactive mean-lives and production rates can reveal complex histories of exposure and burial. Similar methods using chi-square minimization to resolve complex histories are beginning to be used in the terrestrial field today (e.g., Siame et al., 2004; Wolkowinsky and Granger, 2004; Balco et al., 2005).

The summary above is by no means a comprehensive review, but rather a glimpse into the development of multiple-nuclide methods. The point to be made is that the theoretical and mathematical framework for using multiple cosmogenic nuclides has existed since the very beginning of the field. The development of cosmogenic nuclide methods in terrestrial research has closely followed that of its meteoritic predecessor, and in many cases merely elaborates upon these early-established methods. The following section reviews the development of multiple-nuclide measurements in the terrestrial field. There are far fewer applications in this arena, partly because terrestrial cosmogenic nuclides are more difficult to measure, but also because many Earth surface processes occur quickly with respect to the radioactive decay of the cosmogenic nuclides. In these cases, the cosmogenic nuclides behave nearly identically to each other, and little new information can be gleaned from analysis of multiple species. It is only either by analyzing short-lived radionuclides such as ^{14}C or by studying processes that occur over million-year timescales that the strength of multiple-nuclide approaches may be realized.

TERRESTRIAL COSMOGENIC NUCLIDES

Early Measurements

Although the measurement of cosmogenic nuclides in meteorites was in full swing by the 1960s, terrestrial samples were far more difficult to measure by the decay counting methods then in use. Only two early attempts to measure cosmogenic nuclides were successful. The first of these was by Davis and Schaeffer (1955), who detected ^{36}Cl in rock. They were able to measure the accumulation of this nuclide in an unglaciated surface, but had difficulty with glaciated surfaces that had been exposed for less time. Sample treatment was extremely laborious, and the results had high uncertainty, so the method languished. A second measurement of a cosmogenic nuclide was by Hampel et al. (1975), who dissolved nearly 30 kg of SiO_2 to isolate over 400 g of Al_2O_3 for decay counting. These amounts are a thousandfold greater than those employed in modern methods, yet their measurement achieved a precision of only 40%. A combination of difficult chemistry and counting efforts rewarded by imprecise results effectively prohibited terrestrial cosmogenic nuclide geochronology.

With the development of accelerator mass spectrometry (AMS) in the 1980s, the measurement of terrestrial cosmogenic nuclides suddenly became far more tractable. Yiou et al. (1984) were the first to report measurements of terrestrial in situ–produced ^{10}Be and ^{26}Al from tektites and from Libyan Desert Glass, a nearly pure SiO_2 glass of impact origin, but with a terrestrial age of ca. 28 Ma. In a brief but influential paper, Lal and Arnold (1985) then suggested that the nuclide pair ^{26}Al and ^{10}Be would be particularly suitable for measurement in the mineral quartz. It is notable that Lal and Arnold (1985) emphasized the simultaneous measurement of these two radionuclides to take advantage of their radioactive decay. They considered the nuclide pair to be suitable for geologic problems over the timescale of their differential radioactive decay, with a characteristic timescale of ~1 m.y., and suggested a range of applications, including the dating of hominin-bearing sediments.

Interpreting Exposure and Burial Histories

The first two data-intensive papers on terrestrial cosmogenic ^{26}Al and ^{10}Be used these nuclides' radioactive decay to date buried sediments. Klein et al. (1986) were the first to use the terrestrial ^{26}Al-^{10}Be pair in this way, expanding upon the earlier work by Yiou et al. (1984) on Libyan Desert Glass. This glass is an impact melt scattered across a large area of desert. The glass must have been repeatedly buried and reexposed by passing sand dunes and during transport by flowing water. Although Klein et al. (1986) were unable to date the glass absolutely, they were able to show that ^{26}Al and ^{10}Be concentrations fell within the range of allowed ^{26}Al/^{10}Be ratios that is governed by accumulation and radioactive decay. By plotting the nuclide ratio ^{26}Al/^{10}Be against ^{10}Be concentration, Klein et al. (1986) showed that all possible data must fall within an area that they called the "exposure-burial triangle." The exposure-burial triangle is delimited by two lines with an apex at the point of secular equilibrium, where production is balanced by decay. The uppermost line of the triangle is defined by the condition of constant accumulation. Because ^{26}Al decays faster than ^{10}Be, the ratio ^{26}Al/^{10}Be can never exceed the production rate ratio. For constant exposure, this ratio can be predicted by solving equation 2 for both nuclides:

$$N_{26}/N_{10} = (P_{26}/P_{10})(\tau_{26}/\tau_{10})(1 - e^{-t/\tau_{26}})/(1 - e^{-t/\tau_{10}}). \quad (10)$$

The lower line on the exposure-burial triangle is defined by radioactive decay from the secular equilibrium endpoint, assuming that production ceases. In this case, the ^{26}Al/^{10}Be ratio decreases according to

$$N_{26}/N_{10}(t) = P_{26}\tau_{26}/P_{10}\tau_{10}\, e^{-t(1/\tau_{26} - 1/\tau_{10})}. \quad (11)$$

Using the exposure-burial triangle, Klein et al. (1986) demonstrated that earlier estimates of production rates for the two nuclides could not be correct; they estimated production rates from their own measured data, assuming that their highest measured concentrations were near saturation. Although their assumption of saturation was later revealed to be incorrect, the method of constraining erosion rates and half-lives from the exposure-burial triangle can still be useful today, as I will show in a later section on half-life determination.

The exposure-burial triangle was modified by Nishiizumi et al. (1991b) and also by Lal (1991) to distinguish conditions of exposure versus steady erosion. Recall that Whipple and Fireman (1959) pointed out that cosmic-ray exposure ages could be equally well interpreted in terms of erosion rates. This same issue remains for terrestrial exposure; the erosion rate of a rock can be calculated using equation 1. The ratio of two radionuclides can be calculated from this equation, assuming that they have similar production rate profiles. This is a valid assumption for ^{26}Al and ^{10}Be, whose production profiles are nearly identical (Brown et al., 1992); however, this is not the case with many other nuclides. For example, ^{36}Cl may have a different depth dependence than ^{26}Al and ^{10}Be due to its production from thermal-energy neutron capture on ^{35}Cl. For the ^{26}Al-^{10}Be pair, the ratio is given by equation 12:

$$N_{26}/N_{10}(t) = (P_{26}/P_{10})\,[(1/\tau_{10}) + \rho\varepsilon/\Lambda]\,/\,[(1/\tau_{26}) + \rho\varepsilon/\Lambda]. \quad (12)$$

If the ratio ^{26}Al/^{10}Be is plotted versus ^{10}Be, following Klein et al. (1986), then the constant-exposure curve and the steady-erosion curve deviate somewhat from one another, although they must converge under the condition of secular equilibrium. The area between these two curves was denoted the "steady-state erosion island" by Lal (1991). The curves are shown in Figure 1, on logarithmic axes, together with lines showing radioactive decay and burial isochrons. This diagram is sometimes referred to as a "banana diagram" due to the shape of the steady-state erosion island. I will refer to it as the exposure-burial diagram following the usage of Klein et al. (1986).

The exposure-burial diagram has played an important role in multiple-nuclide methods; it will be presented in several configurations later in this paper. Although the fundamentals of the diagram have not varied since its inception, some details have varied among users. For example, Klein et al. (1986) used linear axes to show values near saturation; however, most terrestrial values have low ^{10}Be concentrations and thus group tightly to the left. Nishiizumi et al. (1991b) resolved this issue by expressing the abscissa on a logarithmic scale. Lal (1991) later expressed both axes logarithmically. Although opinions differ on the best way to express this diagram, the log-log axes have two advantages for geochronology: (1) Radioactive decay lines are straight on this diagram, making it easier to visualize changes over time, and (2) although the ordinate is logarithmic with respect to ^{26}Al/^{10}Be, this implies that it is linear with respect to burial time. The data and their uncertainties can thus be interpreted straightforwardly in terms of burial time rather than the less-intuitive ^{26}Al/^{10}Be ratio.

One modification to the exposure-burial diagram that is useful when comparing samples from multiple sites is to normalize the abscissa by the local production rates, so that the ^{10}Be concentrations are scaled to sea level, high latitude (SLHL). This can be done in two ways: Ivy-Ochs et al. (1995) scaled by a dimensionless factor of P_{SLHL}/P_{local}, so that the abscissa maintained units of concentration. On the other hand, Stone (2000) scaled by $1/P_{local}$, so that the abscissa had units of years. Because it is more intuitive to maintain the ^{10}Be concentration in terms of atoms per gram, and because cosmogenic nuclide practitioners are familiar with scaling production rates to SLHL, I will use the dimensionless normalization factor in this paper.

The exposure-burial diagram has several important characteristics. First, if production rates are constant through time, then all rocks with continuous exposure should plot along the exposure line. Their position on the exposure line depends on the exposure time. Likewise, all rocks that have been steadily eroding for long enough to erode several penetration lengths (Λ) should plot along the erosion line, with their position dependent on the erosion rate. Rocks that were suddenly exposed, and subsequently began to erode but have not yet eroded through many penetration lengths, plot within the steady-state erosion island. The position within the steady-state erosion island can in principle be used to determine both how long a rock has been exposed and how fast it is eroding (Nishiizumi et al., 1991b; Lal, 1991), thus resolving the conundrum first posed by Whipple and Fireman (1959). In practice, it is often difficult to distinguish erosion from exposure

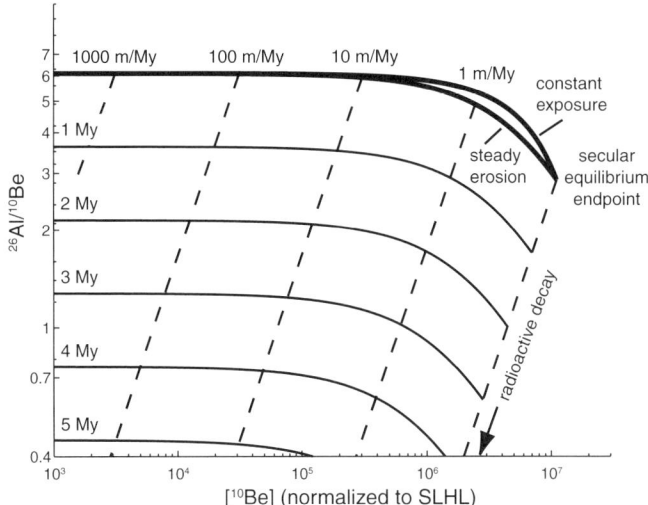

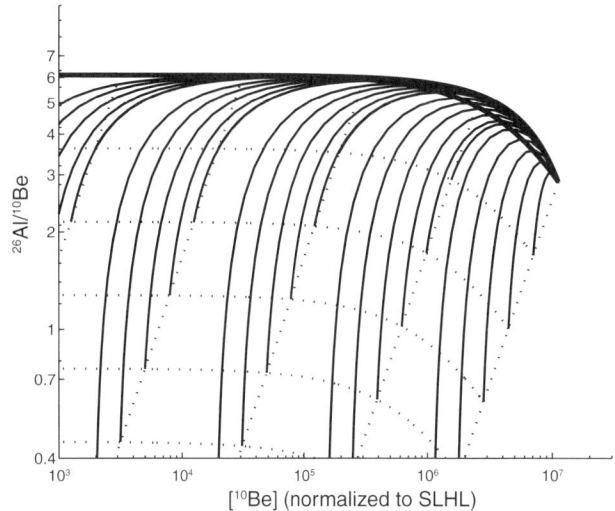

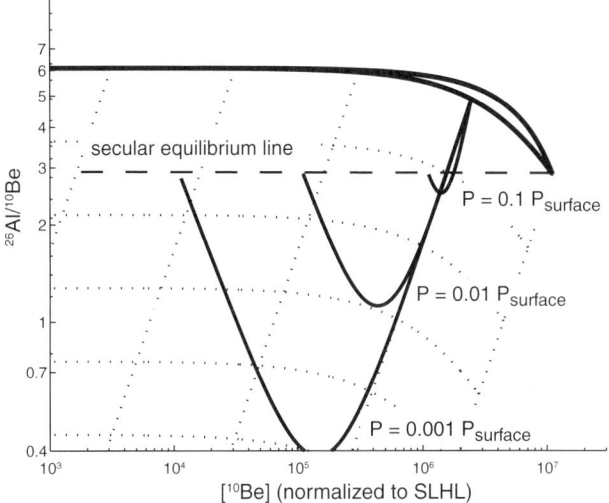

Figure 1. Exposure-burial diagrams illustrating the evolution of ^{26}Al and ^{10}Be in quartz. The top left diagram shows conditions of steady erosion or constant exposure followed by burial. A sample that has experienced constant exposure will plot on the uppermost line. A sample that has experienced long-term steady-state erosion will plot on the lower line, at a position indicated by the erosion rates shown. Samples at the surface that were exposed and then eroded will plot between these curves, in the steady-state erosion island (Lal, 1991). Upon deep burial, the ^{26}Al/^{10}Be ratio evolves parallel to the dashed radioactive decay lines. Million-year decay isochrons are shown. The bottom left diagram illustrates the evolution of a sample whose production rate decreases, but does not stop. Three curves show how the ^{26}Al/^{10}Be ratio evolves under reduced production rates. Note that for long exposure times, the ^{26}Al/^{10}Be ratio always approaches the secular equilibrium line. The top right diagram includes a family of curves illustrating the growth of ^{26}Al/^{10}Be ratios in response to reexposure at the surface. Each growth curve on the diagram begins at a node corresponding to the uppermost diagram, and evolves over time to approach the secular equilibrium endpoint. SLHL—sea level, high latitude; P—production rate.

with the ^{26}Al-^{10}Be pair because measurement uncertainties may be comparable to the difference between the exposure and erosion lines.

A second important characteristic of the exposure-burial diagram is that all samples must plot below and to the left of the exposure line, and to the left of radioactive decay from secular equilibrium. This is the original exposure-burial triangle of Klein et al. (1986). All other space on this diagram is forbidden. A sample may fall to the right of the erosion line if it recently experienced production at a higher rate than its present location, but it is never allowed to have a ^{26}Al/^{10}Be ratio higher than their production ratio. Samples that plot below the exposure or erosion line have experienced at least one episode of burial, in which production rates decrease.

Radioactive decay lines may take two forms, depending on the depth of burial. If production ceases entirely, under the conditions of deep burial, then the ^{26}Al/^{10}Be ratio decreases in a straight path on Figure 1, governed only by radioactive decay. If production rates slow, but do not cease, then the ^{26}Al/^{10}Be ratio initially decreases, but then approaches secular equilibrium for the new production rate (Fig. 1, bottom left; cf. Lal, 1991, his Fig. 7). For deeply buried samples, the radioactive decay lines on the exposure-burial diagram can be traced back to the exposure or erosion line to determine the inheritance, or the amount of cosmogenic nuclides present prior to burial.

If a sample is buried, and subsequently reexposed at the surface, then its ^{26}Al/^{10}Be ratio reapproaches the exposure curve asymptotically, as shown in Figure 1 (top right). A sample that experiences multiple episodes of exposure and burial may thus evolve along a zigzag path on the exposure-burial diagram, moving parallel to the burial and reexposure curves (Fig. 2).

It is important to realize that the position of a sample on the exposure-burial diagram alone cannot uniquely resolve the sample's exposure and burial history. There are many paths that a sample can take, involving repeated episodes of exposure, erosion, shallow burial, or deep burial, that may yield identical

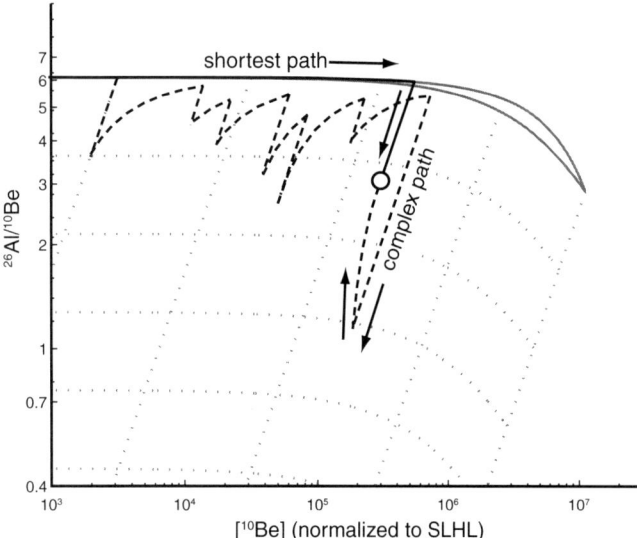

Figure 2. Exposure-burial diagram showing two different pathways leading to the same final point. The shortest path, shown as a solid black line, represents the minimum time required, while the complex path, shown as a dashed line, represents one of many possible scenarios involving repeated exposure and burial.

^{26}Al and ^{10}Be concentrations. Only a minimum age can be established with absolute confidence. It is only through the judicious choice of a sampling site, and the addition of geologic or stratigraphic information, that a sample's exposure and burial history can be reconstructed.

APPLICATIONS IN GEOLOGY AND GEOMORPHOLOGY

In a list that presaged the next two decades of terrestrial cosmogenic nuclide research, Lal (1987, 1988) proposed an array of geologic applications for cosmogenic nuclides. Many of these applications implicitly involve the use of multiple nuclides, in particular the ^{26}Al-^{10}Be pair in quartz, but also stable nuclides and shorter-lived species such as ^{36}Cl, ^{14}C, and ^{39}Ar. Several of these proposed applications can be classified into two groups that are discussed here in more detail: (1) constraining complex burial histories of glacial sediments and sand dunes, and (2) dating single-stage burial histories of deeply and rapidly buried sediments, such as found in caves, river terraces, and potentially marine or lake deposits.

Complex Burial Histories

The fate of sediments on the surface is often complicated, with repeated episodes of burial and exposure as the sediment is transported down hillslopes and rivers, and into basins. It is often impossible to say with certainty how much time any individual grain has spent buried in a sedimentary deposit versus exposed while in transport. The fate of sand grains in deserts or in longshore transport is equally complex, with individual grains being recycled many times to the surface. Likewise, rocks exposed at the surface may be buried and exhumed many times, either by sediments or by glacial ice. In these situations, the use of multiple cosmogenic nuclides such as ^{26}Al and ^{10}Be can provide useful information, such as whether sediment or rock has been buried, and a minimum estimate of the burial and exposure times. Although these are not discrete dates, they can nevertheless be critical for geologic and geomorphic interpretation.

Sand Dunes, Littoral Sediments, and Desert Pavements

Case 1: Libyan Desert Glass. The first use of the ^{26}Al-^{10}Be pair in Libyan Desert Glass, discussed above, was just such a case of complex exposure and burial (Klein et al., 1986). In this case, 12 individual clasts were found to have ^{26}Al/^{10}Be ratios depressed below the production ratio, indicating significant burial. It is likely that each clast had experienced different histories as they were repeatedly buried and exhumed by sand dunes, and transported by flowing water. Klein et al. (1986) recognized that such a complex history was beyond the resolution of the two-nuclide approach, but they also recognized that the measurements provided a minimum estimate of the burial time. In other words, the shortest path to a particular point on the exposure-burial diagram is taken by a single episode of exposure followed by a single episode of burial (Fig. 2). The total exposure and burial history of the clast must exceed the sum of these two ages. The Libyan Desert Glass was found to be at least several million years old, consistent with its fission-track age of 28 Ma.

Case 2: Littoral Sediments. Sediment on shorelines or marine terraces may have complicated histories of burial and exposure during longshore transport or due to eustasy. Nishiizumi et al. (1993) presented data from marine terraces in Peru and South Carolina, USA, with low ^{26}Al/^{10}Be ratios, suggesting either that the sediment experienced multiple episodes of exposure and burial during transport and deposition, or that older material was incorporated into the marine sands. A similar situation was observed by Boaretto et al. (2000), who attempted to date beach sediments that had been blown or washed into Tabun Cave, Israel, a prehistoric site containing Paleolithic artifacts. The sand was derived from dunes along the shore, which were in turn likely derived from longshore transport from the mouth of the Nile River. They found depressed ^{26}Al/^{10}Be ratios in the sand dunes, indicating long transport times or an older source area. Unfortunately, the inherited ^{26}Al/^{10}Be ratios were comparable to those in the cave, thus complicating their interpretation of the data.

Case 3: Pediments. Pediments are a common desert landform where bedrock is beveled to a nearly flat surface at the base of mountains. They are thought to form by weathering and erosion as a thin layer of gravel is transported over the surface by sheetwash. Bierman and Caffee (2001) measured ^{26}Al and ^{10}Be in gravelly pavements on piedmonts, and found that individual quartz clasts have greatly varying exposure and burial histories.

Using the same scheme as Klein et al. (1986) to estimate minimum exposure plus burial times, the clasts were found to have minimum near-surface ages ranging from 0.4 to 2.8 Ma, indicating a great age for the landform.

Case 4: Soil Cover. The ^{26}Al-^{10}Be pair can also be used to detect burial by accumulated soil. Albrecht et al. (1993) reported measurements from samples of the Bandelier Tuff, New Mexico. Nearly all of the samples showed depressed ^{26}Al/^{10}Be ratios, indicative of burial; two samples collected from beneath a thin soil cover indicated burial times of nearly 1 m.y. The authors suggested that a much thicker soil once covered the tuff, thus burying the surface. The soil was later stripped from much of the landscape, reexposing the original surface.

In a very different environment, Braucher et al. (2000) used ^{26}Al and ^{10}Be to show that quartz pebbles incorporated into buried stone lines in tropical laterites were first exposed at the surface and later buried. The depressed ^{26}Al/^{10}Be ratios in stone lines in Cameroon and Gabon indicate minimum burial times ranging from 0.7 to 1.15 m.y. These data help to resolve discrepant views of whether stone lines represent a former surface now buried or if they developed by in situ differentiation.

All of these examples share similar characteristics. The samples, whether individual clasts or bulk sediment, experienced repeated episodes of burial and exposure in a complex history of erosion, transport, and sedimentation. It is impossible to distinguish a detailed history from two radionuclides alone, although a minimum age for the grains can be inferred from the most direct exposure-burial path on the exposure-burial diagram (Fig. 2). An important lesson to be learned from these measurements is that sediment samples must be interpreted with caution, particularly when their maturity suggests a long history of weathering and transport. Even deeply buried sediments may have a history of earlier burial from their source area that is difficult to detect with cosmogenic nuclides alone, and impossible to detect with a single nuclide such as ^{10}Be.

Burial of Glacial Erratics and Bedrock, and Evidence of Cold-Based Glaciation

Sediments and rocks can be buried not only by other sediments, but also by snow and ice. Such is the case beneath cold-based glaciers. Under this scenario, the rocks themselves are not eroded because the ice is frozen to its bed and does not slide, but they are shielded from secondary cosmic rays by the glacier itself. After deglaciation, there may be little or no geomorphic evidence that the ice was ever there. Two different efforts have detected burial of rocks beneath cold-based glaciers using ^{26}Al and ^{10}Be.

Case 1: Laurentide Ice Sheet. Bierman et al. (1999) measured ^{26}Al and ^{10}Be in two different areas near the margins of the former Laurentide Ice Sheet, and found evidence of burial at both sites. The northern site on Baffin Island was characterized by a high degree of weathering on upland surfaces. Previous interpretations of this area had suggested that the area remained ice-free during the last glaciation. Bierman et al. (1999), however, found that the bedrock contained high concentrations of ^{26}Al and ^{10}Be, with ^{26}Al/^{10}Be ratios well below the steady-erosion island. Moreover, erratics on this surface showed young exposure ages consistent with deposition by ice during the Last Glacial Maximum. They concluded that the area must have been buried at least once beneath a thick cover of ice. Following the procedure of Klein et al. (1986), they determined that the minimum age for the bedrock surface must include at least an exposure time of ~200,000 yr and a burial time of ~400,000 yr. The fact that this total near-surface time exceeds the glacial-interglacial period suggests that the bedrock surface could have been buried multiple times.

Bierman et al. (1999) made an important methodological step beyond the interpretation of Klein et al. (1986) by modeling the trajectories of ^{26}Al and ^{10}Be on the exposure-burial diagram under conditions of repeated exposure and burial such as might be expected during glacial-interglacial cycles. They also considered the condition of partial shielding under a thin cover, so that production rates are lowered but not halted. Under the partial shielding scenario, the samples could be exposed for a long period of time. If the cover were suddenly and recently removed, then the samples would lie under the steady-erosion island as drawn for the present-day production rates. Gosse and Phillips (2001) considered a similar scenario in which glacial plucking rapidly removed an overburden in a rock with a very long exposure history, causing the sample to shift to the left on the exposure-burial diagram and thus fall beneath the steady-erosion island. Bierman et al. (1999) concluded that the most likely scenario to produce their observations included multiple burial episodes, but because many different burial scenarios could produce the same final ^{26}Al and ^{10}Be concentrations, they were cautious to limit their interpretation to minimum exposure and burial ages.

Case 2: Fennoscandian Ice Sheet. A different research group followed a similar approach for the Fennoscandian Ice Sheet. Stroeven et al. (2002) and Fabel et al. (2002) measured ^{26}Al and ^{10}Be in an area near the center of the Fennoscandian Ice Sheet that retains a rugged, deeply weathered topography characterized by tors, cliffs, and intact grus. Previous work had suggested that this area had been repeatedly buried beneath cold-based ice, and that the tors were part of a relict preglacial landscape. These authors found that ^{26}Al and ^{10}Be plotted consistently beneath the steady-erosion island, demonstrating conclusively that this landscape had been preserved for at least several glacial-interglacial cycles. They carried the analysis one step further by using a specific model for the timing of glaciation over the past 2.75 m.y., previously constructed from the marine benthic foraminiferal oxygen isotope record from DSDP 607 (Kleman et al., 1997). The modeled history slightly overestimated the final ^{10}Be concentration but could be reconciled with a subdued production rate due to thin cover, or perhaps due to surface erosion during interglacials.

These applications of ^{26}Al and ^{10}Be to bedrock beneath the Fennoscandian Ice Sheet represent a further advance in methodology, in that the exposure-burial histories are expressly modeled. Although a detailed history is still not attainable from the data alone, the regularity of glacial-interglacial cycles lends itself to constructing a credible model of periodic exposure and burial. This leads to more information than can be gleaned from more poorly constrained or stochastic processes such as sand dune migration or longshore transport. However, it is still difficult to reconstruct an exact exposure and burial history, because many different scenarios could lead to the same ^{26}Al and ^{10}Be concentrations. In the next section, I will discuss a different sort of application, in which the ^{26}Al-^{10}Be pair is used to date discrete events that are clearly evident from the sedimentary or geologic record. In these cases, because there is geologic and stratigraphic information that supports a simple exposure and burial history, the burial histories can be dated with much higher confidence.

Discrete Burial Histories

Although the cases discussed so far emphasize the stochastic nature of sediment transport and burial, it is also common for sediment to experience a simple history of a single exposure followed by burial. For example, sediment can be laid at the bottom of a thick deposit, perhaps on an alluvial terrace that accumulates over hundreds or thousands of years. Sediment can be buried beneath volcanic ash or lava that is laid down in days. Or stream sediment can be carried into a cave, where it is instantly shielded by up to hundreds of meters of rock as it passes beneath the surface. In these cases, cosmogenic nuclide production practically ceases, and the sedimentation event can be dated with confidence by the relative radioactive decay of multiple nuclides. The methodology of this type of burial dating has been recently reviewed by Granger and Muzikar (2001); I will not review the methodology here, but will instead focus on several geologic examples.

Dating Cave Sediments

Case 1: New River, Virginia. The first use of the ^{26}Al-^{10}Be pair to date sediment burial events was by Granger et al. (1997), who realized that cave sediments are ideal for this method. Quartz-bearing sediments in caves are almost always derived from outside of the cave itself, which is generally formed in a soluble rock such as limestone. Prior to deposition, the quartz was eroded from its parent rock and transported over the ground surface, accumulating cosmogenic nuclides. Once washed into the cave, it is instantly shielded by tens to hundreds of meters of solid rock, so cosmogenic nuclide production practically stops. This simple, single-stage history of exposure followed by burial allows the burial age to be readily calculated using equation 13 for both ^{26}Al and ^{10}Be, using the simplifying assumption that sediment begins under a condition of steady-state erosion (equation 12). (For different initial conditions, see Granger and Muzikar, 2001.)

$$N_i = P_i/[(1/\tau_i) + \rho\varepsilon/\Lambda]\, e^{-t/\tau_i}. \quad (13)$$

Granger et al. (1997) applied this method to date quartz pebbles found in caves alongside the New River, Virginia, in the Valley and Ridge province of the southern Appalachian Mountains. These quartz pebbles were derived from a metamorphic rock more than 75 km upstream; thus the only source of quartz pebbles in this landscape is from the New River itself. Where the New River flows through limestone, caves discharge water as springs at river level. Some of the New River's bed load inevitably spills into the cave openings, particularly if the caves are large. (Divers in the river can observe this happening today.) As the river cuts down over time, the caves and the sediments contained within them are left abandoned above the river. The water finds a new path to the river underground, dissolving a new cave, and the ancient alluvium can remain virtually undisturbed in the cave for millions of years.

In many ways the caves are similar to terraces. The caves form at river level, just as terraces do. When the river level is stable for a long time, the caves enlarge, in the same way that terraces widen. When the river incises, the horizontal cave passages are abandoned; sometimes caves form narrow sinuous canyons that match the river's incision rate, or just as often the groundwater chooses a new flow path and abandons the old cave entirely. If the river aggrades, then the caves can fill with sediment, just as terraces are capped with alluvium. However, caves offer an advantage over terraces, in that erosion and weathering is negligible underground. The cave passage, because it is formed in bedrock, leaves a faithful record of river level that is not lowered or degraded over time. Although caves are sometimes more difficult to find than terraces, the record of river incision and aggradation contained within them may offer much higher resolution than from terraces alone. Moreover, caves offer an additional advantage, in that their sediments can be readily dated with cosmogenic nuclides.

In a survey of 55 caves along the New River, Granger et al. (1997) found five that contained in situ quartz-pebble alluvium. Using the burial ages calculated from equation 13, they showed that the New River is incising at ~27 m/m.y., at a pace similar to regional erosion rates.

One nagging problem remains unresolved with the New River study. When Granger et al. (1997) measured ^{26}Al and ^{10}Be in modern river sediments, they found evidence of prior burial. The quartz sediments have an apparent burial age of ~100,000 yr. It is possible that this inherited burial age reflects the travel time of quartz down the New River, as it was deposited in point bars and later reworked. If this is the case, then the cave sediments should have 100,000 yr subtracted from their age. A similar problem of prior burial, discussed above, was found for the beach sediments studied by Boaretto et al. (2000) in Israel. However, Granger et al. (1997) point out that sediment transport and sediment sources in the modern river have been seriously modified by land use. A dam now prevents quartz pebbles from the mountain source area from reaching the limestone in the Valley and Ridge. Moreover, entire cities are built on old river terraces, and erosion due to construction has introduced old gravels into the river. It is not clear that the modern river sediments are a suitable analog for those from the past. A better analog in this case would be sediments deposited only 100–200 yr ago, prior to major land use but recent enough to be little affected by cosmogenic nuclide production and decay.

Case 2: Mammoth Cave, Kentucky. The Mammoth Cave system is the largest, and arguably the most studied, cave system in the world (Granger et al., 2001). Mammoth Cave has developed in a ridge of nearly horizontally bedded limestones capped by a protective cover of sandstone and quartz-pebble conglomerate. The cave formed by carrying water from a karst plain to discharge as springs on the Green River. Sinking streams carry abundant quartz gravel into passages containing underground streams many kilometers long. In this way, Mammoth Cave is different from the caves studied on the New River, Virginia. There, the small caves were primarily repositories of river sediment that washed into the caves through happenstance. At Mammoth Cave, the quartz sediments were carried by the cave-forming waters themselves. The Mammoth Cave system is an example of a multi-level cave. Each level corresponds to a time when the Green River was stable, and the water table remained at the same elevation for a long period of time. The cave morphology shows that these long episodes of river stability were punctuated by rapid incision, when the Green River cut to a lower elevation. Sometimes, river aggradation would fill the cave passages with thick sediment packages. This is an ideal setting for a geologic application of burial dating, because the timing of river incision is thought to be linked to climate history (Miotke and Palmer, 1972).

Granger et al. (2001) collected sediment from the entire vertical extent of the Mammoth Cave system. They were able to identify seven major events in the cave's development, and to link most of these with the expansion of the Laurentide Ice Sheet and associated drainage rearrangements within the Ohio River system. Major river incision events were associated with the earliest glaciations at ca. 2 Ma, and with a particularly extensive glaciation that was responsible for integrating the Ohio River system along the ice margin at ca. 1.5 Ma. Importantly, sediment from outside the cave had a burial age of zero, as expected. In contrast to the quartz pebbles in the New River, these sediments came from a local source and thus had little opportunity for long-term deposition and remobilization. The data from this study are shown in Figure 3 and fall clearly along a radioactive decay line, indicating that the sediment entering the cave maintained a uniform inheritance over millions of years. Similar results were obtained by Anthony and Granger (2004), who dated sediments in large caves along the Cumberland River, Tennessee. These large caves are associated with periods of river stability and were abandoned in sequence as the Cumberland River incised in response to climate change and drainage rearrangement.

The Mammoth Cave study made several advances in the development of the burial dating methodology. First, it demonstrated that burial ages are reproducible to within analytical uncertainty, helping to confirm that a simple sediment burial scenario is appropriate. Sediment was not remobilized from one cave level to another, and the sediment entered the cave with little or no history of prior burial. Second, it extended the burial dating record to beyond 4 Ma. At these ages, the remaining concentrations of ^{26}Al and ^{10}Be are small, and they are particularly susceptible to postburial production of cosmogenic nuclides. Third, it demonstrated that ^{26}Al and ^{10}Be can be used to determine a record of paleoerosion rates. Erosion rates in the Mammoth Cave area had remained slow and unvaried for millions of years. Despite these apparent successes, it must be emphasized that cave sediments should be collected and interpreted cautiously; the sediment source area should be identified, and the possibility of sediment remobilization in the cave should be minimized.

Case 3: Sierra Nevada, California. Other cave sediment studies soon followed. The most notable of these is that of Stock et al. (2004), who measured the incision rate of five different rivers in the Sierra Nevada, California. Although the Sierra Nevada are a batholith, remnants of marble country rock are scattered through the range. Stock et al. (2004) dated granitic river sediment found in ten caves now perched high in canyon walls. These caves could be used to reconstruct the river incision history in much the same way as the work of Granger et al. (1997) on the New River. Stock et al. (2004) found that river incision rates were rapid from 3 to 1.5 Ma, but were drastically reduced after that time. They posited that rapid incision was a transient response to tectonic uplift, and that the incision later slowed as rivers were choked with sediment due to glaciation in the headwaters.

Case 4: Sterkfontein, South Africa. All of the above cave-dating studies used caves to infer long-term river incision rates. This is a natural application of caves in geomorphology, because cave formation is linked to the regional water table and is controlled by river incision and aggradation. However, caves are often interesting for other reasons as well. For example, caves often contain fossils, including those of our own ancestors. As suggested by Lal and Arnold (1985), the burial dating method is applicable over a timescale that spans hominin evolution, and is

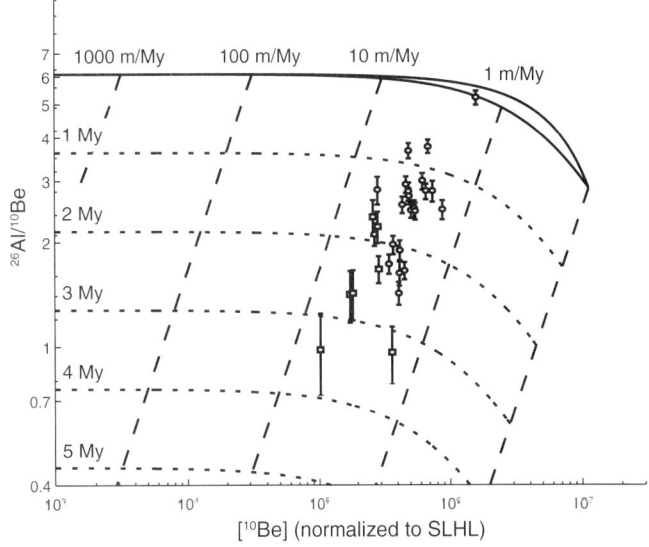

Figure 3. Data from Mammoth Cave, Kentucky, show burial ages ranging from 0 to 4 Ma. The data spread parallel to the radioactive decay lines, indicating that local erosion rates remained nearly constant over this time. Modified from Granger et al. (2001).

thus ideal for dating hominin-bearing caves. The caves at Sterkfontein, South Africa, offer a good example of this application.

Sterkfontein is the source of more hominin fossils than any other site in the world. It is the source of most *Australopithecus africanus* fossils ever found and thus holds an important place in paleoanthropology. The only previous dating of the cave sediments, however, was by paleomagnetic stratigraphy of intercalated flowstone (Partridge et al., 1999). Because paleomagnetic stratigraphy does not provide absolute dates due to the possibility of missing sections, it was important to determine radiometric ages for the sediments as well. The sediments and the fossils originated on the surface, and fell into a vertical cave entrance, forming a debris cone on the cave floor (Partridge, 1978). The soil at the site contains quartz that is derived from a nearby sandstone; this quartz thus forms much of the sediment that encases the hominin fossils and can be dated using ^{26}Al and ^{10}Be.

Partridge et al. (2003) measured ^{26}Al and ^{10}Be in five samples from the caves at Sterkfontein, including two from a new fossil find, plus an additional sample from the surface. These samples indicated that the cave sediments were approximately four million years old, substantially older than previously suspected.

Dating with Cosmogenic Nuclide Profiles

Caves provide the simplest possible scenario for burial dating, because the sediments are very deeply buried and cosmogenic nuclide production effectively ceases. However, sediment is often buried beneath only several meters of overburden. In this case, the evolution of the ^{26}Al and ^{10}Be concentrations is not as simple as that of deep burial. In addition to radioactive decay, one must also account for postburial production, which is depth- and time-dependent. The relevant equation to be solved, for the case of a steadily eroding deposit, becomes

$$N(z,t) = N_{inh}e^{-t/\tau} + \int_0^t P(z+\rho\varepsilon t')e^{-t'/\tau}dt', \qquad (14)$$

where N_{inh} is the inherited concentration prior to deposition, and z is depth. Equation 14 includes at least three unknowns; these are the depositional age, the inherited cosmogenic nuclide concentration, and the time dependence of the postburial production rate, which is equivalent to the erosion rate in steady state. This equation cannot be solved using only two cosmogenic nuclide measurements from a single sample. However, the equation can be solved using multiple samples in a vertical profile.

If all of the sediment in a profile was deposited quickly with respect to cosmogenic nuclide production and radioactive decay, then every sediment sample from that profile must have the same age as well as the same erosion rate. On the other hand, it is not necessary that each sample have the same inheritance. Equation 14 can be solved exactly for two samples from the same profile, because there are four nuclide measurements and four unknowns (these are the depositional age, erosion rate, and each sample's inheritance). It is better to collect several samples to overconstrain the model solution. It is necessary that the samples be taken from sufficiently different depths that there is significantly different postburial production. In practice, this requires depths of at least 5–10 m (Wolkowinsky and Granger, 2004). An age determination is then made by using equation 14 to reproduce the measured concentrations of ^{26}Al and ^{10}Be in every sample. The best-fit age may be determined by chi-square minimization, in much the same manner as previously used to constrain complex exposure histories of meteorites (e.g., Nishiizumi et al., 1991a, 1996).

Case 1: Kentucky River. The first use of a cosmogenic nuclide profile to date sediment burial was by Granger and Smith (2000), who measured ^{26}Al and ^{10}Be in nine samples from a 10 m profile in sandy quartzose sediments on a terrace of the Kentucky River. The terrace sediments were deposited prior to a major river incision event that had been tied to Laurentide glaciation. Granger and Smith (2000) realized that postburial production due to deeply penetrating muons could be significant even at depths of 10 m. They parameterized the production rate profiles of Stone et al. (1998) and Heisinger (1998) in order to calculate the muogenic contribution to each sample. The importance of this correction is revealed by the model results, which reveal that even for the most shielded sample at 10 m depth, fully 1/3 of the measured ^{26}Al is due to postburial muogenic production. Nevertheless, by fitting a model to the observations, they were able to date the terrace's deposition to ca. 1.5 ± 0.3 Ma. This age agrees very well with the timing of incision inferred for a similar event at Mammoth Cave, Kentucky (Granger et al., 2001).

Case 2: San Juan River. A second example of dating terrace sediments with a cosmogenic nuclide profile is that of Wolkowinsky and Granger (2004), who dated a high-level alluvial deposit above the San Juan River, Utah. The San Juan River is a tributary of the Colorado River, which it joins upstream of Grand Canyon at Glen Canyon. Wolkowinsky and Granger (2004) measured ^{26}Al and ^{10}Be in two profiles. The first was a profile of seven samples nearly 12 m deep; the second was a profile of five samples over 5 m deep. In retrospect it was found that the 5 m profile was insufficiently deep to constrain the terrace's age, because too large a fraction of the cosmogenic nuclides at depth was due to postburial production; however, the 12 m profile was sufficiently deep to obtain a burial age. The terrace age was found to be $1.36^{+0.20}_{-0.15}$ Ma.

Although the methodology of Wolkowinsky and Granger (2004) was nearly identical to that of Granger and Smith (2000), they used a different graphical illustration of uncertainty that is worthy of discussion. It is difficult to visualize the goodness of fit for a profile using the exposure-burial diagram, for two reasons. First, each sample in a profile follows a different path after burial due to differences in postburial production. Second, each sample may be deposited with a different cosmogenic nuclide inheritance. It is therefore not possible to uniquely determine a burial age using the exposure-burial diagram alone. The goodness of fit between the model and the data is revealed by the chi-square statistic. Wolkowinsky and Granger (2004) plotted the chi-square surface as a function of the model terrace age and the model terrace erosion rate (Fig. 4). The chi-square approach has several advantages. (1) The best-fit age and erosion rate are apparent by the minimum on the chi-square surface. Trade-offs between age and erosion rate are immediately apparent. (2) The reduced chi-square value can be used to quantitatively estimate the likelihood that the model fully explains the data. If a poor fit is obtained, then it is likely that the profile does not match the assumptions in the model. For example, the profile may consist of two separate deposits of different age. (3) The curvature of the chi-square surface can be used to reveal the uncertainty in the modeled age and erosion rate. Following Bevington and Robinson (2003), one standard error of uncertainty is circumscribed on the chi-square surface where it increases by one from its lowest value. The uncertainty in a depositional age obtained in this way is very similar to that obtained by Monte Carlo methods. Siame et al. (2004) used a similar chi-square method to constrain erosion rates and exposure ages from profiles of a single nuclide.

Case 3: Paleosol Profile Beneath Glacial Till. The examples of dating river terraces using cosmogenic nuclide profiles each assumed that the sediment was deposited quickly and in a single episode (although the episode could have occurred over hundreds or thousands of years). However, more complex situations can be easily modeled, as long as they are well constrained by stratigraphy. For example, multiple sedimentary packages may be deposited in a sequence. Each sedimentary package will have its own depositional age and a burial history that is determined by the depositional age of the overlying packages. The problem is reminiscent of the complex meteorite exposure history modeled by Nishiizumi et al. (1991a), in that the samples have experienced a history of discrete events.

An excellent example of dating a sedimentary sequence is the recent work of Balco et al. (2005). These authors dated a paleosol buried beneath till of the Laurentide Ice Sheet. The paleosol developed over a long period of exposure prior to glaciation, so it contains exceptionally high cosmogenic nuclide concentrations. Because the paleosol indicated bedrock weathering in place, it was assumed that the cosmogenic nuclides were in equilibrium with erosion prior to burial (equation 12). It was buried beneath a stack of two tills capped by a loess deposit. Both tills retained paleosols as well, indicating that each was exposed for a long period of surface exposure and that the there was little erosion associated with successive till or loess deposition. Balco et al. (2005) constructed a three-step burial history, with each step corresponding to rapid burial followed by slow surface erosion. They then used a Monte Carlo approach, minimizing the chi-square deviation between model and measurements to determine the age of the paleosol most consistent with the data. They found that the lowermost till was emplaced at 2.41 ± 0.14 Ma. This age demonstrates that the earliest Laurentide glaciations inferred from the marine oxygen isotope record extended equally as far south as more recent ice sheet advances.

REMAINING ISSUES: THE HALF-LIFE PROBLEM

There remain several uncertainties that seriously affect burial dating. Chief among these is the choice of half-lives for ^{26}Al and ^{10}Be. In particular, the half-life of ^{10}Be remains ambiguous. This section reviews the half-life of each radionuclide, but special attention is given to that of ^{10}Be.

The half-life of ^{26}Al was first determined by Rightmire et al. (1959) from mass spectrometric and positron activity measurements of ^{26}Al produced by cyclotron radiation. The value was later adjusted by Samworth et al. (1972), who revised the branching ratio of ^{26}Al decay, yielding a value of $(7.16 \pm 0.32) \times 10^5$ yr. This half-life was later brought into question by Nishiizumi et al. (1980), who compared ^{21}Ne and ^{26}Al in ordinary chondrites with exposure ages less than 3 Ma, and found a significant deficit of ^{26}Al. Nishiizumi et al. (1980) suggested that the ^{26}Al half-life may be in error. In response to this proposal, three independent research groups undertook the measurement of the ^{26}Al half-life.

A half-life is best determined by simultaneous measurement of activity and the radionuclide concentration in a single sample. Norris et al. (1983) remeasured the half-life of ^{26}Al to $(7.05 \pm 0.24) \times 10^5$ yr, by measuring specific activity and ^{26}Al concentrations using mass spectrometric isotope dilution in a solution derived from proton-irradiated silicon. Thomas et al. (1984) measured $(7.8 \pm 0.5) \times 10^5$ yr by producing a known amount of ^{26}Al

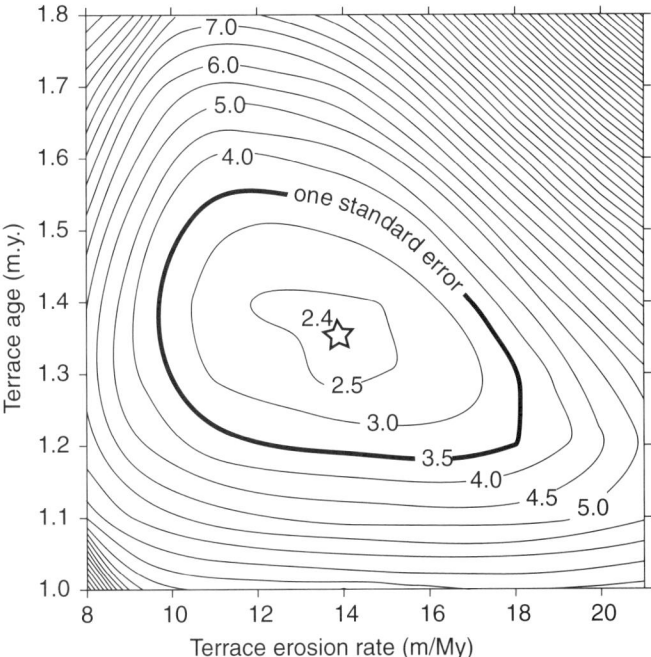

Figure 4. Chi-square contours shown as a function of terrace age and terrace erosion rate, for high-level alluvium above the San Juan River, Utah. The minimum chi-square value represents the best model fit and is shown as a star. The bold contour at a value of one higher than the minimum circumscribes one standard error in the model estimation. Modified from Wolkowinsky and Granger (2004).

from proton bombardment of ^{26}Mg and measuring the subsequent activity. Middleton et al. (1983) measured the half-life to be (7.02 ± 0.56) × 10^5 yr, using absolute AMS to determine the ^{26}Al/^{27}Al ratio in standards of independently certified activity. All three of these values agree with the earlier estimate of Samworth et al. (1972); they have a weighted average of (7.17 ± 0.17) × 10^5 yr. This value was further tested by Wallner et al. (2000), who evaluated AMS standards based on activity measurements, and found agreement to within the 5% experimental error. The half-life of ^{26}Al thus seems to be firmly established to within 3%.

The half-life of ^{10}Be remains far more questionable. It was first measured by Hughes et al. (1947), who analyzed neutron-irradiated BeO and reported a value of 2.9 m.y. That same year, McMillan (1947) reported a similar value of 2.5 ± 0.5 m.y. from deuteron-irradiated beryllium. Both of these values were later lowered by nearly half. Emery et al. (1972) used revised cross sections to lower the half-life of Hughes et al. (1947) to 1.6 m.y. These authors also reported additional specific activity measurements yielding half-lives of 1.3 and 1.9 m.y., with an average of 1.6 ± 0.2 m.y. McMillan (1972) later discovered that he had inadvertently reported a mean-life rather than a half-life, so his previous value was lowered by a factor of ln(2) to (1.7 ± 0.4) × 10^6 yr. Previously that same year, Yiou and Raisbeck (1972) had determined a half-life of 1.5 ± 0.3 m.y. by irradiating carbon and producing both ^{7}Be and ^{10}Be. They calculated the half-life from the known half-life of ^{7}Be and the relative cross sections of the two nuclides. Later, Makino et al. (1975, in Japanese, cited in Hofmann et al., 1987) measured a half-life of (1.48 ± 0.15) × 10^6 yr, although I am not familiar with the details of their method. By the mid-1970s, most estimates had converged on a ^{10}Be half-life near 1.5 m.y., although considerable scatter remained.

Prompted by the need for a better AMS standard, a new effort was begun in the mid-1980s to redetermine the half-life of ^{10}Be. This effort led to disparate estimates of the ^{10}Be half-life, so the procedures are recounted in some detail. A sample of neutron-irradiated beryllium shim-rod from Idaho National Engineering Laboratory was dissolved and then enriched at Oak Ridge National Laboratory. The solution was diluted with water to produce a solution known as ORNL-A. Half of this solution was then spiked with beryllium to produce a solution for isotope dilution known as ORNL-MASTER (Inn et al., 1987). The ORNL-MASTER solution was then split, with 20% of the solution going to a separate research group (Hofmann et al., 1987) for dilution and analysis. The remaining half of ORNL-A was again diluted with pure beryllium, and its isotopic ratio measured by a combination of secondary ion mass spectrometry (SIMS) and AMS. A series of four dilutions led ultimately to an AMS standard available as standard reference material (SRM) 4325. It has a certified ^{10}Be/^{9}Be ratio of 2.68 × 10^{-11}. The published dilutions for this standard (Inn et al., 1987; Hofmann et al., 1987) are in perfect agreement with this value (although the ratio originally reported by Inn et al., 1987, is different). The ^{10}Be half-life was determined from activity measurements on both the ORNL-A and ORNL-MASTER solutions.

The ORNL-MASTER aliquot was diluted by Hofmann et al. (1987), and the activity of the diluted solution measured by liquid scintillation counting. Using the beryllium concentration and isotope ratio derived from the isotope dilution from ORNL-A (Inn et al., 1987), they determined the half-life of ^{10}Be to be (1.51 ± 0.06) × 10^6 yr.

The activity of the ORNL-A solution was determined by Inn et al. (1987) using liquid scintillation counting. They determined the half-life of ^{10}Be to be (1.34 ± 0.07) × 10^6 yr (as reported in documentation for SRM 4325), significantly different from the determination by Hofmann et al. (1987).

Two measurements of activity from the same parent solution led inexplicably to two disparate determinations of the ^{10}Be half-life. In an attempt to resolve this discrepancy, Middleton (1993) estimated the ^{10}Be/^{9}Be ratio in SRM 4325 using absolute AMS. The accuracy of this method is difficult to evaluate, given the possibility of undetected systematic error. Nevertheless, Middleton (1993) determined the ^{10}Be/^{9}Be ratio in the standard to be (3.06 ± 0.14) × 10^{-11}, in agreement with the measurements reported by Hofmann et al. (1987). Using this isotopic ratio and the activities inferred from SRM 4325, Middleton (1993) derived a ^{10}Be half-life of (1.53 ± 0.07) × 10^6 yr, in very good agreement with the earlier estimate of Hofmann et al. (1987). It was suggested that the isotopic ratio of SRM 4325 is in error, despite the documented dilution series. The community has largely adopted this interpretation by normalizing all measurements to the higher value when reporting ages and production rates, although some AMS laboratories continue to use the SRM 4325 standard.

It is important to consider an additional line of evidence when evaluating the ^{10}Be half-life. Widely used and accepted AMS standards have been prepared by K. Nishiizumi from dilutions of ICN solutions of known activity (e.g., Nishiizumi et al., 1991b). The isotopic ratios of these standards are based upon an assumed ^{10}Be half-life of 1.5 × 10^6 yr. A comparison between the Nishiizumi standards and standards diluted from SRM 4325 show a discrepancy of 14%, a perfect mirror of the discrepancy in the half-life (Middleton, 1993; unpublished standards calibration at PRIME Lab). The coincidence in this offset persuasively suggests that the activity of SRM 4325 was correctly measured. There remain two possible resolutions to the half-life discrepancy. (1) The reported isotopic ratio of SRM 4325 could be too low by 14%, despite the documented dilution series. This would imply that the correct half-life is 1.5 m.y., but it would also call into question the measurements of Hofmann et al. (1987), who used the isotopic ratio of the parent solution to calculate the half-life. (2) The isotopic ratio and half-life of SRM 4325 could be correct. The ^{10}Be half-life remains disputed, with two very different estimates, each reported to high precision.

Apart from remeasurement of the ^{10}Be half-life from a new parent solution, there are geologic tests that can be used to discriminate between the two competing half-lives. Recall that Klein et al. (1986) used the exposure-burial triangle to constrain production rates, assuming that half-lives were known. The same approach can be used to constrain half-lives, assuming that production rates are known. Figure 5 shows two versions of the exposure-burial diagram with measurements compiled from published values of ^{26}Al and ^{10}Be near saturation. The figure contains all published ^{26}Al-^{10}Be pairs that I know of with ^{10}Be concentrations higher than 4×10^6 atoms/g. The ^{26}Al and ^{10}Be concentrations primarily come from Antarctica; they have been scaled to sea level, high latitude (SLHL) using the algorithm of Stone (2000), so that all of the values can be plotted on the same graph. The first graph (Fig. 5, top) is drawn using the currently accepted ^{10}Be half-life of 1.51 m.y. The second graph (Fig. 5, bottom) is drawn using the lower half-life of 1.34 m.y. It is important to note that when the half-life is changed, all of the measurements must be changed as well, because they were measured against standards calibrated by activity. Thus, all ^{10}Be concentrations must be lowered by 14%, and likewise the ^{26}Al/^{10}Be ratios must be raised by 14%.

The data in Figure 5 are consistent with either half-life of ^{10}Be, although they are more tightly bound by the lower half-life. In fact, all of the data are consistent to within one standard error with the shorter half-life. Moreover, assuming that the scaling algorithm of Stone (2000) is correct, the longer half-life is inconsistent with any of the samples being at the saturation endpoint. Although it is not possible to rule out either half-life at this point, the saturation endpoint shown on Figure 5 lends some support to the shorter half-life.

It is possible to carry this analysis one step further, by using the data to calculate a half-life. If any sample is independently known to be saturated with respect to both ^{26}Al and ^{10}Be, then its ratio can be used to calculate the relative half-lives of these two nuclides. Equations 15 and 16 can be used to determine the half-life, given a measured ratio R_{meas}:

$$R_{meas} (2.18 \text{ m.y.}/\tau_{new}) = (P_{26}/P_{10}) (2.18 \text{ m.y.}/\tau_{new}) (\tau_{26}/\tau_{new}), \quad (15)$$

$$\tau_{new} = (P_{26}\tau_{26}/P_{10}R_{meas}). \quad (16)$$

For example, one can use the ^{26}Al/^{10}Be ratio closest to saturation on Figure 5, determined for a sample from the Atacama Desert (Nishiizumi et al., 2005). This sample has a ratio of 3.15 ± 0.10. Using equation 16, this value yields a ^{10}Be half-life of 1.37 ± 0.04 m.y., in agreement with the value determined for SRM 4325. Although it is difficult to prove that any single sample is at saturation, it is compelling that the closest value ever published corresponds within error to the half-life of 1.34 m.y.

Other geologic tests have been used to estimate the ^{10}Be half-life, although it must be emphasized that geologic uncertainties are difficult to evaluate. Bourlès et al. (1989) examined authigenic ^{10}Be/^{9}Be ratios in a marine sediment core whose age was determined independently by magnetostratigraphy. The core spanned ~9 m.y. of deposition, giving sufficient time for ^{10}Be to decay by two orders of magnitude. Assuming that the ^{10}Be/^{9}Be ratio in the authigenic component remains constant through time, the half-life can be calculated against the magnetic polarity timescale.

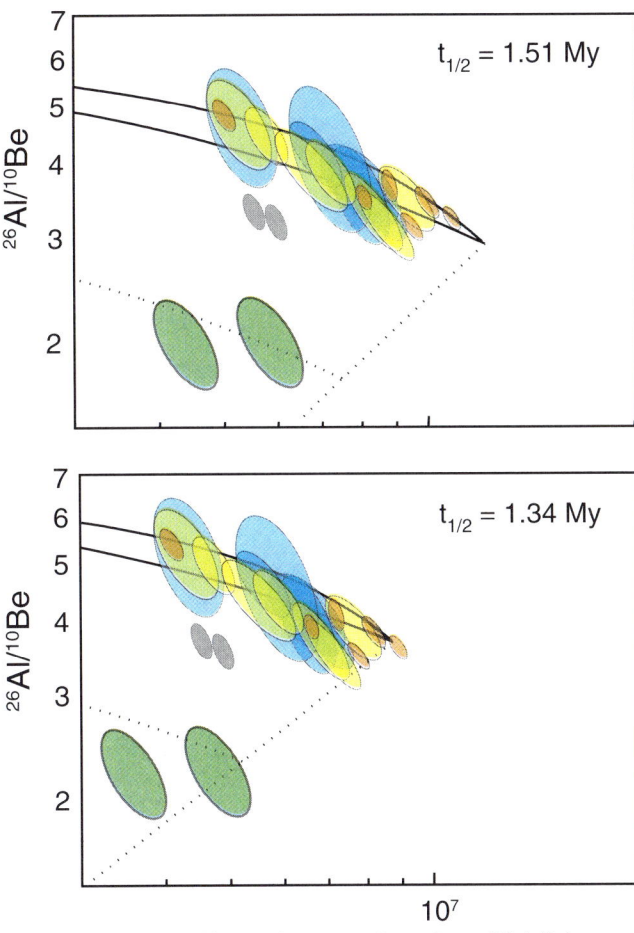

Figure 5. Exposure-burial diagrams showing published data with [^{10}Be] (SLHL) greater than 4×10^6 atoms/g. The upper diagram is constructed using a ^{10}Be half-life of 1.51 m.y., and the lower diagram is constructed using a half-life of 1.34 m.y. All data have been scaled to sea level, high latitude (SLHL) using the algorithm of Stone (2000). All data should lie within the exposure-burial triangle, and it is expected that the data will approach the secular equilibrium endpoint. The data are consistent with either half-life, although they are more tightly bound by the shorter half-life as shown in the lower diagram. Red—Nishiizumi et al. (2005); gray—Bierman and Caffee (2001); green—Klein et al. (1986); yellow—Ivy-Ochs et al. (1995); blue—Brook et al. (1995).

The magnetic stratigraphy of the core is clear for the Pliocene and Pleistocene; specific correlations beyond that time are more uncertain. Bourlès et al. (1989) separated the data into two groups by age and regressed the ^{10}Be/^{9}Be ratio against sedimentation time to infer the half-life. Using the accepted magnetic timescale at the time, they obtained a half-life of 1.26 ± 0.02 m.y. over the range 0–5.8 Ma; for the range 5.8–8.9 Ma, they obtained a half-life of 1.56 ± 0.19 m.y. A second data set from the same core gave a best-fit half-life of 1.43 ± 0.03 m.y. over the entire range. The magnetic timescale has improved significantly since Bourlès et al. (1989) constructed their sedimentation curve. Adopting the orbitally tuned magnetic polarity timescale of Ogg and Smith (2004), and using only polarity chrons 1, 2, and 2A for which correlations are very clear, I obtain a ^{10}Be half-life of 1.33 ± 0.03 m.y. using the data of Bourlès et al. (1989). Using the data set up to 8.2 Ma, for which the paleomagnetic stratigraphy is more difficult to reliably correlate, leads to a probable half-life of 1.34 ± 0.07 m.y. Considered together, the half-lives estimated from marine sediments are more consistent with 1.34 m.y. than 1.51 m.y.

Geologic tests such as the ones discussed here can never substitute for physical measurements of the ^{10}Be half-life. Until the half-life issue is resolved, it is recommended that dates determined using the radioactive decay of ^{10}Be be published with two values, corresponding to the two disparate half-lives of 1.51 ± 0.06 m.y. and 1.34 ± 0.07 m.y.

SUMMARY

The measurement of multiple cosmogenic nuclides with different production rates, half-lives, and depth dependencies has been an integral part of cosmogenic nuclide research since the very beginnings of the field. These methodologies were initially developed to study meteorites, which by virtue of their exposure in space contain a host of cosmogenic nuclides in concentrations far exceeding those of terrestrial rocks. The study of meteorites led naturally to the use of multiple nuclides, because the events being dated occurred over timescales long with respect to radioactive decay. Thus, multiple nuclides with different half-lives could yield information about changes in production rate through time.

The measurement of terrestrial cosmogenic nuclides began in the 1980s following the advent of accelerator mass spectrometry. Early papers in this new field reflected the mind-set, cultivated by experience in meteoritics, that multiple cosmogenic nuclides measured in the same rock or mineral would best yield information about the exposure and burial histories. Proposed applications included dating buried sediments, and dating hominin evolution over millions of years with the ^{26}Al-^{10}Be pair in quartz. However, the early measurements of buried sediments revealed that exposure and burial histories in the terrestrial environment were often complex, and difficult to interpret in terms of absolute dates.

The utility of cosmogenic nuclides for surface exposure dating, rather than burial dating, quickly led to its application in younger environments where radioactive decay is unimportant. In these cases, interpreting exposure ages or erosion rates is relatively straightforward. Multiple cosmogenic nuclides such as ^{26}Al and ^{10}Be were still analyzed, but usually only to confirm that a sample had a simple exposure history. The presence of a depressed ^{26}Al/^{10}Be ratio was used to indicate that a rock or sediment had been buried, but could only constrain a minimum burial age. Thus, the use of multiple cosmogenic nuclides for explicitly dating sediment burial histories languished for a decade, while the geologic community focused instead on measuring and interpreting exposure ages.

The past ten years have seen a renewal of cosmogenic nuclide applications to dating buried sediments. This was led by the realization that quartz-bearing cave sediments represent a simple history of exposure followed by deep, instantaneous burial. Thus, the interpretation of burial ages in caves is seldom complicated by considerations of sediment recycling or partial shielding. The sediment ages represent a clear geomorphic event: the deposition of sediment in the cave. Moreover, because caves are formed by groundwater, the sediment ages can often be interpreted in terms of local river incision histories. The method has even proven useful for dating hominin- or other fossil-bearing deposits in caves, which were previously difficult or impossible to date.

More recently, multiple cosmogenic nuclides have been used to date sediments buried in other environments, such as alluvial terraces or beneath glacial till. In these applications, cosmogenic nuclide production does not completely cease after burial, but must be accounted for by explicitly modeling depth profiles and long-term production by muons. The dating of buried paleosols and the recognition of paleosurfaces represents an important advance, and will undoubtedly play an important role in the near future.

The future of dating with multiple cosmogenic nuclides was anticipated by the first practitioners in the terrestrial field, who envisioned a path similar to that forged in the meteoritic field. Under this vision, an increasingly large array of both stable and radioactive cosmogenic nuclides will be measured at single sites or in single samples. These multiple cosmogenic nuclides may take advantage of variations in radioactive decay, variations in production rates with depth, or even variations in production rates with altitude to explore the exposure, burial, and uplift history of rocks and sediment. Although future applications will undoubtedly benefit from advances in measurement and detection, as well as in constraining production rates with time, altitude, and depth, they are just as dependent on the recognition of geologic scenarios that are amenable to dating. Nearly every rock on the surface of Earth contains multiple cosmogenic nuclides; the future lies not only in the development of analytical methods to measure these nuclides, but in the creativity of geologists and geomorphologists to discover new ways to interpret them in terms of meaningful exposure and burial histories, as well as rates of surface processes.

ACKNOWLEDGMENTS

This work was partly supported by National Science Foundation grants EAR-0092459 and EAR-0242797. I thank two anonymous reviewers for their suggestions, and David Fink for his advice.

REFERENCES CITED

Albrecht, A., Herzog, G.F., Klein, J., Dezfouly-Arjomandy, B., and Goff, F., 1993, Quaternary erosion and cosmic-ray-exposure history derived from ^{10}Be and ^{26}Al produced in situ—An example from Pajarito Plateau, Valles caldera region: Geology, v. 21, p. 551–554, doi: 10.1130/0091-7613(1993)021<0551:QEACRE>2.3.CO;2.

Anthony, D.M., and Granger, D.E., 2004, A Late Tertiary origin for multi-level caves along the western escarpment of the Cumberland Plateau, Tennessee and Kentucky, established by cosmogenic ^{26}Al and ^{10}Be: Journal of Cave and Karst Studies, v. 66, p. 46–55.

Arnold, J.R., 1961, Nuclear effects of cosmic rays in meteorites: Annual Reviews of Earth and Planetary Sciences, p. 349–370.

Arnold, J.R., Honda, M., and Lal, D., 1961, Record of cosmic ray intensity in the meteorites: Journal of Geophysical Research, v. 66, no. 10, p. 3519–3531.

Arrol, W.J., Jacobi, R.B., and Paneth, F.A., 1942, Meteorites and the age of the solar system: Nature, v. 149, p. 235–238.

Balco, G., Rovey, C.W., II, and Stone, J.O.H., 2005, The first glacial maximum in North America: Science, v. 307, p. 222, doi: 10.1126/science.1103406.

Bauer, C.A., 1947, Production of helium in meteorites by cosmic radiation: Physical Review, v. 72, p. 354–355, doi: 10.1103/PhysRev.72.354.

Begemann, F., Geiss, J., and Hess, D.C., 1957, Radiation age of a meteorite from cosmic-ray-produced He-3 and H-3: Physical Review, v. 107, no. 2, p. 540–542, doi: 10.1103/PhysRev.107.540.

Bevington, P.R., and Robinson, D.K., 2003, Data reduction and error analysis for the physical sciences: New York, McGraw-Hill Higher Education, 336 p.

Bierman, P.R., and Caffee, M.W., 2001, Slow rates of rock surface erosion and sediment production across the Namib Desert and escarpment, southern Africa, in Pazzaglia, F.J., and Knuepfer, P.L.K., eds., The steady-state orogen: Concepts, field observations, and models: New Haven, Connecticut, Yale University, Kline Geology Laboratory, American Journal of Science, v. 301 (Special Issue), p. 326–358.

Bierman, P.R., Marsella, K.A., Patterson, C., Davis, P.T., and Caffee, M., 1999, Mid-Pleistocene cosmogenic minimum-age limits for pre-Wisconsinan glacial surfaces in southwestern Minnesota and southern Baffin Island: A multiple nuclide approach: Geomorphology, v. 27, p. 25–39, doi: 10.1016/S0169-555X(98)00088-9.

Boaretto, E., Berkovits, D., Hass, M., Hui, S.K., Kaufman, A., Paul, M., and Weiner, S., 2000, Dating of prehistoric caves sediments and flints using ^{10}Be and ^{26}Al in quartz from Tabun Cave (Israel): Progress report: Nuclear Instruments and Methods in Physics Research, Section B, Beam Interactions with Materials and Atoms, v. 172, p. 767–771, doi: 10.1016/S0168-583X(00)00349-9.

Bourlès, D., Raisbeck, G.M., and Yiou, F., 1989, ^{10}Be and ^{9}Be in marine sediments and their potential for dating: Geochimica et Cosmochimica Acta, v. 53, p. 443–452, doi: 10.1016/0016-7037(89)90395-5.

Braucher, R., Bourlès, D.L., Brown, E.T., Colin, F., Muller, J.-P., Braun, J.-J., Delaune, M., Edou Minko, A., Lescouet, C., Raisbeck, G.M., and Yiou, F., 2000, Application of in situ-produced cosmogenic ^{10}Be and ^{26}Al to the study of lateritic soil development in tropical forest: theory and examples from Cameroon and Gabon: Chemical Geology, v. 170, p. 95–111.

Brook, E.J., Brown, E.T., Kurz, M.D., Ackert, R.E., Jr., Raisbeck, G.M., and Yiou, F., 1995, Constraints on age, erosion, and uplift of Neogene glacial deposits in the Transantarctic Mountains determined from in situ cosmogenic ^{10}Be and ^{26}Al: Geology, v. 23, p. 1063–1066, doi: 10.1130/0091-7613(1995)023<1063:COAEAU>2.3.CO;2.

Brown, E.T., Brook, E.J., Raisbeck, G.M., Yiou, F., and Kurz, M.D., 1992, Effective attenuation lengths of cosmic rays producing ^{10}Be and ^{26}Al in quartz: Implications for exposure age dating: Geophysical Research Letters, v. 19, no. 4, p. 369–372.

Davis, R.J., and Schaeffer, O.A., 1955, Chlorine-36 in nature: Annals of the New York Academy of Sciences, v. 62, no. 5, p. 105–122.

Emery, J.F., Reynolds, S.A., Wyatt, E.I., and Gleason, G.I., 1972, Half-lives of radionuclides, IV: Nuclear Science and Engineering, v. 48, p. 319–323.

Fabel, D., Stroeven, A.P., Harbor, J., Kleman, J., Elmore, D., and Fink, D., 2002, Landscape preservation under Fennoscandian ice sheets determined from in situ produced ^{10}Be and ^{26}Al: Earth and Planetary Science Letters, v. 201, p. 397–406, doi: 10.1016/S0012-821X(02)00714-8.

Fireman, E.L., and Schwarzer, D., 1957, Measurement of Li-6, He-3, and H-3 in meteorites and its relation to cosmic radiation: Geochimica et Cosmochimica Acta, v. 11, p. 252–262, doi: 10.1016/0016-7037(57)90098-4.

Gosse, J.C., and Phillips, F.M., 2001, Terrestrial in situ cosmogenic nuclides: Theory and application: Quaternary Science Reviews, v. 20, p. 1475–1560, doi: 10.1016/S0277-3791(00)00171-2.

Granger, D.E., and Muzikar, P.F., 2001, Dating sediment burial with cosmogenic nuclides: Theory, techniques, and limitations: Earth and Planetary Science Letters, v. 188, no. 1-2, p. 269–281, doi: 10.1016/S0012-821X(01)00309-0.

Granger, D.E., and Smith, A.L., 2000, Dating buried sediments using radioactive decay and muogenic production of ^{26}Al and ^{10}Be: Nuclear Instruments and Methods in Physics Research, v. B172, p. 822–826.

Granger, D.E., Kirchner, J.W., and Finkel, R.C., 1997, Quaternary downcutting rate of the New River, Virginia, measured from differential decay of cosmogenic ^{26}Al and ^{10}Be in cave-deposited alluvium: Geology, v. 25, p. 107–110, doi: 10.1130/0091-7613(1997)025<0107:QDROTN>2.3.CO;2.

Granger, D.E., Fabel, D., and Palmer, A.N., 2001, Pliocene-Pleistocene incision of the Green River, Kentucky, determined from radioactive decay of cosmogenic ^{26}Al and ^{10}Be in Mammoth Cave sediments: Geological Society of America Bulletin, v. 113, p. 825–836, doi: 10.1130/0016-7606(2001)113<0825:PPIOTG>2.0.CO;2.

Hampel, W., Tagaki, J., Sakamoto, K., and Tanaka, S., 1975, Measurement of muon-induced ^{26}Al in terrestrial silicate rock: Journal of Geophysical Research, v. 80, no. 26, p. 3757–3760.

Heisinger, B.P., 1998, Myonen-induzierte produktion von radionukliden [Ph.D. thesis]: Technischen Universität Munchen, 153 p.

Hofmann, H.J., Beer, J., Bonani, G., Von Gunten, H.R., Raman, S., Suter, M., Walker, R.L., Wolfli, W., and Zimmerman, D., 1987, ^{10}Be: Half-life and AMS standards: Nuclear Instruments and Methods in Physics Research, v. B29, p. 32–36.

Honda, M., Shedlovsky, J.P., and Arnold, J.R., 1961, Radioactive species produced by cosmic rays in iron meteorites: Geochimica et Cosmochimica Acta, v. 22, p. 133–154, doi: 10.1016/0016-7037(61)90112-0.

Hughes, D.J., Eggler, C., and Huddleston, C.M., 1947, The half-life of Be^{10}: Physical Review, v. 71, p. 269, doi: 10.1103/PhysRev.71.269.

Inn, K.G.W., Raman, S., Coursey, B.M., Fassett, J.D., and Walker, R.L., 1987, Development of the NBS Be-10/Be-9 Isotopic Standard Reference Material: Nuclear Instruments and Methods in Physics Research, Section B, Beam Interactions with Materials and Atoms, v. 29, no. 1-2, p. 27–31, doi: 10.1016/0168-583X(87)90197-2.

Ivy-Ochs, S., Schluchter, C., Kubik, P.W., Dittrich-Hannen, B., and Beer, J., 1995, Minimum ^{10}Be exposure ages of early Pliocene for the Table Mountain Plateau and the Sirius Group at Mount Fleming, Dry Valleys, Antarctica: Geology, v. 23, p. 1007–1010, doi: 10.1130/0091-7613(1995)023<1007:MBEAOE>2.3.CO;2.

Klein, J., Giegengack, R., Middleton, R., Sharma, P., Underwood, J.R.J., and Weeks, R.A., 1986, Revealing histories of exposure using in situ produced ^{26}Al and ^{10}Be in Libyan Desert Glass: Radiocarbon, v. 28, no. 2A, p. 547–555.

Kleman, J., Hattestrand, C., Borgstrom, I., and Stroeven, A.P., 1997, Fennoscandian paleoglaciology reconstructed using a glacial geological inversion model: Journal of Glaciology, v. 43, p. 283–289.

Lal, D., 1987, Cosmogenic nuclides produced in situ in terrestrial solids: Nuclear Instruments and Methods in Physics Research, v. B29, p. 238–245.

Lal, D., 1988, In situ-produced cosmogenic isotopes in terrestrial rocks: Annual Review of Earth and Planetary Sciences, v. 16, p. 355–388.

Lal, D., 1991, Cosmic ray labeling of erosion surfaces: In situ nuclide production rates and erosion models: Earth and Planetary Science Letters, v. 104, p. 424–439, doi: 10.1016/0012-821X(91)90220-C.

Lal, D., and Arnold, J.R., 1985, Tracing quartz through the environment: Proceedings of Indian Academy of Science: Earth and Planetary Science Letters, v. 94, p. 1–5.

Makino, T., Gensho, R., and Honda, N., 1975, Preparation of ^{10}Be and the Measurement of Isotopic Ratio of Beryllium: Shitsuryo Bunseki, v. 23, p. 33.

McMillan, E.M., 1947, Energy and half-life of the Be^{10} radioactivity: Physical Review, v. 72, p. 591–593, doi: 10.1103/PhysRev.72.591.

McMillan, E.M., 1972, Half-life of ^{10}Be: A correction: Physical Review C, v. 6, no. 6, p. 2296, doi: 10.1103/PhysRevC.6.2296.

Middleton, R., 1993, On Be-10 standards and the half-life of Be-10: Nuclear Instruments and Methods in Physics Research, v. B82, p. 399–403.

Middleton, R., Klein, J., Raisbeck, G.M., and Yiou, F., 1983, Accelerator mass spectroscopy with ^{26}Al: Nuclear Instruments and Methods in Physics Research, v. 218, p. 430–438, doi: 10.1016/0167-5087(83)91017-7.

Miotke, F.-D., and Palmer, A.N., 1972, Genetic relationship between caves and landforms in the Mammoth Cave National Park area: Wurtzburg, Germany, Bohler Verlag, 69 p.

Nishiizumi, K., Imamura, M., and Honda, M., 1979, Cosmic ray produced radionuclides in Antarctic meteorites: Memoirs of the National Institute of Polar Research (Japan), Special Issue 12, p. 161–177.

Nishiizumi, K., Regnier, S., and Marti, K., 1980, Cosmic ray exposure ages of chondrites, pre-irradiation and constancy of cosmic ray flux in the past:

Earth and Planetary Science Letters, v. 50, p. 156–170, doi: 10.1016/0012-821X(80)90126-0.

Nishiizumi, K., Arnold, J.R., Klein, J., Fink, D., Middleton, R., Kubik, P.W., Sharma, P., Elmore, D., and Reedy, R.C., 1991a, Exposure histories of lunar meteorites: ALHA81005, MAC88104, MAC88105, and Y791197: Geochimica et Cosmochimica Acta, v. 55, p. 3149–3155, doi: 10.1016/0016-7037(91)90479-O.

Nishiizumi, K., Kohl, C.P., Arnold, J.R., Klein, J., Fink, D., and Middleton, R., 1991b, Cosmic ray produced ^{10}Be and ^{26}Al in Antarctic rocks: Exposure and erosion history: Earth and Planetary Science Letters, v. 104, p. 450–454.

Nishiizumi, K., Kohl, C.P., Arnold, J.R., Dorn, R., Klein, J., Fink, D., Middleton, R., and Lal, D., 1993, Role of in situ cosmogenic nuclides ^{10}Be and ^{26}Al in the study of diverse geomorphic processes: Earth Surface Processes and Landforms, v. 18, p. 407–425.

Nishiizumi, K., Caffee, M.W., Jull, A.J.T., and Reedy, R.C., 1996, Exposure history of lunar meteorites Queen Alexandra Range 93069 and 94269: Meteoritics and Planetary Science, v. 31, p. 893–896.

Nishiizumi, K., Caffee, M.W., Finkel, R.C., Brimhall, G., and Mote, T., 2005, Remnants of a fossil alluvial fan landscape of Miocene age in the Atacama Desert of northern Chile, using cosmogenic nuclide exposure age dating: Earth and Planetary Science Letters, v. 237, p. 499–507, doi: 10.1016/j.epsl.2005.05.032.

Norris, T.L., Gancarz, A.J., Rokop, D.J., and Thomas, K.W., 1983, Half-life of ^{26}Al: Proceedings of the Fourteenth Lunar and Planetary Science Conference, Part I: Journal of Geophysical Research, v. 88, p. B331–B333.

Ogg, J.G., and Smith, A.G., 2004, The geomagnetic polarity time scale, in Gradstein, F.M., et al., eds., A geologic Time Scale 2004: Cambridge, UK, Cambridge University Press, p. 63–86.

Partridge, T.C., 1978, Re-appraisal of lithostratigraphy of Sterkfontein hominid site: Nature, v. 275, p. 282–287, doi: 10.1038/275282a0.

Partridge, T.C., Shaw, J., Heslop, D., and Clarke, R.J., 1999, The new hominid skeleton from Sterkfontein, South Africa: Age and preliminary assessment: Journal of Quaternary Science, v. 14, no. 4, p. 293–298, doi: 10.1002/(SICI)1099-1417(199907)14:4<293::AID-JQS471>3.0.CO;2-X.

Partridge, T.C., Granger, D.E., Caffee, M.W., and Clarke, R.J., 2003, Lower Pliocene hominid remains from Sterkfontein: Science, v. 300, p. 607–612, doi: 10.1126/science.1081651.

Rightmire, R.A., Simanton, J.R., and Kohman, T.P., 1959, Disintegration scheme of long-lived aluminum-26: Physical Review, v. 113, p. 1069, doi: 10.1103/PhysRev.113.1069.

Samworth, E.A., Warburton, E.K., and Engelbertink, G.A.P., 1972, Beta decay of the ^{26}Al ground state: Physical Review C, v. 5, no. 1, p. 138–142, doi: 10.1103/PhysRevC.5.138.

Siame, L., Bellier, O., Braucher, R., Sébrier, M., Cushing, M., Bourlès, D., Hamelin, B., Baroux, E., de Voogd, B., Raisbeck, G., and Yiou, F., 2004, Local erosion rates versus active tectonics: Cosmic ray exposure modeling in Provence (south-east France): Earth and Planetary Science Letters, v. 220, p. 345–364, doi: 10.1016/S0012-821X(04)00061-5.

Stock, G.S., Anderson, R.S., and Finkel, R.C., 2004, Pace of landscape evolution in the Sierra Nevada, California, revealed by cosmogenic dating of cave sediments: Geology, v. 32, p. 193–196, doi: 10.1130/G20197.1.

Stone, J.O., 2000, Air pressure and cosmogenic isotope production: Journal of Geophysical Research, v. 105, no. B10, p. 23,753–23,759, doi: 10.1029/2000JB900181.

Stone, J.O.H., Evans, J.M., Fifield, L.K., Allan, G.L., and Cresswell, R.G., 1998, Cosmogenic chlorine-36 production in calcite by muons: Geochimica et Cosmochimica Acta, v. 62, no. 3, p. 433–454, doi: 10.1016/S0016-7037(97)00369-4.

Stroeven, A.P., Fabel, D., Hattestrand, C., and Harbor, J., 2002, A relict landscape in the centre of Fennoscandian glaciation: Cosmogenic radionuclide evidence of tors preserved through multiple glacial cycles: Geomorphology, v. 44, no. 1-2, p. 145–154, doi: 10.1016/S0169-555X(01)00150-7.

Thomas, J.H., Rau, R.L., Skelton, R.T., and Kavanagh, R.W., 1984, Half-life of ^{26}Al: Physical Review C, v. 30, p. 385–387, doi: 10.1103/PhysRevC.30.385.

Wallner, A., Ikeda, Y., Kutschera, W., Priller, A., Steier, P., Vonach, H., and Wild, E., 2000, Precision and accuracy of ^{26}Al measurements at VERA: Nuclear Instruments and Methods in Physics Research, Section B, Beam Interactions with Materials and Atoms, v. 172, p. 382–387, doi: 10.1016/S0168-583X(00)00136-1.

Whipple, F.L., and Fireman, E.L., 1959, Calculation of erosion in space from the cosmic-ray exposure ages of meteorites: Nature, v. 183, p. 1315.

Wolkowinsky, A.J., and Granger, D.E., 2004, Early Pleistocene incision of the San Juan River, Utah, dated with ^{26}Al and ^{10}Be: Geology, v. 32, p. 749–752, doi: 10.1130/G20541.1.

Yiou, F., and Raisbeck, G.M., 1972, Half-life of ^{10}Be: Physical Review Letters, v. 29, no. 6, p. 372–375, doi: 10.1103/PhysRevLett.29.372.

Yiou, F., Raisbeck, G.M., Klein, J., and Middleton, R., 1984, ^{26}Al/^{10}Be in terrestrial impact glasses: Journal of Non-Crystalline Solids, v. 67, p. 503–509, doi: 10.1016/0022-3093(84)90172-8.

MANUSCRIPT ACCEPTED BY THE SOCIETY 11 APRIL 2006

Extending ¹⁰Be applications to carbonate-rich and mafic environments

Régis Braucher[†]
Pierre-Henri Blard[‡]
Lucilla Benedetti[§]
Didier L. Bourlès[#]

CEREGE: UMR 6635–CNRS, IRD, Université Paul Cézanne, Plateau de l'Arbois, BP 80, 13545 Aix-en-Provence, France

ABSTRACT

Quantitative geomorphologic studies using cosmogenic nuclides in carbonate-rich and mafic environments have up to now been restricted to the cosmogenic radionuclide ^{36}Cl ($T_{1/2}$ = 0.31 m.y.), and to the stable ^{3}He and ^{21}Ne cosmogenic nuclides, respectively. To extend the time span and erosion rate range quantifiable in carbonate-rich environments, and to provide the possibility to decipher complex exposure histories by differential radioactive decay over several Ma in mafic environments, the in situ production rate of ^{10}Be ($T_{1/2}$ = 1.5 m.y.), the nuclide with the longest half-life of the well-established terrestrial cosmogenic radionuclides, has been determined in calcite and clinopyroxenes.

The development of new chemical decontamination procedures efficiently removing meteoric ^{10}Be from carbonates and altered clinopyroxenes allows determining ^{10}Be production rates. A ^{10}Be production rate in clinopyroxenes of 3.1 ± 0.8 atoms/g/yr at sea level and high latitude is proposed from measurements of ^{10}Be and ^{3}He concentrations in K-Ar-dated Quaternary basaltic flows of Etna volcano.

Through measurements of ^{10}Be and ^{36}Cl concentrations in the same calcite samples and of ^{10}Be concentrations in depth profiles of flint from the same erosional surface, a value of 37.9 ± 6.0 atoms/g/yr has been determined for the ^{10}Be production rate in calcite at sea level and high latitude. Approximately sixfold higher than production in the coexisting flint, this higher rate of production may be due to high production cross sections for C spallation by cosmic rays with energies below 50 MeV. These results also open the possibility of dating burial events in carbonate-rich environments by differential radioactive decay of ^{10}Be and ^{36}Cl.

Keywords: cosmogenic, ^{10}Be, ^{3}He, ^{36}Cl, calcite, clinopyroxene, olivine.

[†]E-mail: braucher@cerege.fr.
[‡]E-mail: blard@cerege.fr.
[§]E-mail: benedetti@cerege.fr.
[#]E-mail: bourles@cerege.fr.

INTRODUCTION

Cosmogenic nuclides produced in terrestrial rocks provide an efficient way to quantify Earth surface processes governing landscape evolution and to date geological events that exhume material from depth, such as glaciation, mass wasting, volcanic and tectonic activity, ... (see review in Gosse and Phillips, 2001). The peculiar physical responses to impinging particles of each of the various targets involved in the production mechanisms, as well as the differing chemical properties, imply that measurement of in situ–produced nuclides is easier and more reliable for specific pure mineral phases. The selected phase must indeed be able to be efficiently decontaminated from the atmospheric variety (if it exists) (Brown et al., 1991), it must fully retain the studied cosmogenic nuclide (Brook et al., 1995), and the production rate in that phase must be particularly well constrained (Masarik and Reedy, 1995). Over the past 15 yr, these considerations led to the use of the following stable and radioactive in situ cosmogenic nuclide systems: (1) ^{10}Be ($T_{1/2}$ = 1.5 Ma), ^{26}Al ($T_{1/2}$ = 0.73 Ma), and ^{21}Ne (stable) in quartz (e.g., Brown et al., 1991; Staudacher and Allègre, 1991), (2) ^{36}Cl ($T_{1/2}$ = 0.31 Ma) in carbonates (Stone et al., 1996), and (3) ^{3}He (stable) and ^{21}Ne in olivines and clinopyroxenes (e.g., Kurz, 1986; Marti and Craig, 1987).

However, the time span and erosion rate range that can be evaluated using a given radioactive cosmogenic nuclide increase with its half-life. Constraining the in situ production rate of ^{10}Be within carbonates can thus significantly extend the range of cosmic-ray exposure ages and erosion rates that can be determined in carbonate-rich environments, given the half-life of this radionuclide ($T_{1/2}$ = 1.5 Ma). Furthermore, the ability to measure couples of cosmogenic nuclides with different half-lives in the same mineral phase allows complex exposure histories to be deciphered by differential radioactive decay. This approach, successfully applied using ^{26}Al and ^{10}Be in quartz-rich environments, will have thus the potential to be extended to carbonate-rich environments using ^{10}Be and ^{36}Cl. Similarly, studies using cosmogenic nuclides in mafic environments are limited to the stable isotopes ^{3}He and ^{21}Ne. Developing the ability to measure in situ–produced ^{10}Be in olivines or clinopyroxenes might therefore have important implications on the characterization and dating of burial events over several m.y. in these environments.

However, cosmogenic ^{10}Be is also produced in the atmosphere through cosmic-ray particle reactions with atmospheric ^{14}N and ^{16}O with an average flux of atmospheric ^{10}Be orders of magnitude higher than the integrated rate of in situ ^{10}Be produced in 1 cm^2 of surficial rock (Monaghan et al., 1986). Atmospheric ^{10}Be may then be adsorbed on minerals and/or incorporated into weathered mineral zones via superficial circulation of meteoric waters. Reliable measurements of the in situ–produced ^{10}Be concentration within carbonates, weathered olivines, or clinopyroxenes thus require an efficient decontamination procedure of this meteoric ^{10}Be variety. For quartz, chemical cleaning based on hydrofluoric acid (HF) step dissolutions proved to remove efficiently this undesirable meteoric component (Brown et al., 1991). A similar step-dissolution method showed efficiency to decontaminate olivines (Nishiizumi et al., 1990; Shimaoka et al., 2002) but appeared to be unsuitable for clinopyroxenes (Ivy-Ochs et al., 1998). New specific decontamination procedures have been thus developed for removing meteoric ^{10}Be contamination from calcite and altered clinopyroxenes. They were tested on Tortonian calcite from southeastern France (43°N) and on variously altered clinopyroxenes from exposed Quaternary lava flows of Etna volcano (Sicily, 38°N).

Cosmogenic ^{36}Cl measured in the same calcite samples and in situ–produced ^{10}Be measured within associated flints allow validating the cleaning procedure developed on carbonates and proposing an empirical ^{10}Be production rate in CaCO$_3$. Similarly, the clinopyroxenes' ^{10}Be decontamination procedure was assessed both by measuring the cosmogenic ^{3}He concentrations in the same samples and by K-Ar dating of the sampled lava flows. Once the procedure had been validated, empirical ^{10}Be production rates were computed for clinopyroxenes and compared to the modeled production rate (Masarik, 2002).

METEORIC ^{10}Be DECONTAMINATION PROCEDURES

Because the average atmospheric ^{10}Be flux to Earth's surface is orders of magnitude higher than the integrated total rate of in situ ^{10}Be produced in a 1 cm^2 column of surficial rock, measurement of the accumulated in situ–produced ^{10}Be concentration requires elimination of the meteoric ^{10}Be. Specific cleaning procedures to remove meteoric ^{10}Be were thus developed on carbonates and variously altered clinopyroxenes.

Carbonates

A cleaning procedure to remove meteoric ^{10}Be was developed on Oligocene carbonates sampled at Limans, France (43°58′N, 5°43′E; altitude 671 m) (LIM02, LIM04, LIM08, and LIM09 in Table 1). X-ray diffraction indicates that all these carbonate samples are pure calcite. Their measured bulk density is 2.61 g cm^{-3}.

TABLE 1. ^{10}Be CONCENTRATIONS IN THE LIMANS CALCITE SAMPLES SUBJECTED TO VARIOUS CLEANING TREATMENTS

Samples[†]	Depth (cm)	Dissolved material (g)	^{10}Be (× 10^6 atoms/g)	^{10}Be (× 10^6 error atoms/g)
LIM02 W, D	65	7.46	1.05	0.33
LIM02 W, 1N, D	65	1.52	0.89	0.18
LIM02 W, 4N, D	65	8.52	1.21	0.13
LIM04 W, D	110	11.26	0.20	0.02
LIM04 W, 1N, D	110	1.66	0.27	0.07
LIM04 W, 4N, D	110	11.26	0.23	0.05
LIM08 W, D	30	6.42	0.22	0.06
LIM08 W, 2N, D	30	5.91	0.21	0.04
LIM08 W, 4N, D	30	5.29	0.19	0.04
LIM09 W, D	65	10.48	1.33	0.60
LIM09 W, 4N, D	65	6.34	1.34	0.35

[†]W—24 h water rinsing; xN—x partial dissolution(s) in 1.5 M HNO$_3$; D—total dissolution in 15 M HNO$_3$.

To test different cleaning procedures, each sample was crushed, sieved (250–500 µm), and divided into five subsamples. The first group was stirred for 24 h in ultrapure water (18 MΩ-cm), dried, and weighed. This cleaned material was prepared for ^{10}Be measurement by complete dissolution in 15 M HNO$_3$ followed by addition of 300 µg ^{9}Be carrier (Merck 1000 mg/L Be standard) and subsequent purification by solvent extractions and alkaline precipitations (Bourlès, 1988). The second group, after being subjected to the same deionized water treatment, was partially dissolved by gradual addition of sufficient 1.5 M HNO$_3$ to dissolve ~10% of the calcite. The pH was monitored and maintained below 5 throughout this step. The cleaned calcite was then dried, weighed, and prepared for ^{10}Be measurement as described above. The other three groups were treated similarly, but were subjected to two, three, and four repetitions of the partial dissolution treatment in 1.5 M HNO$_3$. ^{10}Be/^{9}Be ratios were measured by accelerator mass spectrometry (AMS) at the Tandétron facility of Gif-sur-Yvette, France. Measured ratios were calibrated directly against the National Institute of Standards and Technology (NIST) standard reference material SRM 4325 using its certified ^{10}Be/^{9}Be ratio of $(26.8 \pm 1.4) \times 10^{-12}$. It has been noted that the ratio reported by NIST is incompatible with the ICN standards used at the University of Pennsylvania and at the University of California (Middleton et al., 1993). We have thus normalized our ^{10}Be concentrations by a factor of 1.143 ± 0.039 to make them directly comparable to those based on ICN standards. ^{10}Be uncertainties (1σ) include a 3% contribution conservatively estimated from observed standard variations during the runs, a 1σ statistical uncertainty in the number of ^{10}Be events counted, and the uncertainty of the blank correction (associated ^{10}Be/^{9}Be blank ratio was $(2.4 \pm 1.2) \times 10^{-15}$).

The constant ^{10}Be concentration of the stepwise dissolutions, even after the first water washing step, indicates that meteoric ^{10}Be has been removed by the cleaning procedure (Table 1; Fig. 1), even by the mildest treatment (deionized water). These results are in agreement with previous observations in marine environments (Bourlès et al., 1989; Southon et al., 1987) showing that atmospherically produced ^{10}Be is not adsorbed on carbonates.

Clinopyroxenes

A cleaning procedure to remove the meteoric ^{10}Be contamination was developed on variously altered clinopyroxenes from surficial samples of exposed Quaternary basaltic flows of Etna

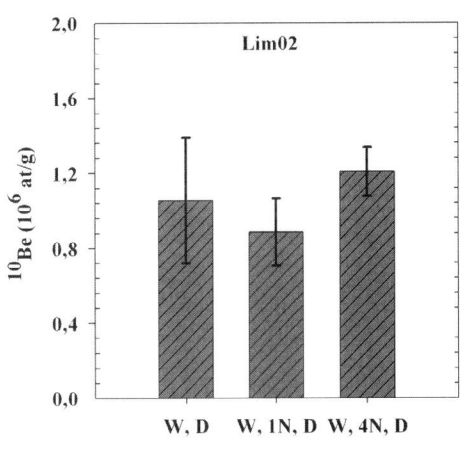
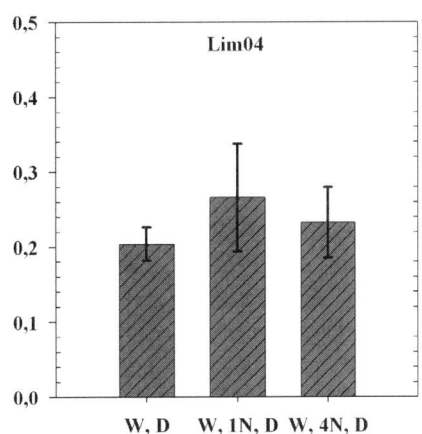
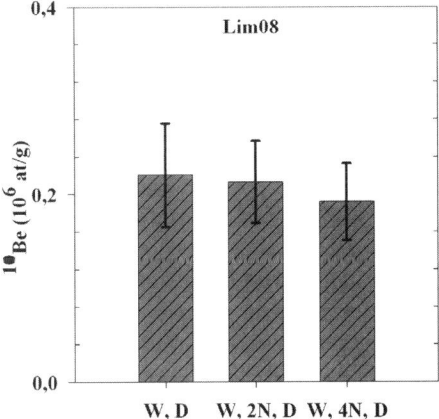
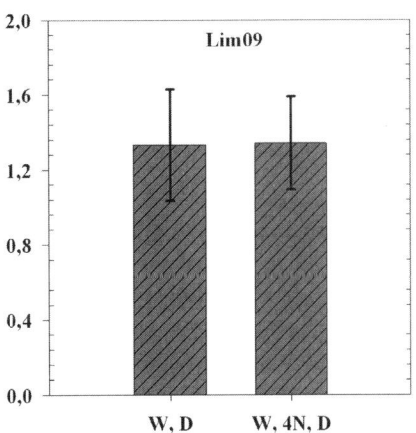

Figure 1. ^{10}Be in cleaning fractions of calcite.

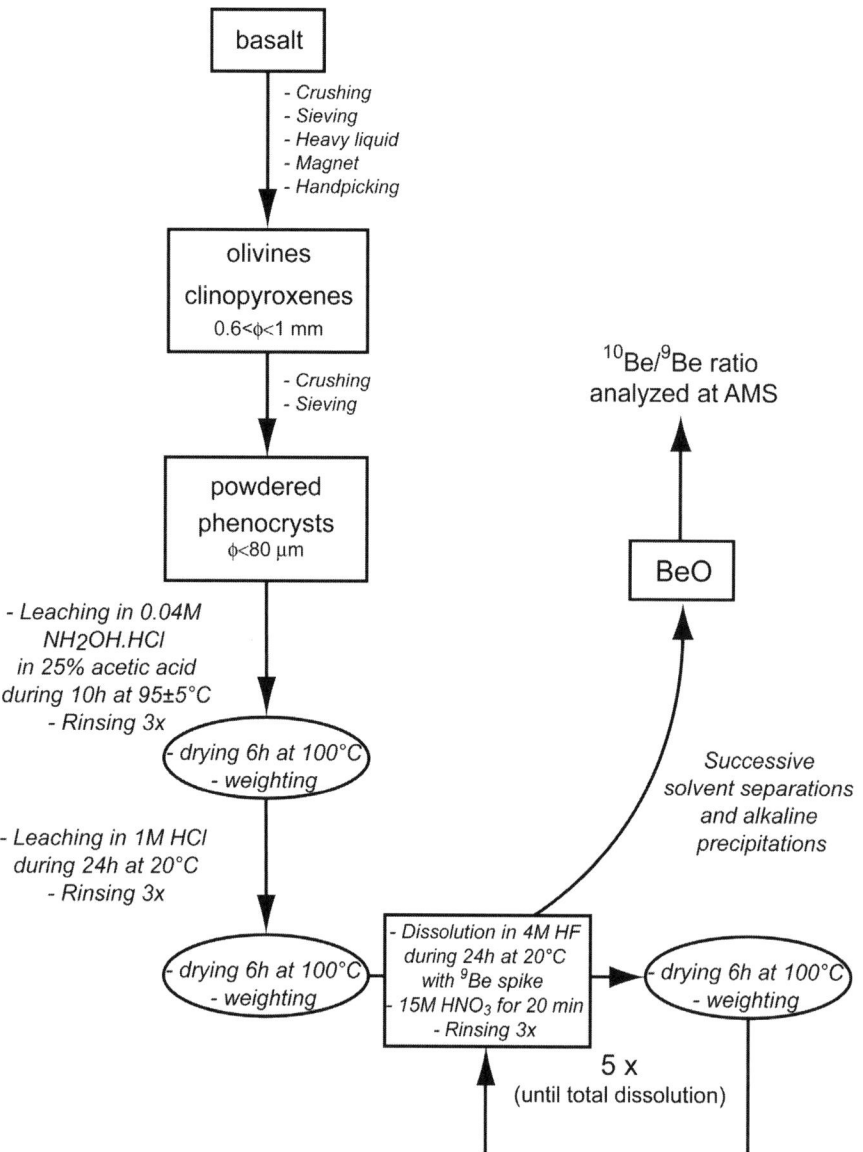

Figure 2. Schematic representation of the atmospheric decontamination procedure developed for in situ cosmogenic ^{10}Be measurement in mafic phenocrysts.

volcano (Sicily, 38°N) (see Fig. 2). Two surficial samples were collected on the pahoehoe Nave flow (SI41, 830 m, and SI27a, 1190 m) at the top of plurimetric tumuli (Table 2). SI43, from the Piano Della Lepre site (2070 m) was collected at the top of a pahoehoe flow, today shielded by the emplacement of an overlying flow (Table 2).

In order to decontaminate from meteoric ^{10}Be, which can be hosted both into mineral overgrowth (Friedmann and Weed, 1987) and into secondary clay minerals (Kabata-Pendias and Pendias, 2000; Wedepohl, 1974), we had to deal with a specific drawback linked to the clinopyroxenes crystallographic structure. Ivy-Ochs et al. (1998) indeed highlighted that sequential dissolution with HF of monomineralic grains (0.2–0.5 mm) was unable to efficiently remove this contamination. This result was interpreted as a proof of the meteoric ^{10}Be accumulation within the grains, in micrometric locked pits where weathering generates clay formation. As a consequence, the originality of the proposed cleaning method is to overcome this difficulty through a preliminary crushing step releasing most of the contaminated zones.

Basalt samples from Etna were crushed and sieved, the 0.6–1 mm fraction being processed. Extraction of mafic phenocrysts from the basaltic groundmass was performed by magnetic separation and density partition in heavy liquids. Since olivines and clinopyroxenes have close densities (~3.3 g cm^{-3}) and magnetic susceptibilities, handpicking under binocular microscope improved the separation of both species. Approximately 25 g of >95%-pure clinopyroxenes (<5% olivines) were isolated for each sample, except for the SI27a sample (65% clinopyroxenes, 35% olivines). These phenocrysts were then crushed in an agate mortar and sieved to keep the fraction smaller than 80 μm.

TABLE 2. MOUNT ETNA BASALTIC FLOWS SAMPLES DESCRIPTION AND LOCATION—K-Ar AND COSMOGENIC ^{3}He DATA

Sample	Altitude (m)	Latitude (°N)	Longitude (°E)	Description	Sampling depth correction†	K-Ar age/ exposure duration (k.y. ± 1σ)	^{3}He$_c$‡ (10^6 atoms/g ± 1σ)	^{3}He$_c$ exposure ages (k.y. ± 1σ)§
Nave flow (top)								
SI27b	1195	37°49'39	14°54'96	Pahoehoe flow	K-Ar dating	32 ± 4	-	-
SI27a	1195	37°49'39	14°54'96	Cords potentially eroded	0.96	32 ± 4	6.25 ± 0.19	24 ± 3
Nave flow (bottom)								
SI28	860	37°51'60	14°51'17	True column basalt	K-Ar dating	33 ± 2	-	-
SI41	820	37°50'91	14°50'10	Cords	0.89	33 ± 2	6.11 ± 0.19	32 ± 3
Piano Della Lepre								
SI14	2070	37°42'57	15°1'55	Overlying flow—true column	K-Ar dating	10 ± 3	-	-
SI16	2070	37°42'57	15°1'55	Underlying flow—true column	K-Ar dating	20 ± 1	-	-
SI43	2070	37°42'57	15°1'55	Cords of the underlying flow	0.89	10 ± 3	5.04 ± 0.21	11 ± 1

Note: K-Ar and ^{3}He$_c$ are from (Blard et al., 2005).
†Depth corrections are performed assuming a basalt density of 2.5 g cm^{-3} and an attenuation length of 150 g cm^{-2} in basalt (Gosse and Phillips, 2001).
‡Concentrations corrected from sampling depth. For SI43, post exposition through the overlying flow is less than 2% of the measured ^{3}He$_c$.
§Cosmogenic exposure durations are deduced from the ^{3}He$_c$ data using SLHL P_3 of 108 ± 11 atoms/g/yr (Masarik and Reedy, 1996) scaled with the Stone (2000) factors.

To optimize the meteoric ^{10}Be removal without dissolving unaltered minerals, lattice powder samples were leached at 95 ± 5 °C during 10 h in a 0.04 M solution of hydroxylammonium chloride (NH$_2$OH.HCl) in 25% acetic acid. This solution dissolves metallic oxides (Bourlès et al., 1989) and allows the grain-adsorbed ^{10}Be to be released in the aqueous phase. After centrifugation, the leachate solution is poured into a Teflon bottle as well as the three rinses performed using the same leaching solution. The remaining solid phase is oven-dried (6 h at 100 °C) and precisely weighed (±1 mg) to determine the mass of material eliminated by the leaching step.

Powder is next laden within a 60 mL Teflon jar containing 25 mL of 1 M HCl and shaken for 24 h. After separation of the supernatant by centrifugation, the remaining solid phase is rinsed three times with ultrapure deionized water that is added to the supernatant.

Oven-dried and weighed, the remaining powder sample undergoes five sequential dissolution steps in 4 M HF in a 60 mL Teflon jar shaken during 24 h. After each sequential HF dissolution, ~10 mL of concentrated 15 M nitric acid is added to dissolve coprecipitated fluorides (MgF$_2$, CaF$_2$). Twenty minutes later the acid floating is extracted, the solid phase is rinsed three times using ultrapure water that is added to the supernatant, and the remaining powder sample is dried and weighed.

The observed mean value of 0.48 ± 0.14 g of clinopyroxenes dissolved per mL of concentrated (48%) HF added is consistent with the stoichiometric value of ~0.5 g mL^{-1}, suggesting that the dissolving reactions are close to completion (Table 3).

The solutions obtained were spiked with 300 µL of a 1 × 10^{-3} g g^{-1} ^{9}Be Merck solution for ^{10}Be measurement. Iron is extracted from the aqueous phase by organic separation in diisopropylether. Beryllium is then separated following the chemical extraction procedure described in (Bourlès et al., 1989). ^{10}Be/^{9}Be ratios were also measured at the AMS facility of Gif-sur-Yvette, France. During this run, the associated ^{10}Be/^{9}Be blank value was (4.9 ± 2.8) × 10^{-15}.

The K-Ar dating and cosmogenic ^{3}He (^{3}He$_c$) data are presented in Table 2. The K-Ar ages are 32 ± 4 ka and 33 ± 2 ka for the Nave flow samples SI27a and SI41, respectively, and the past exposure interval is 10 ± 3 ka for the Piano Della Lepre site (SI43). Scaled for latitude and altitude with the Stone (2000) factors, the ^{3}He$_c$ production rate of 108 ± 10 atoms/g/yr (Masarik and Reedy, 1996) yields ^{3}He$_c$ cosmic-ray exposure ages in agreement with the K-Ar ages for SI41 (Nave bottom) and SI43 (Piano Della Lepre) (Table 2).

However, the 24 ± 3 ka exposure age calculated from ^{3}He$_c$ measured on SI27a (Nave top) is significantly lower than the 32 ± 4 ka K-Ar age of this flow, suggesting that a mean erosion rate of 11 ± 3 m/m.y. has affected this surface since the flow emplacement.

Table 3 presents the ^{10}Be concentrations measured in aqueous phases obtained during the cleaning procedure for SI27a and in the aqueous phases resulting from the last HF dissolution for all the samples. Uncertainties, reported as 1σ on the ^{10}Be concentrations take into account counting statistics, observed external reproducibility of the AMS (3%), and blank correction.

Assuming a ^{3}He$_c$/^{10}Be$_c$ production ratio of ~24 (Table 4), the SI27a ^{3}He$_c$ concentration (Table 2) thus implies a theoretical in situ–produced ^{10}Be concentration of (2.54 ± 0.08) × 10^5 atoms/g. This theoretical value can be used to check the cleaning procedure efficiency.

The leaching step performed on SI27a yields a ^{10}Be concentration of (6.34 ± 0.43) × 10^7 atoms/g. This concentration is two orders of magnitude higher than the theoretical in situ–produced concentration, evidencing a contamination by atmospherically produced ^{10}Be. The hydroxylamine leaching thus removes a significant part of the meteoric contamination from the crushed clinopyroxenes. The solution resulting from the second cleaning step in HCl has a concentration of $(4.3 \pm 0.6) \times 10^6$ atoms/g, lower than that of the leaching solution but nevertheless higher than the theoretical in situ–produced concentration. Such a ^{10}Be release may result from the dissolution of remaining secondary minerals (oxides, clays) by HCl.

The three analyzed HF dissolution steps agree within 1σ and have ^{10}Be concentrations averaging $(2.55 \pm 0.29) \times 10^5$ atoms/g (weighted mean using relative uncertainties) (Fig. 5). These concentrations agree within uncertainties with the $(2.54 \pm 0.08) \times 10^5$ atoms/g theoretical in situ–produced ^{10}Be concentration. Such an agreement strongly suggests that the atmospherically produced ^{10}Be was efficiently removed and that the concentrations measured from HF dissolutions represent the in situ–produced

TABLE 3. ^{10}Be DATA FROM THE MOUNT ETNA MAFIC PHENOCRYSTS AND COMPUTED SEA LEVEL, HIGH LATITUDE P_{10} PRODUCTION RATES

Sample	Chemical treatment	Mass dissolved (g)	^{10}Be concentration[†] ($10^5 \pm 1\sigma$ atoms/g)	SLHL P_{10} (atoms/g/yr ± 1σ)[‡] Calculated from ^{10}Be data		Simulated value[§]
				Assuming ε = 0 m/m.y.[††]	ε computed from ^{3}He$_c$ data[‡‡]	
Nave flow (top) 65% cpx 35% oli						
SI27a	Leaching	0.77	638 ± 43			
SI27a	Dissolution in 1 M HCl	0.99	43 ± 6			
SI27a	Dissolution in 4 M HF	2.49	Non analyzed			
SI27a	Dissolution in 4 M HF	2.02	2.52 ± 0.42			
SI27a	Dissolution in 4 M HF	4.29	Non analyzed			
SI27a	Dissolution in 4 M HF	3.93	2.53 ± 0.54			
SI27a	Dissolution in 4 M HF	4.99	2.65 ± 0.64			
Weighted mean ± 1σ	3 HF dissolutions		2.55 ± 0.29	3.5 ± 0.6	4.6 ± 0.9	4.58
Nave flow (bottom) 95% cpx						
SI41	Last 4 M HF dissolution	3.78	1.14 ± 0.28	2.2 ± 0.6	2.3 ± 0.6	4.58
Piano Della Lepre 95% cpx						
SI43	Last 4 M HF dissolution	3.03	1.02 ± 0.25	2.5 ± 1.0	2.5 ± 1.0	4.60
Weighted mean[#] ± 1σ				2.7 ± 0.5	3.1 ± 0.8	4.6 ± 0.5

[†]^{10}Be measurements have been normalized to ICN standard.
[‡]^{10}Be concentrations are corrected from sampling depth to calculate P_{10}. Calculated P_{10} is scaled to sea level high, latitude using the Stone (2000) factors.
[§]From Masarik (2002).
[#]Means are calculated weighting the data over relative uncertainties.
[††]Calculated from K-Ar data using Equation (3).
[‡‡]Calculated from K-Ar and ^{3}He$_c$ data combining Equations (3) and (4).

TABLE 4. MAJOR ELEMENT COMPOSITION (wt%) AND THEORETICAL SIMULATED SEA LEVEL, HIGH LATITUDE ^{10}Be PRODUCTION RATES

Sample	Mg	Ti	Cr	Mn	Fe	Ni	K	Ca	Al	Na	Si	O	Modeled ^{10}Be[†] production rate (atoms/g/yr)	Modeled ^{3}He[‡] production rate (atoms/g/yr)
Nave (bottom) cpx														
SI41	9.43	0.84	0.01	0.17	8.11	0.03	0.01	13.77	2.34	0.42	22.44	42.43	4.58	107
Nave (top) cpx														
SI27a	8.08	1.07	0.01	0.15	6.73	0.02	0.01	15.34	2.79	0.47	22.69	42.63	4.58	107
Piano Della Lepre cpx														
SI43	8.55	0.87	0.02	0.16	6.71	0.02	0.01	15.06	2.37	0.47	23.05	42.72	4.60	108
Elemental ^{10}Be production rate[†] (atoms/g/yr)	1.74				0.35				1.03		0.89	9.82		
Elemental ^{3}He production rate[‡] (atoms/g/yr)	116				40			61	107		111	135		

[†]From Masarik (2002).
[‡]From Masarik and Reedy (1996).

component. The tested chemical cleaning procedure therefore appears suitable to allow measurement of in situ–produced ^{10}Be in clinopyroxenes.

^{10}Be IN SITU PRODUCTION RATES

Carbonates

To estimate the ^{10}Be in situ production rate in carbonates, coexisting depth profiles of calcite and flint from the vertical southern limb of the Trevaresse anticline were sampled (43°37′N, 5°25′E; altitude 375 m; Tortonian lacustrine series; Chardon and Bellier, 2003) (Fig. 3). The calcite and flint samples were collected along separate vertical profiles less than 1 m apart within a single trench.

In situ–produced ^{10}Be concentrations were measured in both flint and calcite profile samples. The flint samples were decontaminated using sequential HF dissolutions (Brown et al., 1991), and the calcite samples as described above. All solutions obtained after total dissolution of the material remaining after the specific decontamination procedures were spiked with 300 μL of a 1 × 10^{-3} g g^{-1} ^{9}Be Merck solution for ^{10}Be measurement. Beryllium was then extracted and purified by successive solvent separations and alkaline precipitations (Bourlès et al., 1989).

In addition, in situ–produced ^{36}Cl concentrations were measured in the calcite profile samples. After grinding, leaching, and chemical separation of chlorine by precipitation of silver chloride, the ^{36}Cl and chloride concentrations in the carbonate were determined for all samples by isotope dilution accelerator mass spectrometry at the Lawrence Livermore National Laboratory CAMS (Center for Accelerator Mass Spectrometry) facility. Blanks were two orders of magnitude lower than the samples. Uncertainties of ^{36}Cl concentration include the statistical uncertainties of measurements and the blank correction. ^{36}Cl production rates from calcium of Stone et al. (1998) were used. Those production rates were calculated for our site latitude and altitude using Stone (2000) coefficients.

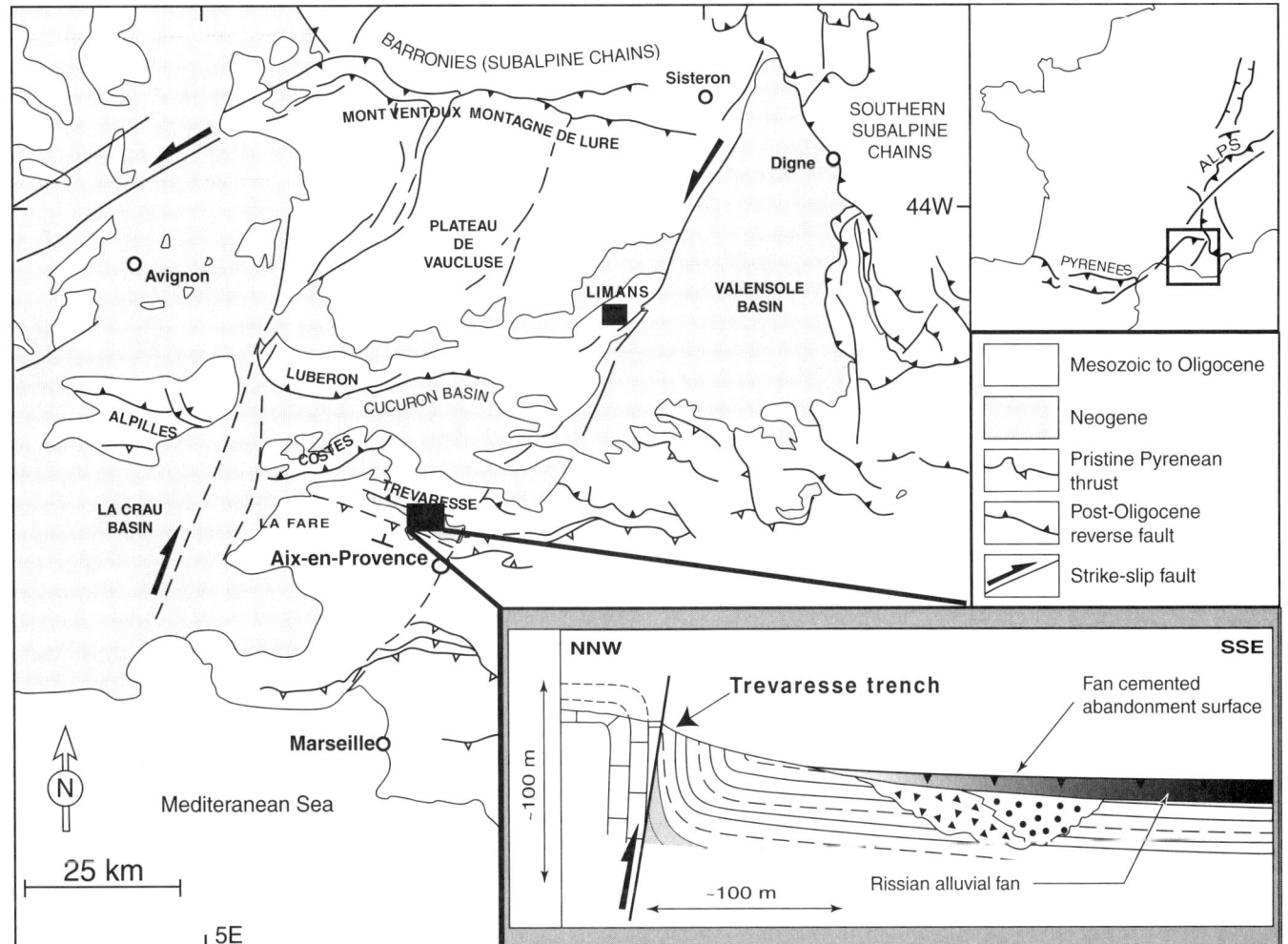

Figure 3. Site map of the sampled flints and carbonates (modified after Chardon and Bellier, 2003).

The ^{10}Be concentrations in the flint (Table 5) and calcite (Table 6) samples from the Trevaresse profiles decrease exponentially with depth, consistent with in situ production by cosmic rays (Fig. 4).

Furthermore, ratios of ^{10}Be to in situ ^{36}Cl concentrations (Table 6) measured within the decontaminated calcite samples are constant

$$2.14\,^{+0.13}_{-0.13},\ 2.16\,^{+0.15}_{-0.15},\ 2.15\,^{+0.19}_{-0.18},\ 1.91\,^{+0.14}_{-0.13},$$

at depths of 104, 141, 183, 235 g/cm², respectively). Virtually all of the ^{36}Cl in these samples is likely to have been produced by Ca spallation; their low Cl content and their sampling depths preclude significant production by thermal neutron capture. Since the ^{36}Cl analytical procedure is well established (Stone et al., 1996), this independently suggests that the ^{10}Be concentrations measured in the decontaminated calcite correspond to the in situ–produced ^{10}Be.

To model the ^{10}Be production rate within calcite, the following equation is used:

$$C(x;t) = \frac{P_0 \cdot p_n}{\frac{\varepsilon}{\Lambda_n} + \lambda} \cdot \exp\left(-\frac{x}{\Lambda_n}\right) \cdot \left[1 - \exp\left(-t \cdot \left(\frac{\varepsilon}{\Lambda_n} + \lambda\right)\right)\right]$$

$$+ \frac{P_0 \cdot p_{\mu s}}{\frac{\varepsilon}{\Lambda_{\mu s}} + \lambda} \cdot \exp\left(-\frac{x}{\Lambda_{\mu s}}\right) \cdot \left[1 - \exp\left(-t \cdot \left(\frac{\varepsilon}{\Lambda_{\mu s}} + \lambda\right)\right)\right]$$

$$+ \frac{P_0 \cdot p_{\mu f}}{\frac{\varepsilon}{\Lambda_{\mu f}} + \lambda} \cdot \exp\left(-\frac{x}{\Lambda_{\mu f}}\right) \cdot \left[1 - \exp\left(-t \cdot \left(\frac{\varepsilon}{\Lambda_{\mu f}} + \lambda\right)\right)\right], \quad (1)$$

where P_0 is the production rate; p_n, $p_{\mu s}$, and $p_{\mu f}$ refer to the neutron, slow muon, and fast muon contributions (these are 97.85%, 1.50%, and 0.65%, respectively, in quartz, but unknown in calcite); Λ_n, $\Lambda_{\mu s}$, and $\Lambda_{\mu f}$ are the neutron, slow muon, and fast muon attenuation lengths, which are 150, 1500, and 5300 g/cm², respectively (Braucher et al., 2003); λ is the radioactive decay constant; and ε is the erosion rate. The erosion rate can be determined from the depth profiles of ^{10}Be in flint and ^{36}Cl in calcite. These indicate steady-state erosion rates of (13.5 ± 1.9) m/m.y. and (13.4 ± 1.4) m/m.y., respectively, following calculations of Braucher et al. (2003) and Stone et al. (1996).

The ^{10}Be profile in calcite indicates an apparent attenuation length of ~325 g/cm². The profile is not deep enough for quantitative differentiation between fast and slow muons. Nevertheless, at steady state under the erosional conditions of the site, this apparent attenuation length may correspond to various combinations of both types of muon contributions within ranges of 0%–6% for the slow muons and of 0%–2% for the fast muons.

Using the well-constrained local erosion rate, and assuming contributions for fast and slow muons that yield an apparent attenuation length of ~325 g/cm², equation 1 can be solved for the production rate. This yields a ^{10}Be production within calcite of 50.5 ± 8.0 atoms/g/yr, 6.3 ± 1.3 times higher than that of 7.9 ± 1.2 atoms/g/yr within flint based on the polynomials of Stone (2000). This corresponds to a value of 37.9 ± 6.0 atoms/g/yr (sea level and high latitude). The calculated uncertainty includes propagation of a 6% uncertainty on the ^{10}Be production rate used to estimate the erosion rate within flint and the analytical uncertainties (~5%) of ^{10}Be in calcite.

Elemental compositions of minerals play a major role in inducing higher in situ ^{10}Be production in calcite relative to quartz. In quartz, the targets responsible for ^{10}Be production are O, through the reactions ^{16}O(n, 4p3n)^{10}Be and ^{16}O(μ^-, αpn)^{10}Be, and Si, through the reactions ^{28}Si(n, x)^{10}Be and ^{28}Si(μ^-, x)^{10}Be. Relative elemental production rates can be estimated by dividing the product of the production cross section and the mass fraction

TABLE 5. IN SITU–PRODUCED ^{10}Be CONCENTRATIONS IN THE TREVARESSE FLINT SAMPLES

Samples	Depth (cm)	^{10}Be† (× 10⁵ atoms/g)	^{10}Be error (× 10⁵ atoms/g)
TR03-01	40	3.43	0.48
TR03-05	50	3.11	0.39
TR03-02	54	2.23	0.26
TR03-03	64	2.80	0.47
TR03-04	90	2.08	0.38

Note: density = 2.57 g cm⁻³.
†^{10}Be concentrations have been normalized to ICN standard.

TABLE 6. IN SITU–PRODUCED ^{10}Be AND ^{36}Cl CONCENTRATIONS IN THE TREVARESSE CALCITE SAMPLES

Samples	Depth (cm)	Dissolved material (g)	^{10}Be† (× 10⁶ atoms/g)	^{10}Be error (× 10⁶ atoms/g)	Cl (ppm) (error < 3%)	Ca (weigh %) (error < 3%)	^{36}Cl (× 10⁶ atoms/g)	^{36}Cl error (× 10⁶ atoms/g)
TR0306	40	25.13	2.47	0.10	7.60	40.09	1.15	0.02
TR0307	54	24.17	2.25	0.10	7.20	39.48	1.04	0.02
TR0308	70	14.33	1.99	0.08	8.65	37.66	0.93	0.04
TR0309	90	13.05	1.66	0.07	10.46	38.16	0.87	0.03

Note: density = 2.61 g cm⁻³.
†^{10}Be concentrations have been normalized to ICN standard.

of the relevant element by its atomic weight. For example, in SiO_2, O represents ~53.3% of the mass, so ~92% of ^{10}Be in SiO_2 is produced from O (Leya et al., 2000b). In calcite, O represents ~47.5% of the mass, so production by O spallation should produce almost the same amount of ^{10}Be as in SiO_2. Another target must be considered. As spallogenic reactions favor product masses that are either slightly less than the target or much less, such as protons and neutrons (Gosse and Phillips, 2001), the most favorable target in carbonate is C.

The occurrence of ^{10}Be production from carbon has been demonstrated in meteorites (Nagai et al., 1993) and studied in diamonds (Lal et al., 1987). ^{10}Be is produced from C by protons mainly via the reaction $^{12}C(p, 3p)^{10}Be$. Secondary neutrons produce ^{10}Be from carbon via the reaction $^{12}C(n, 2pn)^{10}Be$. To determine production of nuclides by galactic cosmic ray (GCR) or solar cosmic ray (SCR) particles, models based on physical approaches have been developed (Leya et al., 2000a; Masarik and Reedy, 1995; Michel et al., 1995). Amounts of produced nuclides or production cross sections (Leya et al., 2000b; Raisbeck and Yiou, 1977) have also been estimated after irradiation of pure or composite targets. These studies indicate that, for particles having energy higher than 500 MeV, production of ^{10}Be from C is 2–4 times higher than that from O (Leya et al., 2000a; Nagai et al., 1990).

More recently, from experimental cross sections in natural carbon targets, Kim et al. (2002) reported that the production rate of ^{10}Be in carbon at 40 MeV is 22 times higher than that in oxygen, while, in agreement with studies in meteorites, it is only ~2 times higher at 400 MeV. As the energy of the particles (mainly secondary neutrons) that impinge on Earth's surface is lower than 500 MeV (Lal, 1958, 1988), this altogether may well explain the ratio found in the Trevaresse samples. By compiling production cross sections from Michel et al. (1995), Sisterson et al. (1997), Leya et al. (2000b), and Kim et al. (2002), the production ratio of carbon over oxygen can be estimated for spallation by protons with energies ranging from 30 to 600 MeV. This indicates systematically higher production of ^{10}Be from C than O, the highest production ratio values being associated with low energies.

Production ratios can be evaluated by integrating the excitation functions for oxygen and carbon (Kim et al., 2002) over the ground-level energy spectra for neutrons and protons (Ashton et al., 1971; Lal, 1958, 1988). This can be expressed by the following equation:

$$\frac{\int_{E_1}^{E_2}\sigma_C(E)\times 0.12\times dE}{\int_{E_1}^{E_2}\sigma_O(E)\times 0.48\times dE}+\frac{\int_{E_1}^{E_2}\sigma_C(E)\times 0.12\times dE}{\int_{E_1}^{E_2}\sigma_O(E)\times 0.48\times dE}, \quad (2)$$

where σ_c and σ_o are the excitation functions for C and O, respectively (we consider cross sections for neutrons equal to those for protons); $n(E)$ and $p(E)$ are the ground energy spectra for neutrons and protons, respectively; E is the energy; and 0.12

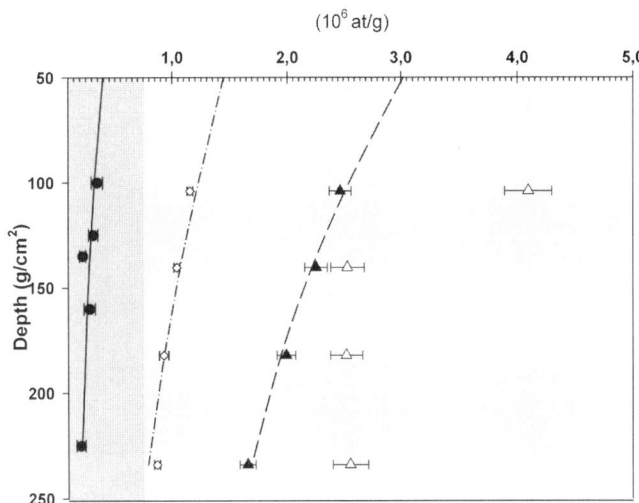

Figure 4. In situ–produced ^{10}Be and ^{36}Cl concentrations as a function of depth for the Trevaresse samples (flint and calcite). Best fits (solid and dash-dotted lines) with erosion rate as free parameter were determined using theoretical depth distributions from the experimental data of ^{10}Be concentrations measured within flint samples (circles) and from the ^{36}Cl concentrations measured within calcite samples (diamonds). They yield identical local erosion rates of 13.5 ± 1.9 m/m.y. and 13.4 ± 1.4 m/m.y., respectively. Best fit (dashed line) to experimental data with production rate as free parameter, performed on decontaminated calcite samples (filled triangles) using the well-constrained local erosion rate, yields in situ ^{10}Be production rate of 50.5 ± 7.5 atoms/g/yr within calcite. Open triangles correspond to untreated calcite samples (samples were crushed then totally dissolved).

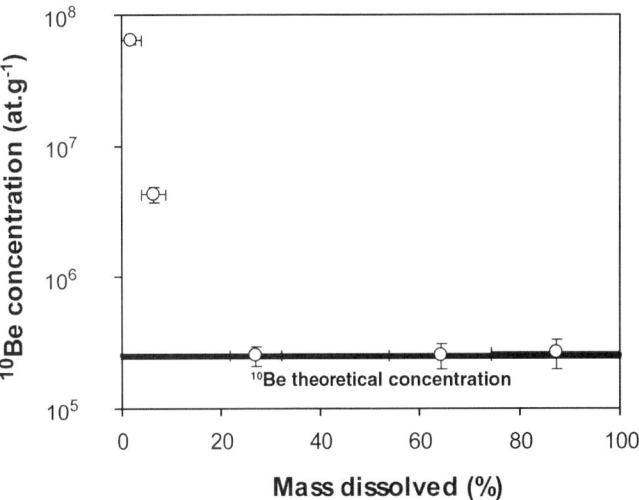

Figure 5. Meteoric ^{10}Be cleaning of olivines and clinopyroxenes by step dissolution of sample SI27a. The cleaning experiment yields a decrease of the ^{10}Be concentration to a constant plateau value in agreement with the theoretical in situ ^{10}Be concentration. This theoretical $^{10}Be_c$ level of (2.54 ± 0.08) × 10^5 atoms/g is deduced from the ^3He_c concentration ((5.99 ± 0.18) × 10^6 atoms/g, value uncorrected from sampling depth), assuming that (1) the $^{10}Be_c/^3He_c$ production ratio is ~24 in clinopyroxenes (Masarik, 2002), (2) both ^3He_c and $^{10}Be_c$ have the same attenuation length in basalt, and (3) the $^{10}Be_c$ loss from radioactive decay ($T_{1/2}$ = 1.5 m.y.) is negligible (<1%) for this 32 ± 4 ka flow (K-Ar age).

and 0.48 are the relative mass proportions of C and O within calcite. Integration over an energy domain ranging from 30 to 600 MeV shows a higher ^{10}Be production rate from C throughout and yields a production rate of ^{10}Be in calcite ~1.9 times greater than in flint. This is insufficient to explain the factor of 6.3 we observe.

We propose two potential sources for this inconsistency. A significantly lower threshold for ^{10}Be production by C spallation than that for O spallation, coupled with the high ground-level particle fluxes at low energies (i.e., <25 MeV where few production cross sections data are available), could lead to significantly higher production in calcite than we calculate above. Secondly, experimental cross sections have generally been determined for protons, whereas neutrons dominate ground-level production of cosmogenic nuclides. The fact that we consider cross sections for neutrons equal to those for protons may underestimate production of ^{10}Be from both C and O. Basic theoretical and statistical principles indicate that the reaction ^{16}O(n, 3n4p)^{10}Be should have higher reaction cross sections than ^{16}O(p, 2n5p)^{10}Be and that this effect is more substantial for ^{12}C(n, n2p)^{10}Be compared to ^{12}C(p, 3p)^{10}Be.

Clinopyroxenes

Similarly, measuring cosmogenic nuclide concentrations in independently dated lava flows for which prior exposure to cosmic rays can be excluded ($N_{^{10}Be}(t_0) = 0$ atoms/g) allows calculating empirical production rates. Equation 3 (Lal, 1991) indeed describes the evolution of the surficial in situ–produced ^{10}Be concentration, $N_{^{10}Be}$ (atoms/g), as a function of the exposure duration, t (yr); of the surficial in situ ^{10}Be production rate at the sampling site, P_{10} (atoms/g/yr); and of the erosion rate, ε (cm yr^{-1}):

$$N_{^{10}Be}(t) = \frac{P_{10}}{\left(\frac{\rho \cdot \varepsilon}{\Lambda} + \lambda\right)} \left[1 - e^{-t\left(\frac{\rho \cdot \varepsilon}{\Lambda} + \lambda\right)}\right], \quad (3)$$

where $\Lambda = 150$ g cm^{-2} is the attenuation length in basalt; $\lambda = 1.621 \times 10^{-7}$ yr^{-1} is the radioactive decay constant of ^{10}Be; and $\rho = 2.5$ g cm^{-3} is the density of the vesicles-rich sampled rocks.

If erosion is negligible ($\rho\varepsilon/\Lambda \ll \lambda$), P_{10} can be calculated knowing only the exposure age, t, and the ^{10}Be concentration, $N_{^{10}Be}(t)$, at the surface. For a lava flow, the exposure age, t, can be assumed to be the same as the K-Ar age of the eruption.

If the flow surface denudation has been significant ($\varepsilon > 1$ m/m.y. for flows younger than 50 ka), erosion has to be determined independently to avoid underestimating the production rates. This can be done by measuring in the same sample the concentration of the stable cosmogenic nuclide ^{3}He$_c$, which is described by

$$N_{^{3}He}(t) = \frac{P_3 \cdot \Lambda}{\rho \cdot \varepsilon}\left[1 - e^{-t\left(\frac{\rho\varepsilon}{\Lambda}\right)}\right], \quad (4)$$

where P_3 (atoms/g/yr) is the surficial ^{3}He$_c$ production rate at the sampling site. Since the sea-level high-latitude (SLHL) value of P_3 is well constrained in clinopyroxenes (108 ± 10 atoms/g/yr from Masarik and Reedy, 1996), equations 3 and 4 can be combined to eliminate the erosion term ε and, therefore, to calculate P_{10}.

Both approaches were used to calculate empirical in situ ^{10}Be production rates (P_{10}). For this, the measured in situ ^{10}Be concentrations (Table 3) were first adjusted for their sampling depth assuming an attenuation length of 150 g cm^{-2} in basalt (Gosse and Phillips, 2001). This correction is less than 15% for the thickest sample (SI41, Nave bottom, 15 cm thick). Topographic mask corrections are negligible. The computed rates were finally adjusted from sampling latitudes and elevations to SLHL applying the Stone (2000) scaling factors (Table 3). Despite various clinopyroxenes' weathering degrees, all the so-calculated empirical values agree within the 1σ level (Table 3), which strongly supports the efficiency of the decontamination procedure. Using the calculation based only on the ^{10}Be and K-Ar data, calculated SLHL P_{10} values range from 2.2 ± 0.6 to 3.5 ± 0.6 atoms/g/yr. When the erosion term is eliminated using the ^{3}He$_c$ data, calculated SLHL P_{10} values then range from 2.3 ± 0.6 to 4.6 ± 0.9 atoms/g/yr. Means weighed to relative uncertainties are 2.7 ± 0.5 and 3.1 ± 0.8 atoms/g/yr, respectively. The effect of erosion on this production rate calibration is thus limited, because underestimates of erosion remain inside the 1σ uncertainty for all the samples.

Table 4 presents the computed SLHL P_{10} using the physical model developed by Masarik (2002). The values of the SLHL P_{10} modeled from the SI41, SI27a, and SI43 clinopyroxene (diopside) compositions, 4.58, 4.58, and 4.60 atoms/g/yr, respectively, are similar. The uncertainties attached to these simulated rates are assumed to be ~10%, leading to a mean (1σ) of 4.6 ± 0.5 atoms/g/yr. Therefore, these simulated values are slightly higher than the weighed mean of the empirical values, 3.1 ± 0.8 atoms/g/yr, but both values nevertheless remain statistically compatible within uncertainties.

Simulated SLHL production rates are, however, valid for the present cosmogenic neutron flux. Discrepancies between simulated and calibrated rates may thus result from the past geomagnetic fluctuations as the natural calibration sites are affected by the so-induced neutron flux variations over the exposure duration. Virtual dipole moment (VDM) variations from Carcaillet et al. (2004) database and dipole position fluctuations from Lanza and Zanella (2003) were used to calculate the past neutron flux according to the equations given in (Dunai, 2001). This signal was thus integrated over the appropriate exposure interval to compute the geomagnetic corrections for each studied flow. This correction has to be considered only for the production rate estimate based on equation 3. Indeed, the calculation based on both equations 3 and 4 necessarily integrates magnetic variations through the ^{3}He$_c$ data input. Nevertheless, given the considered exposure durations (last 33 ± 2 k.y. for the Nave flow and from 20 ± 1 to 10 ± 3 ka for the Piano Della Lepre site) at sampling latitudes and altitudes, geomagnetic variations do not induce shifts larger than 3%. Such a limited effect can therefore be considered as insignificant given the uncertainty attached to the empirical SLHL P_{10}.

Finally, the data published in (Nishiizumi et al., 1990) may also be used to compute empirical SLHL P_{10} in olivines (Forsterite$_{80}$). The authors indeed measured both ^3He_c and ^{10}Be in olivines from eroded Maui lavas erupted ca. 500 ka. Combining equations 3 and 4, and using a SLHL P_3 of 110 ± 10 atoms/g/yr from Masarik and Reedy (1996) scaled with the Stone (2000) factors, these data yield a SLHL P_{10} of 3.2 ± 1.0 atoms/g/yr for olivines. This SLHL P_{10} computed for olivines is similar to that determined for clinopyroxenes in this study, suggesting that the compositional differences between olivines Fo$_{80}$ and diopside clinopyroxenes do not induce significant in situ ^{10}Be production rate discrepancies.

However, further studies could be done in order to strengthen the evidence that in situ ^{10}Be can be measured accurately in mafic phenocrysts. These investigations should particularly focus on (1) the potential loss of in situ ^{10}Be enhanced by the preliminary crushing step and (2) the in situ ^{10}Be overestimation due to the trapping of atmospheric ^{10}Be from the magmatic chamber.

CONCLUSIONS

The new chemical procedures proposed in this study to decontaminate carbonates and clinopyroxenes from atmospheric ^{10}Be have been tested on calcite from a depth profile sampled in the Tortonian lacustrine series from the Trevaresse anticline (France, 43°N), and on K-Ar-dated basaltic flows of Etna volcano (Sicily, 38°N), respectively.

Measurement of ^{10}Be concentrations within both flint and calcite from two depth profiles sampled below the same erosional surface and measurement of ^{36}Cl concentrations within the same calcite samples demonstrate that the cleaning procedure developed to remove meteoric ^{10}Be contamination from calcite is efficient. The effective attenuation length along the calcite profile implies that production induced by muons plays a significant role in the in situ ^{10}Be production in calcite. The in situ ^{10}Be production rate in calcite thus computed at the sampling site is 50.5 ± 7.5 atoms/g/yr, 6.3 ± 0.3 times higher than that in flint. This corresponds to an in situ ^{10}Be production rate in calcite at sea level and high latitude of 37.9 ± 6.0 atoms/g/yr. Leading to a production ratio significantly higher than can be explained by existing production cross sections, this work emphasizes the need for additional measurements at low energies.

The efficiency of the chemical procedure proposed to decontaminate weathered clinopyroxenes from atmospheric ^{10}Be is attested to not only by the ^{10}Be concentrations reaching a plateau during final HF dissolutions, but also by a strong agreement between the measured ^{10}Be concentrations and those inferred from theoretical production rates (Masarik, 2002). The key step of the proposed chemical decontamination procedure relies on a preliminary crushing step of the pyroxenes (<80 μm) performed to maximize the release of the contaminated zones. To limit the undesirable dissolution of clinopyroxenes during the decontamination steps, the crushing step is followed by leachings of the sample in hydroxylamine and HCl solutions.

The combined ^{10}Be and ^3He_c data obtained from K-Ar-dated basaltic flows of Etna volcano allowed in situ ^{10}Be production rates to be calculated for clinopyroxenes. The yielded sea-level high-latitude mean in situ ^{10}Be production rate of 3.1 ± 0.8 atoms/g/yr is similar, within uncertainties, not only to simulated values (Masarik, 2002), but also to the other empirical values.

REFERENCES CITED

Ashton, F., Edwards, H.J., and Kelly, G.N., 1971, The spectrum of cosmic ray neutrons at sea level in the range 0.4–1.2 GeV: Journal of Physics A: General Physics, v. 4, p. 352–366, doi: 10.1088/0305-4470/4/3/013.

Blard, P.-H., Lavé, J., Pik, R., Quidelleur, X., Bourlès, D., and Kieffer, G., 2005, Fossil cosmogenic ^3He record from K-Ar dated basaltic flows of Mount Etna volcano (Sicily, 38°N): Evaluation of a new paleoaltimeter: Earth and Planetary Science Letters, v. 236, p. 613–631.

Bourlès, D.L., 1988, Etude de la géochimie de l'isotope cosmogénique ^{10}Be et de son isotope stable ^9Be en milieu océanique, Application à la datation des sédiments marins [Ph.D. thesis]: Paris-sud Centre d'Orsay, 225 p.

Bourlès, D.L., Raisbeck, G.M., and Yiou, Y., 1989, ^{10}Be and ^9Be in marine sediments and their potential for dating: Geochimica et Cosmochimica Acta, v. 53, p. 443–452, doi: 10.1016/0016-7037(89)90395-5.

Braucher, R., Brown, E.T., Bourles, D.L., and Colin, F., 2003, In situ produced Be-10 measurements at great depths: Implications for production rates by fast muons: Earth and Planetary Science Letters, v. 211, no. 3-4, p. 251–258, doi: 10.1016/S0012-821X(03)00205-X.

Brook, E.J., Kurz, M.D., Ackert, R.P., Raisbeck, G.M., and Yiou, F., 1995, Cosmogenic nuclide exposure ages and glacial history of late Quaternary Ross Sea drift in McMurdo Sound, Antartica: Earth and Planetary Science Letters, v. 131, p. 41–56, doi: 10.1016/0012-821X(95)00006-X.

Brown, E.T., Edmond, J.M., Raisbeck, G.M., Yiou, F., Kurz, M.D., and Brook, E.J., 1991, Examination of surface exposure ages of Antarctic moraines using in situ produced ^{10}Be and ^{26}Al: Geochimica et Cosmochimica Acta, v. 55, p. 2269–2283, doi: 10.1016/0016-7037(91)90103-C.

Carcaillet, J.T., Bourlès, D.L., and Thouveny, N., 2004, Geomagnetic dipole moment and ^{10}Be production rate intercalibration from authigenic $^{10}Be/^9Be$ for the last 1.3 Ma: Geochemistry Geophysics Geosystems, v. 5, no. Q05006.

Chardon, D., and Bellier, O., 2003, Geological boundary conditions of the 1909 Lambesc (Provence, France) earthquake: Structure and evolution of the Trévaresse ridge anticline: Bulletin de la Société Géologique de France, v. 174, no. 5, p. 497–510, doi: 10.2113/174.5.497.

Dunai, T.J., 2001, Influence of secular variation of the geomagnetic field on production rates of in situ produced cosmogenic nuclides: Earth and Planetary Science Letters, v. 193, p. 197–212, doi: 10.1016/S0012-821X(01)00503-9.

Friedmann, E.I., and Weed, R., 1987, Microbial trace-fossil formation, biogenous and abiotic weathering in the Antarctic cold desert: Science, v. 236, p. 703–705.

Gosse, J.C., and Phillips, F.M., 2001, Terrestrial in situ cosmogenic nuclides: Theory and applications: Quaternary Science Reviews, v. 20, no. 14, p. 1475–1560, doi: 10.1016/S0277-3791(00)00171-2.

Ivy-Ochs, S., Kubik, P.W., Masarik, J., Wieler, R., Bruno, L., and Schlüchter, C., 1998, Preliminary results on the use of pyroxene for ^{10}Be surface exposure dating: Schweiz: Mineralogie Petrographie Mitt., v. 78, p. 375–382.

Kabata-Pendias, A., and Pendias, H., 2000, Trace elements in soil and plants: Boca Raton, CRC Press, 413 p.

Kim, K.J., Sisterson, J.M., Englert, P.A.J., Caffee, M.W., Reedy, R.C., Vincent, J., and Castaneda, C., 2002, Experimental cross sections for the production of ^{10}Be from natural carbon targets with 40.6 to 500 MeV protons: Nuclear Instruments and Methods in Physics Research, Section B, Beam Interactions with Materials and Atoms, v. 196, p. 239–244, doi: 10.1016/S0168-583X(02)01297-1.

Kurz, M.D., 1986, Cosmogenic helium in a terrestrial igneous rock: Nature, v. 320, p. 435–439, doi: 10.1038/320435a0.

Lal, D., 1958, Investigations of nuclear interactions produced by cosmic rays [Ph.D. thesis]: Bombay, Bombay University, 90 p.

Lal, D., 1988, In situ produced cosmogenic isotopes in terrestrial rocks: Annual Reviews of Earth and Planetary Sciences, v. 16, p. 355–388.

Lal, D., 1991, Cosmic ray labeling of erosion surfaces: In situ nuclide production rates and erosion models: Earth and Planetary Science Letters, v. 104, p. 424–439, doi: 10.1016/0012-821X(91)90220-C.

Lal, D., Nishiizumi, K., and Arnold, J.R., 1987, In situ cosmogenic ^{3}H ^{14}C and ^{10}Be for determining the net accumulation and ablation rates of ices sheets: Journal of Geophysical Research, v. 92, no. B6, p. 4947–4952.

Lanza, R., and Zanella, E., 2003, Paleomagnetic secular variation at Vulcano (Aeolian Islands) during the last 135 kyr: Earth and Planetary Science Letters, v. 213, p. 321–336, doi: 10.1016/S0012-821X(03)00326-1.

Leya, I., Lange, H.-J., Lüpke, M., Neupert, U., Daunke, R., Fanenbruck, O., Michel, R., Rösel, R., Meltzow, B., Schiekel, T., Sudbrock, F., Herpers, U., Filges, D., Bonani, G., Dittrich-Hannen, B., Suter, M., Kubik, P.W., and Synal, H.-A., 2000a, Simulation of the interaction of galactic cosmic-ray protons with meteoroids: On the production of radionuclides in thick gabbro and iron targets irradiated isotropically with 1.6 GeV protons: Meteoritics and Planetary Science, v. 35, no. 2, p. 287–318.

Leya, I., Lange, H.-J., Neumann, S., Wieler, R., and Michel, R., 2000b, The production of cosmogenic nuclides in stony meteoroids by galactic cosmic-ray particles: Meteoritics and Planetary Science, v. 35, no. 2, p. 259–286.

Marti, K., and Craig, H., 1987, Cosmic-ray-produced neon and helium in the summit lavas of Maui: Nature, v. 325, p. 335–337, doi: 10.1038/325335a0.

Masarik, J., 2002, Numerical simulation of in situ production of cosmogenic nuclides: Geochimica et Cosmochimica Acta, v. 66, no. 15A, p. A491.

Masarik, J., and Reedy, R.C., 1995, Terrestrial cosmogenic-nuclide production systematics calculated from numerical simulations: Earth and Planetary Science Letters, v. 136, p. 381–395, doi: 10.1016/0012-821X(95)00169-D.

Masarik, J., and Reedy, R.C., 1996, Monte Carlo simulation of the in-situ-produced cosmogenic nuclides: Radiocarbon, v. 38, p. 163–164.

Michel, R., Lüpke, M., Herpers, U., Rösel, R., Suter, M., Dittrich-Hannen, B., Kubik, P.W., Filges, D., and Cloth, P., 1995, Simulation and modelling of the interaction of galactic protons with stony meteoroids: Planetary and Space Science, v. 43, no. 3-4, p. 557–572, doi: 10.1016/0032-0633(94)00192-T.

Middleton, R., Brown, L., Dezfouly-Arjomandy, B., and Klein, J., 1993, On ^{10}Be standards and the half-life of ^{10}Be: Nuclear Instruments and Methods in Physics Research, ser. B, v. 52, p. 399–403.

Monaghan, M.C., Krishnaswami, S., and Turekian, K.K., 1986, The global-average production rate of ^{10}Be: Earth and Planetary Science Letters, v. 76, p. 279–287, doi: 10.1016/0012-821X(86)90079-8.

Nagai, H., Imamura, M., Kobayashi, K., Yoshida, K., Ohashi, H., and Honda, M., 1990, High ^{10}Be production rate found in meteoritic carbons: Nuclear Instruments and Methods in Physics Research, Section B, v. 52, p. 568–571, doi: 10.1016/0168-583X(90)90478-D.

Nagai, H., Honda, M., Imamura, M., and Kobayashi, K., 1993, Cosmogenic ^{10}Be and ^{26}Al in metal, carbon, and silicate of meteorites: Geochimica et Cosmochimica Acta, v. 57, p. 3705–3723.

Nishiizumi, K., Klein, J., Middleton, R., and Craig, H., 1990, Cosmogenic ^{10}Be ^{26}Al and ^{3}He in olivine from Maui lavas: Earth and Planetary Science Letters, v. 98, p. 263–266, doi: 10.1016/0012-821X(90)90028-V.

Raisbeck, G.M., and Yiou, F., 1977, The ^{10Be} problem revisited: Plovdiv, Bulgaria, 15th International Cosmic Ray Conference, August 13–26, 1977, Conference Papers, v. 2, (A79-37301 15-93) Sofia, p. 203–207.

Shimaoka, A., Kong, P., Finkel, R.C., Caffee, M.W., and Nishiizumi, K., 2002, The determination of in situ cosmogenic radionuclides in olivine: Geochimica et Cosmochimica Acta, v. 66, no. 15A, p. A709.

Sisterson, J.M., Kim, K., Beverding, A., Englert, P.A.J., Caffee, M., Jull, A.J.T., Donahue, D.J., McHargue, L., Castaneda, C., Vincent, J., and Reedy, R.C., 1997, Measurement of proton production cross sections of ^{10}Be and ^{26}Al from elements found in lunar rocks: Nuclear Instruments and Methods in Physics Research, Section B, Beam Interactions with Materials and Atoms, v. 123, p. 324–329, doi: 10.1016/S0168-583X(96)00409-0.

Southon, J.R., Ku, T.L., Nelson, D.E., Reyss, J.L., Duplessy, J.C., and Vogel, J.S., 1987, ^{10}Be in a deep-sea core: Implications regarding ^{10}Be production changes over the past 420 ka: Earth and Planetary Science Letters, v. 85, p. 356–364, doi: 10.1016/0012-821X(87)90133-6.

Staudacher, T., and Allègre, C.J., 1991, Cosmogenic neon in ultramafic nodules from Asia and in quartzite from Antarctica: Earth and Planetary Science Letters, v. 106, p. 87–102.

Stone, J.O., Allan, G.L., Fifield, L.K., and Cresswell, R.G., 1996, Cosmogenic chlorine-36 from calcium spallation: Geochimica et Cosmochimica Acta, v. 60, no. 4, p. 679–692, doi: 10.1016/0016-7037(95)00429-7.

Stone, J.O., Evans, J.M., Fifield, L.K., Allan, G.L., and Cresswell, R.G., 1998, Cosmogenic chlorine-36 production in calcite by muons: Geochimica et Cosmochimica Acta, v. 62, no. 3, p. 433–454, doi: 10.1016/S0016-7037(97)00369-4.

Stone, J.O., 2000, Air pressure and cosmogenic isotope production: Journal of Geophysical Research, v. 105, no. B10, p. 23753–23759.

Wedepohl, K.H., 1974, Handbook of geochemistry: Berlin, Springer-Verlag, p. 4-A-1–4-G-1.

MANUSCRIPT ACCEPTED BY THE SOCIETY 11 APRIL 2006

Applications of cosmogenic nuclides to Laurentide Ice Sheet history and dynamics

Jason P. Briner[†]
Department of Geology, University at Buffalo, Buffalo, New York 14260, USA

John C. Gosse[‡]
Department of Earth Sciences, Dalhousie University, Halifax, Nova Scotia B3H-4R2, Canada

Paul R. Bierman[§]
Geology Department and School Natural Resources, University of Vermont, Burlington, Vermont 05405, USA

ABSTRACT

Ice sheets play a fundamental role within Earth's climate system and in shaping landscapes. Despite extensive research, the maximum extent and basal dynamics of the Laurentide Ice Sheet (LIS) during the last glacial cycle remain elusive and debated in many areas. Recently, cosmogenic nuclides (e.g., ^{36}Cl, ^{26}Al, ^{10}Be) have played an important role in improving our understanding of LIS extent and behavior. Applications of cosmogenic nuclides to LIS research include surface exposure dating of glacial features, constraining magnitudes of glacial erosion, addressing long-term subaerial exposure and ice sheet burial histories, and burial dating of glacial sediments. These techniques have contributed to the depiction of a more extensive LIS than previously reconstructed for the Last Glacial Maximum. In addition, cosmogenic nuclide research has definitively shown that the LIS covered intensely weathered terrain along its deeply dissected eastern margin, where there were steep gradients in the effectiveness of basal erosion related to basal thermal regime. Cosmogenic nuclide applications, those already employed as well as those yet to be discovered, will undoubtedly continue to contribute to our ever-improving understanding of ice sheet history and dynamics.

Keywords: Laurentide Ice Sheet, cosmogenic nuclide, ice sheet history, ice sheet dynamics, Last Glacial Maximum.

INTRODUCTION

The history and dynamics of the Laurentide Ice Sheet (LIS) have been the focus of research for over a century (e.g., Tyrrell, 1898; Coleman, 1920; Flint, 1943). However, the maximum extent, basal ice dynamics, and pattern of retreat of the LIS during the last glacial cycle have been debated along many of its sectors, in particular in the eastern Arctic (e.g., Miller et al., 2002). The use of in situ–produced cosmogenic nuclides, in conjunction with radiocarbon dating, marine and lake coring efforts, and ice sheet modeling, has led to recent advances in understanding the history and dynamics of glaciers in general (e.g., Phillips et al., 1986; Gosse et al., 1995a) and the LIS in particular (e.g., Gosse

[†]E-mail: jbriner@buffalo.edu.
[‡]E-mail: john.gosse@dal.ca.
[§]E-mail: paul.bierman@uvm.edu.

Briner, J.P., Gosse, J.C., and Bierman, P.R., 2006, Applications of cosmogenic nuclides to Laurentide Ice Sheet history and dynamics, in Siame, L.L., Bourlès, D.L., and Brown, E.T., eds., In Situ–Produced Cosmogenic Nuclides and Quantification of Geological Processes: Geological Society of America Special Paper 415, p. 29–41, doi: 10.1130/2006.2415(03). For permission to copy, contact editing@geosociety.org. © 2006 Geological Society of America. All rights reserved.

et al., 1995b; Bierman et al., 1999; Balco et al., 2002; Briner et al., 2003).

Progress toward understanding the history of the LIS is hampered by limited geochronological tools. Although many different and innovative dating techniques have been applied, radiocarbon dating traditionally has been the most useful and is the basis for much of our current understanding of LIS history during the Last Glacial Maximum (LGM) and deglaciation (Dyke et al., 2002). However, the general scarcity of organic materials associated with glacial deposits, as well as the lag time for vegetation to become established after deglaciation, has limited the application of radiocarbon dating in many areas. An added complication in many sectors of the LIS is the realization that cold-based ice can lead to the preservation of preglacial organic-rich surficial sediments (Davis et al., 2006). It is becoming increasingly clear that while some areas beneath the LIS experienced intense erosion, some landscapes were only slightly modified during the last glacial cycle (e.g., Kleman and Hättestrand, 1999; Bierman et al., 2000; Dredge, 2000; Colgan et al., 2002; Jansson et al., 2003; Briner et al., 2003; Marquette et al., 2004; Staiger et al., 2005), complicating the interpretation of stratigraphic and geomorphic records of the last glaciation.

Cosmogenic exposure dating, based on the buildup of nuclides produced at and near Earth's surface by cosmic-ray bombardment, has recently been added to the geochronological toolbox. Over the past 20 yr, cosmogenic nuclides have filled a niche by making it possible to date directly glacial landforms (Phillips et al., 1986; Nishiizumi et al., 1989; Lal, 1991; Gosse et al., 1995a; Bierman et al., 1999; Briner et al., 2005). Over the last decade, cosmogenic exposure dating has contributed to updated reconstructions of the LIS during the LGM (Dyke et al., 2002; Miller et al., 2002), especially in regions where LGM ice extents have been debated, and contributed to the most recent overall LIS reconstruction (Dyke et al., 2002) (Fig. 1). Dating earlier advances and retreats of the LIS is hampered by the overriding of older deposits by younger ice and the limit of radiocarbon dating to <40 ka. In some cases, marine records distal to the ice sheet record evidence of earlier advances (e.g., Andrews et al., 1998). Recently, cosmogenic nuclide analysis of multiple till sheets (Balco et al., 2005a, 2005b, 2005c), as well as sampling of areas not covered by younger glacial advances (Steig et al., 1998; Kaplan and Miller, 2003), have yielded important chronologic data. In addition, applications that focus on cosmogenic nuclide inheritance (the carryover of cosmogenically produced nuclides from periods of exposure prior to the most recent period of exposure) have been used to address LIS dynamics. Such studies have improved our understanding of ice sheet behavior both in low-relief mid-continent settings (e.g., Colgan et al., 2002) and in the differentially weathered mountain landscapes that fringe eastern Canada (e.g., Bierman et al., 1999; Davis et al., 1999; Briner et al., 2006).

In this paper, we review applications of cosmogenic nuclides that further the understanding of the history and dynamics of the LIS. Whereas Dyke et al. (2002) provide an up-to-date synthesis of the areal extent of the LIS during the LGM, we focus on how our understanding of both the process and history of the LIS has improved from cosmogenic nuclide studies. Although we do not provide a review of the extensive studies of marine records from the continental shelves and slopes, we show how cosmogenic

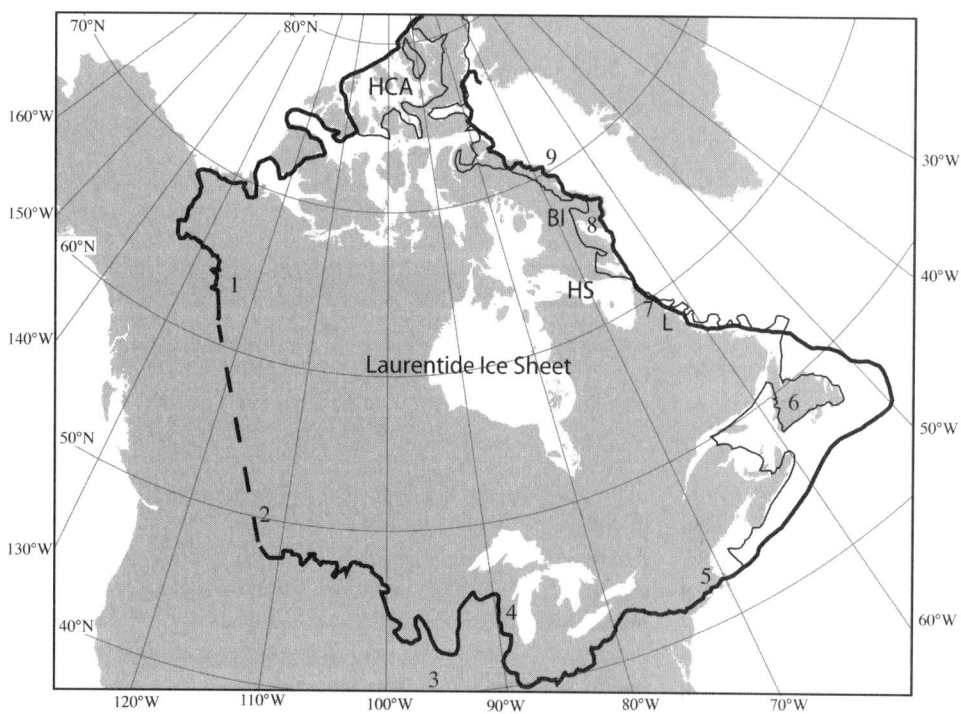

Figure 1. Last Glacial Maximum reconstruction of the Laurentide Ice Sheet (LIS) by Dyke and Prest (1987; thin line) and Dyke et al. (2002; thick line). BI—Baffin Island; HCA—high Canadian Arctic; HS—Hudson Strait; L—Labrador. Numbers refer to sites of cosmogenic nuclide applications on LIS studies. 1—Duk-Rodkin et al. (1996); 2—Jackson et al. (1997, 1999); 3—Balco et al. (2005b); 4—Colgan et al. (2002); 5—Balco et al. (2002); 6—Gosse et al. (1995b, 2006); 7—Clark et al. (2003), Marquette et al. (2004), Staiger et al. (2005); 8—Steig et al. (1998), Bierman et al. (1999), Marsella et al. (2000), Kaplan et al. (2001), Miller et al. (2002), Kaplan and Miller (2003); 9—Briner et al. (2003, 2005, 2006), Davis et al. (2006).

nuclide research has complemented those studies and linked offshore and onshore deglacial chronologies. Following an overview of past and ongoing cosmogenic nuclide applications, we highlight the emerging strengths and challenges of using cosmogenic nuclides to understand ice sheets.

LAURENTIDE ICE SHEET BACKGROUND

The LIS played a dominant role in the climate system during the last glaciation (ca. 120–8 ka; Martinson et al., 1987). Gradually growing to the size of the present-day Antarctic Ice Sheet, perhaps in rapid spurts, and disappearing over a period of ~10 k.y., the LIS increased Earth's albedo, altered atmospheric circulation, sequestered 70–80 m of sea-level equivalent, and significantly altered ocean circulation on millennial timescales (e.g., Imbrie and Imbrie, 1980). Our most complete understanding of the LIS configuration is for the time period of the LGM (22–18 ka) and the subsequent deglaciation (e.g., Dyke et al., 2002).

Reconstructions of the LIS over the past 100 yr have ranged significantly, and often have been cast as minimum versus maximum paradigms. A paradigm shift occurred when Flint (1943) popularized an extensive LIS covering North America during the LGM. Flint's reconstruction had ice extending from the continental shelves in the north to the continental United States in the south. Flint's mode was conceptual, included a single ice dome over Hudson Bay during its maximum phase, and overlooked previous field-based work that depicted relatively restricted ice extent at high latitudes (e.g., Tyrrell, 1898; Coleman, 1920).

The next paradigm shift occurred in the 1960s and 1970s, when the first intensive field-based research was carried out along LIS margins in the north (e.g., Ives and Andrews, 1963; Løken, 1966; Andrews et al., 1970; Grant, 1977). These investigations supported a thinner and less extensive LIS than proposed in Flint's model (Fig. 2), and suggested that the LIS contained multiple domes (Prest et al., 1968; Bryson et al., 1969; Miller and Dyke, 1974; Grant, 1977). These conclusions were based on old radiocarbon ages and the assumption that weathered bedrock terrain indicated ice-free conditions during the LGM. Well into the minimum ice paradigm, some researchers continued to favor a maximum ice model, relying on theoretical ideas based on lessons from Antarctica and the assumption that weathered bedrock terrain could have been covered by ice during the LGM (Fig. 2) (Hughes et al., 1977; Sugden and Watts, 1977; Denton and Hughes, 1981).

Throughout the 1980s, the majority of field evidence continued to point toward relatively restricted ice margins (e.g., ice terminating at the heads of fiords and sounds; Dyke and Prest, 1987), and the controversy regarding the existence of an ice-free corridor between the LIS and the Cordilleran Ice Sheet (CIS) remained unresolved (e.g., Jackson, 1980; Liverman et al., 1989). In the eastern sector of the LIS, field observations and the radiocarbon chronology revealed a far more complicated history of ice cover than could be explained with LIS cover flowing solely from the west (Brookes, 1977; Grant, 1977). With the addition of

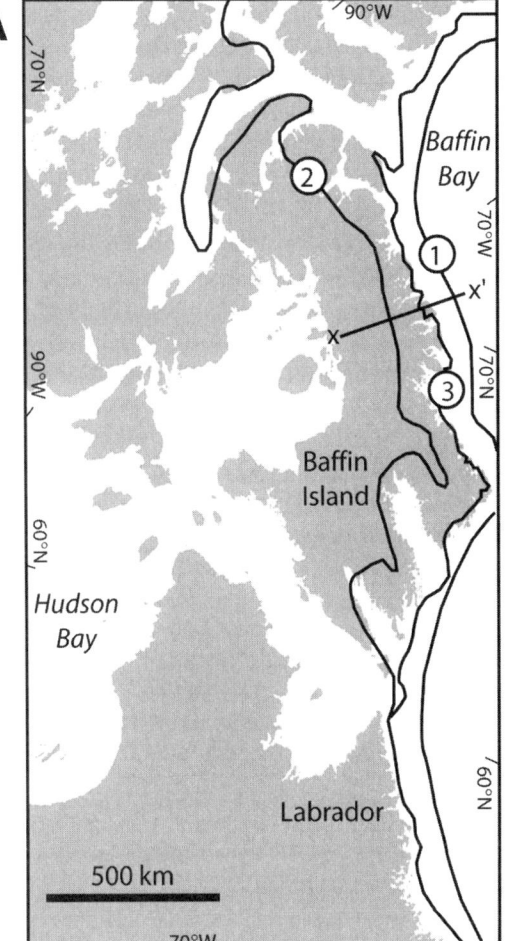

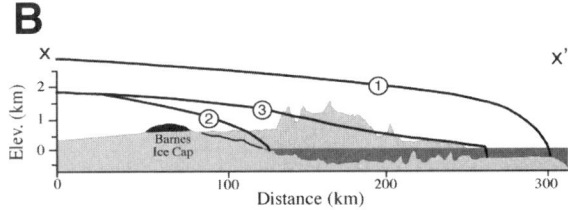

Figure 2. A: Map of eastern Canadian Arctic showing three Last Glacial Maximum reconstructions of the Laurentide Ice Sheet. 1—maximum extent representative of the Flint (1943) paradigm; 2—minimal extent from Dyke and Prest (1987); 3—most recent reconstruction of Dyke et al. (2002). B: Schematic cross section showing ice sheet profiles superposed on topographic profile for the landscape near x–x′ line in A.

subsequent offshore data (Stea et al., 1998; Piper and MacDonald, 2001), it seemed clear that a multidomal LIS covered the entire Atlantic provinces and Maine, or that there were peripheral ice caps over the Atlantic provinces.

Meanwhile, evidence was increasing for a highly dynamic LIS (Andrews and Miller, 1979; Andrews et al., 1985). Research on the Antarctic ice streams revealed that high domes at the center of ice sheets could be drained through low-gradient (~1–2 m/km), very fast-moving (~500 km/yr) ice streams that account for most of an ice sheet's discharge into adjacent seas (Bentley, 1987). Based on the Antarctic Ice Sheet as an analogue (Hughes et al., 1985), research was carried out along the former bed of the LIS, and indicated that it also had low-gradient outlet glaciers and ice streams where it flowed over soft sediments on land (e.g., Clark, 1994; Patterson, 1998) and in marine troughs (e.g., Alley and MacAyeal, 1994). In the North Atlantic deep sea, widespread layers of ice-rafted detritus (Heinrich layers) were found periodically throughout cores of marine sediments deposited during the last glacial cycle (Heinrich, 1988; Bond et al., 1992, 1993). Heinrich-type layers of detrital carbonate were also found in the Labrador Sea and in Baffin Bay (e.g., Andrews et al., 1998), and were traced to Hudson Strait (Andrews and Tedesco, 1992; Andrews et al., 1998; Grousset et al., 2000; Andrews and MacLean, 2003; Hemming and Hajdas, 2003). The combination of terrestrial, marine, and ice sheet modeling studies led to a new understanding of LIS behavior that was far more dynamic than previously envisioned (e.g., Alley and MacAyeal, 1994; Kaplan et al., 1999).

Continued examination of ice extent using new approaches and techniques gradually challenged the minimum LIS model and improved our understanding of ice dynamics (Fig. 2) (Jennings, 1993; Bierman et al., 1999; Marsella et al., 2000; Kaplan et al., 2001; Miller et al., 2002; Marquette et al., 2004). Between earlier (Dyke and Prest, 1987) and more recent compilations of LIS limits during the LGM (Dyke et al., 2002), reconstructed ice margins were expanded in several regions in eastern Canada as a result of both terrestrial (lake sediment and cosmogenic nuclide exposure dating) and marine (coring, imaging) research (e.g., Josenhans et al., 1986; Clark and Josenhans, 1990; Jennings, 1993). Analysis of remotely sensed data revealed mega-scale bedforms across the former footprint of the LIS, and indicated remarkable variation in past ice flow patterns (Boulton and Clark, 1990; Clark, 1997; Veillette et al., 1999; Stokes and Clark, 2002). In Labrador, Klassen's (1994) interpretation of till geochemistry and clast provenance data revealed more than ten large-scale (>200 km) crosscutting flow directions, indicating complex paleoglacier dynamics over the region.

Much of the variation in the LIS ice flow record was thought to have occurred during the last glacial cycle, and its discovery led to three important conclusions: (1) LIS domes that persisted throughout the last glacial cycle were not only deglacial phenomena, but existed and shifted their location throughout the last glacial cycle, (2) the shifting of domes was likely related to ice stream activity (e.g., Stokes and Clark, 2001; Jansson et al., 2003), and (3) in some places erosion was not substantial enough to remove prior bedforms, indicating minimal glacial erosion and landscape preservation near ice sheet centers (Kleman and Hättestrand, 1999). Several studies showed that large portions of former ice sheets were frozen to the bed and preserved ancient (preglacial) terrain upon deglaciation (e.g., Shilts et al., 1979; Dyke, 1993; Kleman, 1994; Dredge, 2000).

COSMOGENIC NUCLIDE BACKGROUND

Isotopes produced in situ at Earth's surface (e.g., the radionuclides ^{36}Cl, ^{26}Al, ^{14}C, ^{10}Be, and the noble gases ^{21}Ne and ^{3}He) via bombardment by cosmic rays are now used routinely in geomorphic studies, including ice sheet research. Because these isotopes are mostly produced in the upper several meters of Earth's surface (Lal and Peters, 1967), a thickness that is readily influenced by glacial erosion, ice sheet processes and their timing can be constrained. Readers are referred to recent comprehensive review papers on cosmogenic nuclide systematics and applications for information on the method (Gosse and Phillips, 2001; Cockburn and Summerfield, 2004).

Cosmogenic nuclides have been used to address LIS history and dynamics in three main ways. First, cosmogenic nuclide exposure dating (Gosse and Phillips, 2001, and references therein) is used to constrain the timing of deglaciation by dating the deposition of erratics and moraine boulders and the exposure of ice-sculpted bedrock surfaces. A second application involves comparing measured nuclide concentrations to independent ages. In this approach, the presence or absence of inheritance (e.g., Briner and Swanson, 1998) is used to address the efficiency of glacial erosion and to elucidate the spatial patterns of erosive versus nonerosive ice. A third application is based on measuring pairs of cosmogenic nuclides with different half-lives in shielded bedrock surfaces or buried sediments on long (>100 k.y.) timescales. This multiple-nuclide approach allows nonunique inferences about surface exposure, burial, and erosion histories over time.

APPLICATIONS

Chronology: Western and Southern LIS Margins

Several cosmogenic nuclide studies over the last decade have dated the timing of LIS deglaciation (Fig. 1). Most of these studies provide age information in areas with previously scant chronology and have made substantial contributions toward solving historical debates concerning the extent of the LIS during the LGM. At the northwestern margin of the LIS (Fig. 1), Duk-Rodkin et al. (1996) report four ^{36}Cl ages that constrain the timing of deglaciation from the LGM advance to 20–28 ka, and two ^{36}Cl ages from a recessional limit to ca. 19 ka. Along the southwestern sector (Fig. 1), Jackson et al. (1997, 1999) provided 19 ^{36}Cl ages from erratics and moraine boulders in an area where the age of the most recent LIS and CIS convergence had remained controversial. All but one of the samples yielded ages between ca. 12 and ca. 18 ka (one anomalously old age is suspected to have resulted from inheritance), indicating an absence of an ice-free corridor between the LIS and CIS during the LGM (Jackson et al., 1997, 1999). At the southeastern margin, where the LIS terminated along the present coast of the northeastern United States (Fig. 1), Balco et al. (2002) provide boulder ^{10}Be and ^{26}Al ages from a terminal moraine and a stratigraphically younger end moraine, respectively. The age of the terminal moraine is 23.2

± 0.5 ka (n = 13); the younger moraine is 18.8 ± 0.4 ka (n = 10). Farther north, Gosse et al. (2006) dated boulders to constrain the timing of deglaciation at 13.5 ± 0.4 ka (n = 4) on the Saint John's Highlands, Great Northern Peninsula of Newfoundland, and 20.9 ± 3.0 ka on a boulder on the summit of Big Level. Both localities were previously believed to be ice-free "nunataks" during at least the last and penultimate glaciations, if not the entire Quaternary (Grant, 1977). These ages helped spawn new interpretations from other regions in western Newfoundland that were also thought to have been ice-free (Brookes, 1977).

Chronology: Northeastern Margins

Several studies have taken place along the northeastern sector of the LIS, an area with historically controversial LGM reconstructions. The glacial history of the Torngat Mountains, Labrador (Fig. 1) was intermittently studied over the past century (e.g., Daly, 1902; Coleman, 1920; Odell, 1933; Ives, 1957, 1978; Clark, 1988; Bell et al., 1989). Owing to the difficulty of access and scale of the range, most studies concentrated in areas with the best-preserved record or where coastal inlets provided access. Most of this mountain range (and the Arctic regions discussed next) is above the tree line, and radiocarbon chronologies are mostly limited to sediments below the marine limit. The principal contributions in Labrador from cosmogenic nuclide studies are the dating of LGM and recessional moraines that could never be dated previously (Clark et al., 2003; Marquette et al., 2004). The prominent Saglek moraine system was shown to consist of two or more components, with an outer ridge dating in the range of 25–18 ka (Marquette et al., 2004), and an inner ridge dating from 13.5 to 11 ka (Clark et al., 2003; Marquette et al., 2004). This new dating built on the only previous age on the Saglek moraine, which was a basal lake sediment date at Square Pond along the outermost and highest moraine ridge (Clark et al., 1989). Younger moraines higher in the fiord and in cirques have been dated to trace the deglacial history through the Holocene (Marquette et al., 2004). However, it was the ubiquitous lack of moraines (with some exceptions in the Nachvak Fiord region) (Bell et al., 1989) for the older glaciations that caused earlier workers to interpret the coastal highlands of the Torngat Mountains to be ice-free (e.g., Koroksoak Zone of Ives, 1957, 1978) and to interpret altitudinally separable zones of differential weathering as trimlines indicating the vertical limit of penultimate glaciations. In lieu of moraines, Marquette et al. (2004) dated boulders exposed on local promontories. The selected boulders were often only subtly different from the underlying bedrock (e.g., in color, grain size, or mineralogy); however, in most instances the boulders were from the Archean gneisses of the Torngat Mountains, which led previous authors to preclude them as erratics. The majority of boulders dated to ca. 11.7 ka (n = 14; Marquette et al., 2004) and were interpreted to indicate that ice persisted through the Younger Dryas chronozone even on the broad summits that were previously interpreted as nunataks. The lack of boulders on the highest steep peaks may indicate they projected through the LIS as nunataks.

Several cosmogenic nuclide studies have shown that glacial features on Cumberland Peninsula, Baffin Island (Fig. 1), previously believed to be pre-LGM in age are in fact LGM features (Steig et al., 1998; Marsella et al., 2000; Kaplan et al., 2001; Miller et al., 2002; Kaplan and Miller, 2003). These studies favor a larger LGM ice extent than previously depicted in the minimum ice model, supporting marine core findings (e.g., Jennings, 1993). On northern Cumberland Peninsula, Steig et al. (1998) provided six exposure ages from surface clasts from three lateral moraine segments. The ages increase on morphostratigraphically older moraines, from 13.4 ± 3.0 ka (n = 3) to 20.6 ± 0.6 ka (n = 1) to ca. 35.7 ± 1.1 ka (n = 2). In the uplands above the highest moraine, bedrock tors have ages of >50 and >65 ka, and lake basins have sediment preserved from prior to the LGM (Steig et al., 1998; Wolfe et al., 2001).

On southern Cumberland Peninsula, the prominent Duval moraines, previously ascribed to pre-LGM based on radiocarbon ages and relative weathering criteria (Dyke, 1979), were part of a comprehensive exposure dating campaign (n = 47) that revealed their LGM age (Marsella et al., 2000). The age interpretation is not without complications, however, even considering the large sample set (n = 17) that was obtained on the Duval and associated recessional moraines. There is a clear bimodal distribution of the ages centered at ca. 22 and ca. 10 ka, which also exists outside of the Duval limit, suggesting that the outer LGM ice margin does not correspond to a moraine. In addition, the 25 ages that Kaplan et al. (2001) reported from scoured bedrock in adjacent Cumberland Sound support LGM ice occupation of the sound to the continental shelf with deglaciation occurring ca. 11–12 ka. However, many of the bedrock samples contain inheritance (Kaplan and Miller, 2003) indicating limited glacial erosion. Finally, six exposure ages from eastern Cumberland Peninsula were reported for the first time in Miller et al.'s (2002) synthesis of the LIS on Baffin Island. Similar to the Steig et al. (1998) study, five of the ages come from a flight of three lateral moraine segments bordering Sunneshine Fiord, with ages of 7.6 ± 0.7 ka (n = 2) to 22.0 ± 2.3 ka (n = 1) to 35.4 ± 3.5 ka (n = 2). An upland bedrock sample is older than 77 ka. The youngest ages are in contradiction with an ocean sediment core taken within the moraines that has a basal age of ca. 14 ka, but appear to be generally compatible with the northern Cumberland Peninsula ages (Miller et al., 2002).

Briner et al. (2005) obtained exposure ages from 103 erratics on the Clyde Foreland to determine the LGM ice limit on northeastern Baffin Island. The foreland was subdivided into zones covered by erosive ice (dominated by moraines and an overall scoured appearance) and nonerosive ice (dominated by lateral meltwater channels and nonglacial topography). Exposure ages from both zones are consistent with LGM ice cover of the foreland, but only a subset of the ages from the unscoured zone fall within the LGM/deglacial period, indicating an overall high amount of inheritance. Davis et al. (2006) obtained 13 exposure ages on erratics on and beyond the raised Aston Delta (Løken, 1966), ~50 km to the south of the Clyde Foreland. Because the intact delta is radiocarbon dated (>54 ka) to prior to the LGM, the

Aston Delta became a cornerstone for limiting LGM ice to fiord and sound heads. The 14.1 ± 0.8 ka average exposure age based on seven erratics that constitute the youngest age cluster reveals that a nonerosive LIS overran the delta during the LGM (Davis et al., 2006). These studies on northeastern Baffin Island, together with the exposure ages from Cumberland Peninsula, clearly illustrate the complicated nature of exposure dating in terrains covered by minimally erosive ice.

Cosmogenic nuclides have been used in other applications aimed at understanding the LIS with mixed success. Erratics from northernmost Baffin Island have ^{10}Be and ^{26}Al ages consistent with restricted LGM ice, contradicting a growing body of evidence from elsewhere; however, the ages are difficult to place in context or evaluate as they have only appeared in one abstract (McCuaig et al., 1994). Raised shorelines provide important indications of the history of ice thickness and rate of deglaciation (Walcott, 1970). Emergence curves reveal the half response time of different regions, and provide insights into mantle rheology (Peltier, 1996). In arctic Canada, shoreline emergence curves have been reconstructed from radiocarbon-dated driftwood and marine and land fauna (Dyke et al., 1992). Uncertainty in the marine radiocarbon reservoir effects and reworking of wood and shell material often limit the accuracy and precision of emergence curves. Cosmogenic nuclide dating of shorelines was done first on Prescott Island in the central Arctic (Gosse et al., 1998). Ages on boulders ($n = 6$) that arose from the reworking of LIS till on beaches from modern to 120 masl allowed the construction of emergence curves indistinguishable from the radiocarbon-dated ($n = 41$) emergence history of Dyke et al. (1992) (Fig. 3). However, shoreline cosmogenic nuclide ages from cobbles, pebbles, and plucked bedrock surfaces ($n = 16$) typically yielded ages older than the radiocarbon data, probably due to nuclide inheritance.

Ice Sheet Dynamics

Cosmogenic nuclides are also used to reconstruct spatial patterns of glacial erosion by the LIS. The origin of differentially weathered mountain landscapes has led to historically contrasting views on ice sheet extent across the Northern Hemisphere. While some workers have argued that highly weathered uplands persisted as nunataks during the last glaciation (e.g., Ives, 1957, 1978; Grant, 1977; Nesje and Dahl, 1990; Steig et al., 1998; Ballantyne, 1998; Rae et al., 2004), others have suggested that different "weathering zones" (Pheasant and Andrews, 1973; Boyer and Pheasant, 1974; Ives, 1978) represent different basal thermal regimes, hence differential glacial erosion by an overriding LGM ice sheet (Sugden, 1977; Sugden and Watts, 1977; Hall and Sugden, 1987; Bierman et al., 2001; Briner et al., 2003; Hall and Glasser, 2003; André, 2004; Marquette et al., 2004; Staiger et al., 2005; Gosse et al., 2006). Until the application of cosmogenic nuclides to these landscapes, their history and relation to ice sheet extent remained enigmatic. Cosmogenic nuclide data from differentially weathered mountain landscapes have several important implications for interpreting nunataks, trimlines, perched erratics, ice sheet thickness, ice sheet extent, spatial patterns of basal thermal regimes and glacial erosion, and ultimately ice sheet dynamics. Thus, all of these topics are covered in this section.

Differentially weathered mountain landscapes have been examined in several areas along the eastern LIS. Several samples from the Marsella et al. (2000) study address differentially weathered landscapes. For example, the exposure ages of erratics both above and below the prominent trimline at the Duval moraines yield the same age distribution, suggesting that the trimline does not represent the LGM ice limit. In addition, Bierman et al. (2001) report a ca. 9 ka upland erratic resting adjacent to >100 ka highly weathered bedrock, indicating the emplacement of the erratic by cold-based ice that did not erode the bedrock. Since the investigations on Cumberland Peninsula and in Newfoundland, more samples have been collected and analyzed both from Baffin Island and from Labrador. On northeastern Baffin Island, Briner et al. (2003) reported three deglacial-aged errat-

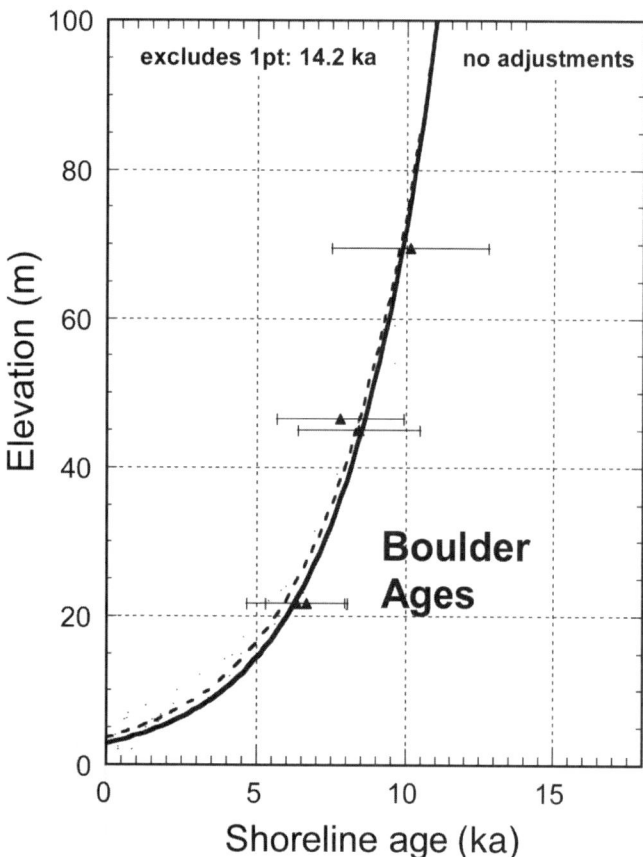

Figure 3. ^{10}Be ages on boulders on beaches of Prescott Island, central Arctic. Ages are not corrected for small effects of snow cover, erosion, or changes in production rate due to uplift. One age (14.2 ka) is not plotted and is considered an outlier based on the regional deglacial history according to Dyke et al. (1992). Errors are 1σ on AMS precisions. Solid curve is exponential fit to ^{10}Be ages. Dashed curve is driftwood-based radiocarbon-dated shoreline emergence curve for adjacent Prince of Wales Island (Dyke et al., 1992).

ics perched on bedrock surfaces (Fig. 4) that have much older exposure ages (older than 60 ka); these data mandate cold-based LIS ice cover on highly weathered interfiord uplands during the LGM. Following their initial study, Briner et al. (2006) obtained cosmogenic nuclide measurements from 27 erratics perched in upland weathered terrain and 33 bedrock samples from three different weathering zones across four fiords on northeastern Baffin Island. These data confirm earlier hypotheses of upland glacierization during the LGM (e.g., Sugden and Watts, 1977; Sugden, 1977), and provide new information on patterns of glacial erosion of a fiord landscape. For example, ^{10}Be and ^{26}Al concentrations increase and their ratios decrease with elevation from fiords to interfiord uplands (Briner et al., 2006). The data indicate that significant glacial erosion (>2 m) occurred in fiords and valleys, between one and two meters of erosion occurred at intermediate elevations, and negligible erosion occurred at the highest elevations during the last glaciation. This pattern of erosion also existed prior to the LGM, as revealed by ^{26}Al/^{10}Be ratios, which provide information on long-term exposure, burial, and glacial erosion histories (see below).

Cosmogenic nuclide measurements on bedrock in the Torngat Mountains, Labrador (Fig. 1), show an increasing trend in concentration with increasing elevation (Staiger et al., 2005). Bedrock in the valleys has cosmogenic nuclide concentrations equivalent to adjacent boulders, whereas bedrock ages are ca. 20–90 ka in the intermediate weathering zone and ca. 80–300 ka in the highly weathered zone. Because ice covered all but the highest peaks, as recently as 11 ka, Staiger et al. (2005) proposed that the ice eroded bedrock more deeply at low elevations than on higher summit plateaus, in accordance with changes in basal thermal regime. The interpretation that erosion rates vary beneath polythermal ice is supported by soils (Marquette et al., 2004) and numerical model simulations of basal sliding and hydrology of the LIS and local ice caps in Labrador (Staiger et al., 2005). Measurements of ^{10}Be inventories in middle Holocene glaciolacustrine deltaic sediment in north-central Baffin Island also reveal a large inheritance component (Gosse et al., 2005). If the deltaic sand ^{10}Be concentrations were interpreted as ages, they would correspond to exposures older than 30 ka. This indicates that even in the glaciofluvial system that fed the delta, most of the sediment is not sourced from meters below the surface, but instead from till and bedrock that has not been significantly eroded by the last ice cover.

A study from the southern margin of the LIS also indicates the spatial variability of glacial erosion. Colgan et al. (2002) provide cosmogenic nuclide measurements on 16 LIS-sculpted bedrock samples near the margin of the Green Bay lobe. A significant amount of inheritance and sample-to-sample variability was found in samples ($n = 7$) that were collected from erosion-resistant lithologies in low-lying areas close to the LGM margin. In contrast, samples collected farther from the ice margin gave tightly clustered model ages consistent with extensive, preexisting radiocarbon estimates for the age of deglaciation. Combined with studies from arctic terrains (e.g., Bierman et al., 1999; Briner et al., 2003, 2005, 2006; Marquette et al., 2004; Davis et al., 2006), these data highlight the spatial variability of glacial erosion.

Cosmogenic Nuclide Ratios

In some cases, cosmogenic nuclides have been used to reconstruct the long-term history of the LIS by exploiting the faster decay of ^{26}Al ($t_{1/2} = 700$ k.y.) with respect to ^{10}Be ($t_{1/2} = 1.5$ m.y.; Lal, 1991; Bierman et al., 1999; Fabel and Harbor, 1999). This isotopic disequilibrium can be used to detect periods of prolonged shielding from cosmic rays following initial exposure. Prolonged burial is implied when the measured ratio of ^{26}Al/^{10}Be is lower than that expected from the measured concentration of ^{10}Be.

Figure 4. Erratics perched on weathered bedrock from Labrador (A) and Baffin Island (B).

Gosse et al. (1993, 1995b) and Bierman et al. (1999) first applied ^{26}Al/^{10}Be data in glacial terrains by sampling upland tors along the eastern margin of the LIS. In Newfoundland, tor samples recorded a minimum of ~100 k.y. of burial, indicating that they had been covered by glacier ice, perhaps intermittently, but not significantly eroded, for that duration (Gosse et al., 1993, 1995b, 2006). The ^{26}Al/^{10}Be ratio for the least weathered (lowest) zone is 5.9 ± 0.5 (n = 1), which is the same as the ratio expected for a continuously exposed surface in the late Pleistocene (~6.0 ± 0.1; Gosse and Phillips, 2001). On the other hand, the ratios of tor-like bedrock knobs in the most weathered (highest) zone are substantially lower at 4.4 ± 0.4 and 5.1 ± 0.2. These ratios show that at least some of the bedrock surfaces within the highest weathering zone have an exposure history that was substantially interrupted by prolonged shielding. Seven tor samples from Cumberland Peninsula on Baffin Island have ^{26}Al and ^{10}Be values consistent with at least 550 k.y. of surface history, including at least 420 k.y. of burial (Bierman et al., 1999). These were some of the first data demonstrating nonerosive cold-based ice cover of weathered uplands, and were suggestive of LGM ice cover, a hypothesis supported by later data from a nearby erratic (Bierman et al., 2001).

Marquette et al. (2004) and Briner et al. (2006) provide additional ratio data for Labrador and northeastern Baffin Island, respectively. In Labrador, ^{26}Al/^{10}Be data from four bedrock samples representing all three weathering zones reveal a lack of long-term burial in the recently eroded zone (n = 1), and increasing burial with elevation from the intermediately weathered (~80 k.y.; n = 1) to highly weathered (~175 k.y.; n = 2) zones. Briner et al. (2006) found similar results, showing no long-term burial in recently sculpted (n = 6) or intermediately weathered (n = 7) bedrock, but an average of ~300 k.y. of burial in upland tors (n = 10).

In midwestern North America, Bierman et al. (1999) used the discordance of ^{26}Al/^{10}Be ratios to place limits of the age of four Sioux Quartzite outcrops exposed outside the LGM margin. The exceptionally hard, striated bedrock had ^{26}Al/^{10}Be ratios varying from 4.3 to 5.2. Although there is no unique interpretation for these data, they allowed Bierman et al. (1999) to set limits on the exposure history and suggest that the bedrock was last covered by ice at least half a million years ago. Balco et al. (2005a, 2005b) have made recent progress toward extending the use of cosmogenic nuclides to LIS deposits farther back in time by developing a technique that utilizes ^{10}Be and ^{26}Al concentrations and ratios in buried tills in the midwestern United States. The ages of several early and middle Pleistocene tills have been constrained (e.g., Balco et al., 2005b), and relatively abundant inheritance in some tills allowed Balco et al. (2005c) to constrain the timing of the oldest maximum ice sheet advances (ca. 2.4 Ma) that overran and reworked pre-Quaternary regolith that had high cosmogenic nuclide concentrations.

EMERGING STRENGTHS AND CHALLENGES

After a decade of cosmogenic nuclide applications to LIS research, we have an improved understanding of LIS chronology, LGM ice extent, and ice sheet dynamics. At the same time, we have a much better appreciation of the complexities associated with cosmogenic nuclide applications, especially those involved with using exposure dating in landscapes formerly covered by cold-based ice.

Because exposure dating studies are complicated by landform instability and inheritance, a relatively large number of samples has the potential to provide a more precise chronology and a more robust means by which to identify outliers. In landscapes covered by cold-based ice, inheritance is prevalent and spatially complicated (Fig. 5). Davis et al. (1999) and Colgan et al. (2002) explicitly deal with inheritance, and together provide some guidelines on minimizing its effects. Of 16 samples dated in the glacially scoured terrain at Cumberland Sound by Kaplan and Miller (2003), at least seven contain nuclides inherited from previous periods of surface or near-surface exposure. Both Marsella et al. (2000) and Briner et al. (2005) report large numbers of moraine and erratic boulders that appear to contain inherited nuclides (50% of Duval moraine boulders, 60% of Clyde Foreland erratics); these studies took place in terrains partially covered by cold-based ice. Boulders with inherited nuclides were either deposited during the most recent ice sheet advance already carrying that inheritance, or deposited on the landscape during a prior advance, then received exposure that was subsequently preserved beneath cold-based ice of the youngest advance. Although it is difficult to separate these scenarios, a cluster of samples with similar inheritance has been interpreted to indicate that boulders can survive beneath younger advances (Marsella et al., 2000; Briner et al., 2005). Davis et al. (2006) report the preservation of a pre-LGM delta (the Aston Delta), along with associated beaches and meltwater channels, beneath cold-based LGM ice on Baffin Island. This finding supports the notion that cold-based ice can overrun preexisting moraines, erratics, and unconsolidated deposits with minimal disturbance. Similarly, clasts isolated from tills in interior Baffin Island contain abundant inheritance, supporting minimal glacial erosion there (Staiger et al., 2004). Inheritance is clearly present and even prevalent within the LGM extent of the LIS (Fig. 5).

Cosmogenic nuclide studies that have taken place in differentially weathered mountain landscapes explicitly address historically debated LGM ice sheet reconstructions. Nuclide concentrations measured in bedrock confirm that weathering zones represent different surface exposure durations, supporting relative weathering studies. The interpretation that differential ice sheet erosion (Fig. 6) gave rise to differentially weathered mountain landscapes is supported by the abundance of bedrock surfaces within the LGM ice limit that have pre-LGM cosmogenic nuclide concentrations and the presence of perched erratics on weathered uplands (Fig. 4). Because ^{10}Be and ^{26}Al half-lives are too long to detect recent burial, ^{26}Al/^{10}Be data only provide information on long-term ice sheet burial history. On the other hand, exposure ages from perched erratics can constrain the age of the most recent period of upland glacierization. Dating upland erratics has the highest potential to address LGM ice sheet reconstructions

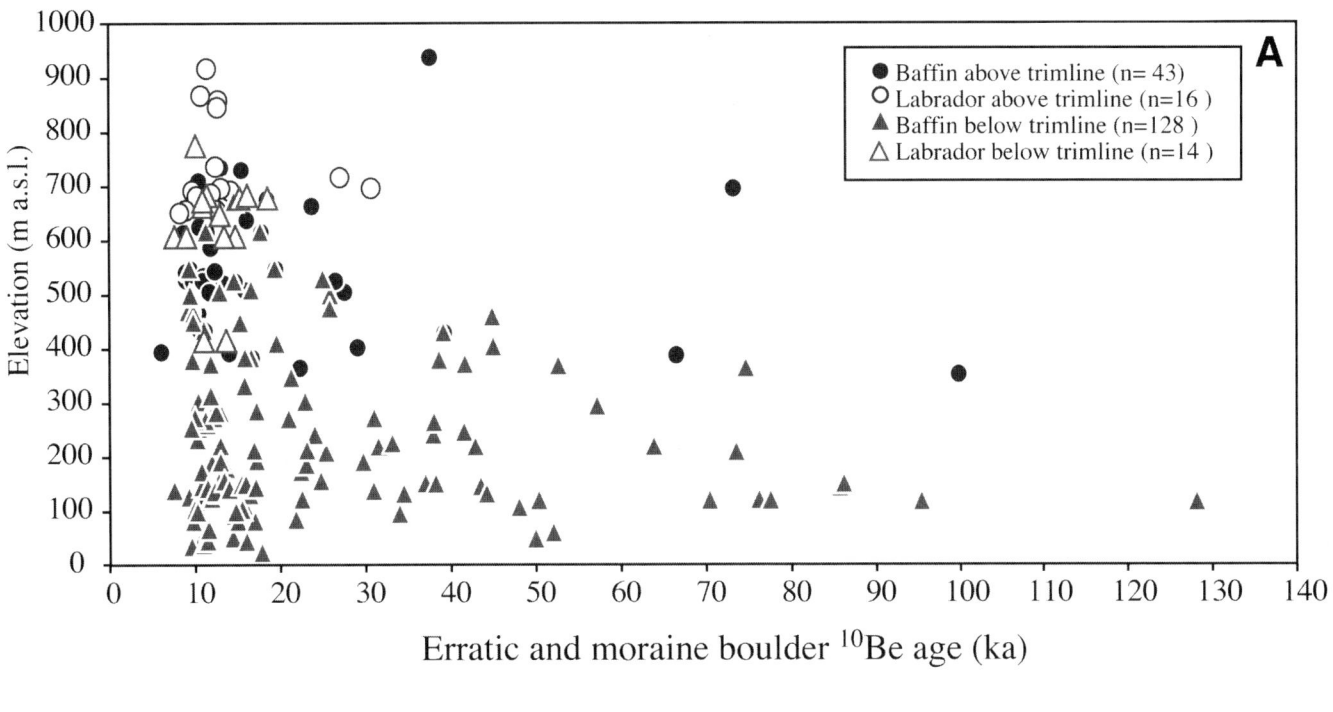
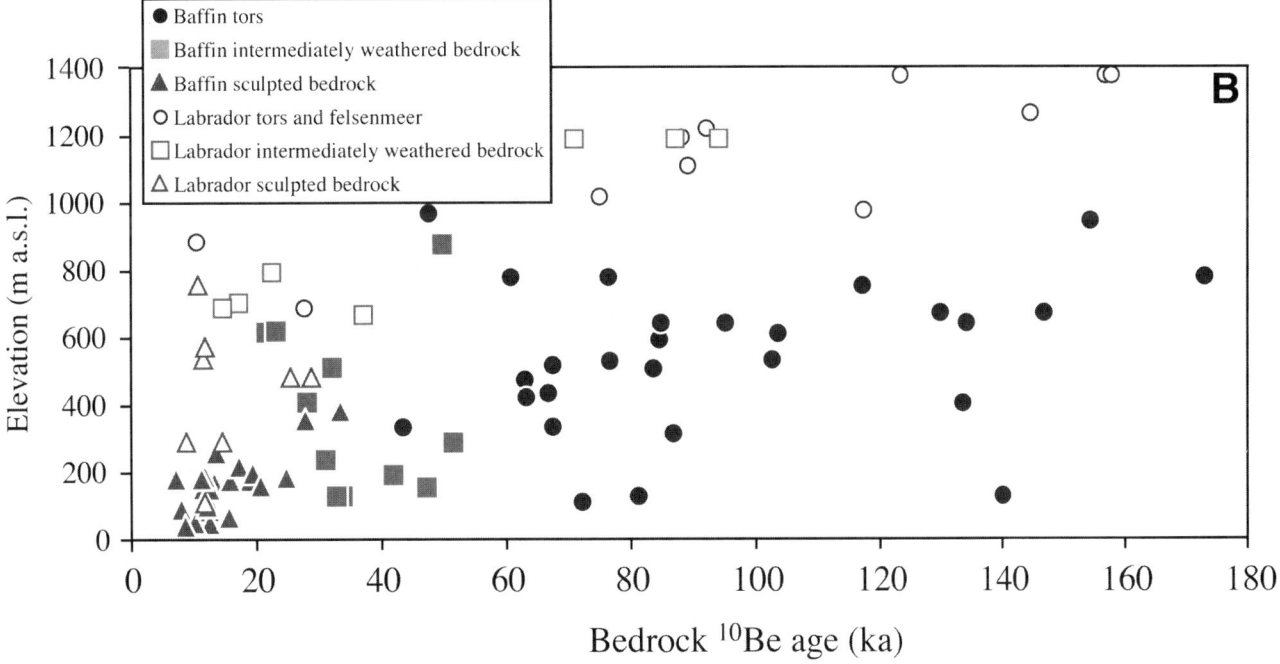

Figure 5. Erratic and moraine boulder (A) and bedrock (B) cosmogenic nuclide ages from the published literature from Labrador and Baffin Island. All ages have been recalculated using the same production rate (5.1 g atoms^{-1} yr^{-1}). A: All samples are interpreted to be from within the LGM ice limit. Samples from above and below regional trimlines have pre-LGM cosmogenic nuclide ages, indicating the abundance of inheritance. B. Cosmogenic nuclide ages are subdivided into sculpted, intermediately weathered, and tor samples, revealing increasing inheritance with surface type/elevation. Because the boundaries (trimlines) between surface type dip seaward, absolute elevation has only a moderate correlation with cosmogenic nuclide age.

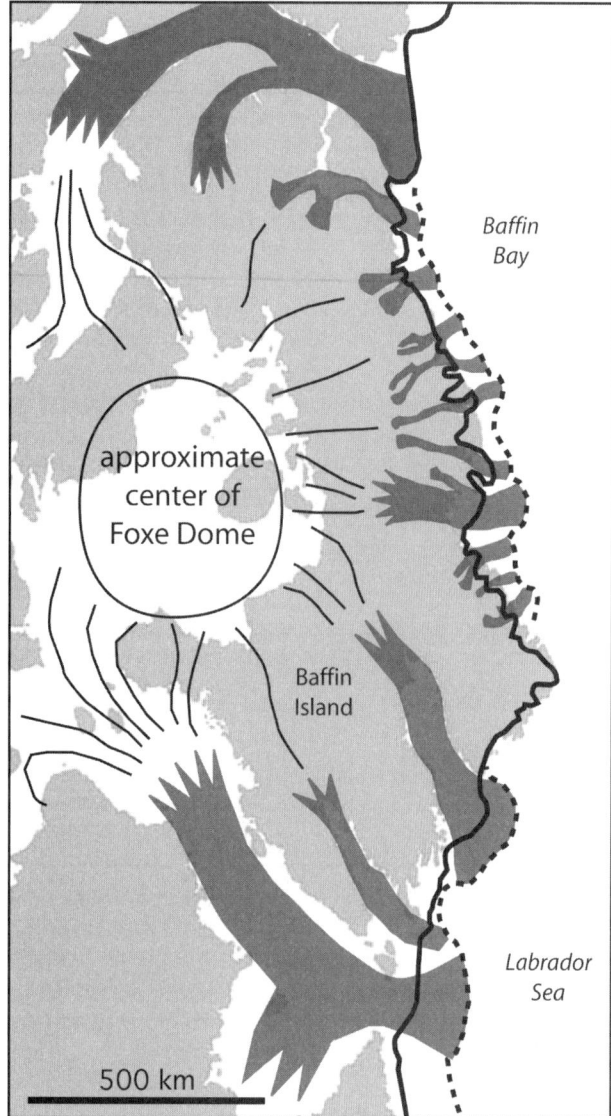

Figure 6. Northeastern sector (Foxe Dome) of the Laurentide Ice Sheet showing the most recently reconstructed Last Glacial Maximum ice limit (thick black line) from Dyke et al. (2002) and hypothesized Last Glacial Maximum ice limit (dashed lines) from Briner et al. (2006). The thin black lines represent ice flow, and the gray areas represent the hypothesized locations of ice streams, whose flow pattern gave rise to differentially weathered mountain landscapes.

nuclide concentrations. In other cases, the preservation of pre-LGM lake sediments in lake basins (e.g., Wolfe et al., 2000) or lateral meltwater channels has provided independent evidence for cold-based ice (e.g., Briner et al., 2005; Davis et al., 2006). Studies in other parts of the world have documented the glacial modification of highly weathered features (e.g., Hall and Glasser, 2003; André, 2004) and glacial erosion by cold-based ice (e.g., Cuffey et al., 2000; Atkins et al., 2002). Combining this type of detailed research with cosmogenic nuclide measurements will add to our growing understanding of ice sheet dynamics.

CONCLUSIONS

The application of cosmogenic nuclides has made substantial contributions toward LIS extent and dynamics. Exposure dating has constrained the age of LIS marginal deposits that previously lacked chronologic control, and has contributed to the resolution of long-standing debates in the Cordillera, eastern Canadian Arctic, and the Maritime provinces. Sediment burial dating has allowed us to date ice sheet deposits much farther back in time than surface exposure techniques. Cosmogenic nuclide studies have proven to be useful in studying LIS dynamics, leading to reinterpretations of trimlines, weathering zones, and enigmatic perched erratics, as well as the subsequent reconstruction of spatial patterns of basal thermal regimes and subglacial erosion. However, many questions remain. For example, although there is now general understanding of LIS dynamics in fiord landscapes, studies have taken place in relatively few regions and it is uncertain whether cosmogenic nuclide results to date can be extrapolated to less-studied regions. Our understanding of glacial erosion processes will likely improve as cosmogenic nuclide concentrations are measured in more samples. In particular, more data should help constrain the frequency and spatial distribution of inheritance from prior exposure periods, at both the local and regional scales. Cosmogenic nuclides also have potential to add to our understanding of ice sheet dynamics in interior regions of the LIS. With the future application of cosmogenic ^{14}C in quartz, difficulties associated with inheritance will be minimized. Continuing improvements in reducing the effect of other sources of error (e.g., snow cover, production rates, moraine erosion) should result in a significantly improved history of the LIS over the past half million years.

ACKNOWLEDGMENTS

The authors have had fruitful collaborations and stimulating discussion with dozens of people, including P.T. Davis, A. Dyke, J. Gray, M. Kaplan, G. Miller, and A. Wolfe. Numerous agencies have contributed to logistical support, including the Nunavut Research Institute, VECO Polar Resources, and Canada-Nunavut Geosciences Office. We are in great debt to our students and technicians (J. Larson, L. Stockli, G. Yang). We thank R. Finkel, M. Caffee, and J. Klein for AMS support. Funding sources include NSF-OPP (J.P.B., J.C.G., P.R.B.), NSF-EAR (P.R.B.), and ACOA-Atlantic Innovation Fund (J.C.G.).

in differentially weathered mountain landscapes, but because nuclides from prior exposure periods remain in many of these boulders, reliably identifying LGM ice presence requires dating many upland erratics.

In most cases, it has been the combination of cosmogenic nuclide data with other information that has provided new information on LIS history and dynamics. For example, relative weathering parameters (soils and bedrock characteristics) from different elevations in mountain landscapes provide an essential context for interpreting the complex pattern of cosmogenic

REFERENCES CITED

Alley, R.B., and MacAyeal, D.R., 1994, Ice-rafted debris associated with binge purge oscillations of the Laurentide Ice Sheet: Paleoceanography, v. 9, p. 503–511, doi: 10.1029/94PA01008.

André, M.F., 2004, The geomorphic impact of glaciers as indicated by tors in North Sweden (Aurivaara, 68 degrees N): Geomorphology, v. 57, p. 403–421, doi: 10.1016/S0169-555X(03)00182-X.

Andrews, J.T., and MacLean, B., 2003, Hudson Strait ice streams: A review of stratigraphy, chronology and links with North Atlantic Heinrich events: Boreas, v. 32, p. 4–17, doi: 10.1080/03009480310001010.

Andrews, J.T., and Miller, G.H., 1979, Glacial erosion and ice sheet divides, northeastern Laurentide Ice Sheet, on the basis of the distribution of limestone erratics: Geology, v. 7, p. 592–596, doi: 10.1130/0091-7613(1979)7<592:GEAISD>2.0.CO;2.

Andrews, J.T., and Tedesco, K., 1992, Detrital carbonate rich sediments, northwestern Labrador Sea—Implications for ice-sheet dynamics and iceberg rafting (Heinrich) events in the North Atlantic: Geology, v. 20, p. 1087–1090, doi: 10.1130/0091-7613(1992)020<1087:DCRSNL>2.3.CO;2.

Andrews, J.T., Buckley, J.T., and England, J.H., 1970, Late-glacial chronology and glacio-isostatic recovery, Home Bay, East Baffin Island, Canada: Geological Society of America Bulletin, v. 81, p. 1123–1148.

Andrews, J.T., Clark, P.U., and Stravers, J.A., 1985, The pattern of glacial erosion across the eastern Canadian Arctic, in Andrews, J.T., ed., Quaternary environments: Eastern Canadian Arctic, Baffin Bay, and West Greenland: Winchester, Allen and Unwin, p. 62–92.

Andrews, J.T., Kirby, M.E., Aksu, A., Barber, D.C., and Meese, D., 1998, Late Quaternary detrital carbonate (DC-) layers in Baffin Bay marine sediments (67 degrees-74 degrees N): Correlation with Heinrich events in the North Atlantic?: Quaternary Science Reviews, v. 17, p. 1125–1137, doi: 10.1016/S0277-3791(97)00064-4.

Atkins, C.B., Barrett, P.J., and Hicock, S.R., 2002, Cold glaciers erode and deposit: Evidence from Alan Hills, Antarctica: Geology, v. 30, p. 659–662, doi: 10.1130/0091-7613(2002)030<0659:CGEADE>2.0.CO;2.

Balco, G., Stone, J.O.H., Porter, S.C., and Caffee, M.W., 2002, Cosmogenic-nuclide ages for New England coastal moraines, Martha's Vineyard and Cape Cod, Massachusetts, USA: Quaternary Science Reviews, v. 21, no. 20-22, p. 2127–2135, doi: 10.1016/S0277-3791(02)00085-9.

Balco, G., Stone, J., Jennings, C., 2005a, Dating Plio-Pleistocene glacial sediments using the cosmic-ray-produced radionuclides ^{10}Be and ^{26}Al: American Journal of Science, v. 305, no. 1, p. 1–41.

Balco, G., Stone, J.O.H., and Mason, J.A., 2005b, Numerical ages for Plio-Pleistocene glacial sediment sequences by ^{26}Al/^{10}Be dating of quartz in buried paleosols: Earth and Planetary Science Letters, v. 232, p. 179–191, doi: 10.1016/j.epsl.2004.12.013.

Balco, G., Rovey, C.W., and Stone, J.O.H., 2005c, The first glacial maximum in North America: Science, v. 307, p. 222, doi: 10.1126/science.1103406.

Ballantyne, C.K., 1998, Age and significance of mountain-top detritus: Permafrost and Periglacial Processes, v. 9, no. 4, p. 327–345, doi: 10.1002/(SICI)1099-1530(199810/12)9:4<327::AID-PPP298>3.0.CO;2-9.

Bell, T., Rogerson, R.J., and Mengel, F., 1989, Reconstructed ice-flow patterns and ice limits using drift pebble lithology, Outer Nachvak Fjord, northern Labrador: Canadian Journal of Earth Sciences, v. 26, p. 577–590.

Bentley, C.R., 1987, Antarctic ice streams: A review: Journal of Geophysical Research, v. 92, p. 8843–8858.

Bierman, P.R., Marsella, K.A., Patterson, C., Davis, P.T., and Caffee, M., 1999, Mid-Pleistocene cosmogenic minimum-age limits for pre-Wisconsinan glacial surfaces in southwestern Minnesota and southern Baffin Island: A multiple nuclide approach: Geomorphology, v. 27, p. 25–39, doi: 10.1016/S0169-555X(98)00088-9.

Bierman, P.R., Davis, P.T., and Caffee, M.W., 2000, Old surfaces on New England summits imply thin Laurentide ice: Geological Society of America Abstracts with Programs, v. 31, no. 7, p. A-330.

Bierman, P.R., Marsella, K.A., Davis, P.T., and Caffee, M.W., 2001, Response to discussion by Wolfe et al. on Bierman et al. (Geomorphology 25 (1999) 25–39): Geomorphology, v. 39, p. 255–260.

Bond, G., Heinrich, H., Broecker, W., Labeyrie, L., McManus, J., Andrews, J., Huon, S., Jantschik, R., Clasen, S., Simet, C., Tedesco, K., Klas, M., Bonani, G., and Ivy, S., 1992, Evidence for massive discharges of icebergs into the North Atlantic Ocean during the Last Glacial Period: Nature, v. 360, p. 245–249, doi: 10.1038/360245a0.

Bond, G., Broecker, W., Johnsen, S., McManus, J., Labeyrie, L., Jouzel, J., and Bonani, G., 1993, Correlations between climate records from North Atlantic sediments and Greenland ice: Nature, v. 365, p. 143–147, doi: 10.1038/365143a0.

Boulton, G.S., and Clark, C.D., 1990, A highly mobile Laurentide Ice Sheet revealed by satellite images of glacial lineations: Nature, v. 346, p. 813–817, doi: 10.1038/346813a0.

Boyer, S.J., and Pheasant, D.R., 1974, Delimitation of weathering zones in the fiord area of eastern Baffin Island, Canada: Geological Society of America Bulletin, v. 85, p. 805–810, doi: 10.1130/0016-7606(1974)85<805:DOWZIT>2.0.CO;2.

Briner, J.P., and Swanson, T.W., 1998, Using inherited cosmogenic Cl-36 to constrain glacial erosion rates of the Cordilleran ice sheet: Geology, v. 26, p. 3–6, doi: 10.1130/0091-7613(1998)026<0003:UICCTC>2.3.CO;2.

Briner, J.P., Miller, G.H., Davis, P.T., Bierman, P.R., and Caffee, M., 2003, Last Glacial Maximum ice sheet dynamics in Arctic Canada inferred from young erratics perched on ancient tors: Quaternary Science Reviews, v. 22, p. 437–444, doi: 10.1016/S0277-3791(03)00003-9.

Briner, J.P., Miller, G.H., Davis, P.T., and Finkel, R., 2005, Cosmogenic exposure dating in Arctic glacial landscapes: Implications for the glacial history of northeastern Baffin Island, Canada: Canadian Journal of Earth Sciences, v. 42, no. 1, p. 67–84.

Briner, J.P., Miller, G.H., Davis, P.T., and Finkel, R.C., 2006, Cosmogenic radionuclides from differentially weathered fjord landscapes support differential erosion by overriding ice sheets: Geological Society of America Bulletin, v. 118, p. 406–420, doi: 10.1130/B25716.1.

Brookes, I.A., 1977, Geomorphology and Quaternary geology of Codroy Lowland and adjacent plateaus, southwest Newfoundland: Canadian Journal of Earth Sciences, v. 14, p. 2101–2120.

Bryson, R.A., Wendland, W.M., Ives, J.D., and Andrews, J.T., 1969, Radiocarbon isochrones on the disintegration of the Laurentide ice sheet: Arctic and Alpine Research, v. 1, p. 1–14, doi: 10.2307/1550356.

Clark, C.D., 1997, Reconstructing the evolutionary dynamics of former ice sheets using multi-temporal evidence, remote sensing and GIS: Quaternary Science Reviews, v. 16, no. 9, p. 1067–1092, doi: 10.1016/S0277-3791(97)00037-1.

Clark, P.U., 1988, Glacial geology of the Torngat Mountains, Labrador: Canadian Journal of Earth Sciences, v. 25, p. 1184–1198.

Clark, P.U., 1994, Unstable behavior of the Laurentide Ice Sheet over deforming sediment and its implications for climate change: Quaternary Research, v. 41, p. 19–25, doi: 10.1006/qres.1994.1002.

Clark, P.U., and Josenhans, H.W., 1990, Reconstructed ice-flow patterns and ice limits using drift pebble lithology, Outer Nachvak Fjord, northern Labrador—Discussion: Canadian Journal of Earth Sciences, v. 27, p. 1002–1006.

Clark, P.U., Short, S.K., Williams, K.M., and Andrews, J.T., 1989, Late Quaternary chronology and environments of Square Lake, Torngat Mountains, Labrador: Canadian Journal of Earth Sciences, v. 26, p. 2130–2144.

Clark, P.U., Brook, E.J., Raisbeck, G.M., Yiou, F., and Clark, J., 2003, Cosmogenic Be-10 ages of the Saglek moraines, Torngat Mountains, Labrador: Geology, v. 31, p. 617–620, doi: 10.1130/0091-7613(2003)031<0617:CBAOTS>2.0.CO;2.

Cockburn, H.A.P., and Summerfield, M.A., 2004, Geomorphological applications of cosmogenic isotope analysis: Progress in Physical Geography, v. 28, p. 1–42, doi: 10.1191/0309133304pp395oa.

Coleman, A.P., 1920, Extent and thickness of the Labrador ice sheet: Geological Society of America Bulletin, v. 31, p. 819–828.

Colgan, P.M., Bierman, P.R., Mickelson, D.M., and Caffee, M., 2002, Variation in glacial erosion near the southern margin of the Laurentide Ice Sheet, south-central Wisconsin, USA: Implications for cosmogenic dating of glacial terrains: Geological Society of America Bulletin, v. 114, p. 1581–1591, doi: 10.1130/0016-7606(2002)114<1581:VIGENT>2.0.CO;2.

Cuffey, K.M., Conway, H., Gades, A.M., Hallet, B., Lorrain, R., Severinghaus, J.P., Steig, E.J., Vaughn, B., and White, J.W.C., 2000, Entrainment at cold glacier beds: Geology, v. 28, p. 351–354, doi: 10.1130/0091-7613(2000)028<0351:EACGB>2.3.CO;2.

Daly, R.A., 1902, The geology of the northeast coast of Labrador: Harbard University Museum of Comparative Zoology Bulletin, v. 38, p. 205–270.

Davis, P.T., Bierman, P.R., Marsella, K.A., Caffee, M.W., and Southon, J.R., 1999, Cosmogenic analysis of glacial terrains in the eastern Canadian Arctic: A test for inherited nuclides and the effectiveness of glacial erosion: Annals of Glaciology, v. 28, p. 181–188.

Davis, P.T., Briner, J.P., Coulthard, R.C., Finkel, R.W., and Miller, G.H., 2006, Preservation of arctic landscapes overridden by cold-based ice sheets: Quaternary Research, v. 62, p. 156–163.

Denton, G.H., and Hughes, T., 1981, The last great ice sheets: New York, Wiley and Sons, 484 p.

Dredge, L.A., 2000, Age and origin of upland block fields on Melville Peninsula, eastern Canadian Arctic: Geografiska Annaler, ser. A, Physical Geography, v. 82A, p. 443–454.

Duk-Rodkin, A., Barendregt, R.W., Tarnocai, C., and Phillips, F.M., 1996, Late Tertiary to late Quaternary record in the Mackenzie Mountains, Northwest Territories, Canada: Stratigraphy, paleosols, paleomagnetism, and chlorine-36: Canadian Journal of Earth Sciences, v. 33, p. 875–895.

Dyke, A.S., 1979, Glacial and sea-level history of southwestern Cumberland Peninsula, Baffin Island, N.W.T., Canada: Arctic and Alpine Research, v. 11, p. 179–202, doi: 10.2307/1550644.

Dyke, A.S., 1993, Landscapes of cold-centered Late Wisconsinan ice caps, arctic Canada: Progress in Physical Geography, v. 17, p. 223–247.

Dyke, A.S., and Prest, V.K., 1987, Late Wisconsin and Holocene history of the Laurentide Ice Sheet: Geographie Physique et Quaternaire, v. 41, p. 237–263.

Dyke, A.S., Morris, T.F., Green, D.E.C., and England, J., 1992, Quaternary geology of Prince of Wales Island, Arctic Canada: Geological Survey of Canada Memoir 433, 142 p.

Dyke, A.S., Andrews, J.T., Clark, P.U., England, J.H., Miller, G.H., Shaw, J., and Veillette, J.J., 2002, The Laurentide and Innuitian ice sheets during the Last Glacial Maximum: Quaternary Science Reviews, v. 21, p. 9–31, doi: 10.1016/S0277-3791(01)00095-6.

Fabel, D., and Harbor, J., 1999, The use of in situ produced cosmogenic radionuclides in glaciology and glacial geomorphology: Annals of Glaciology, v. 28, p. 103–110.

Flint, R.F., 1943, Growth of the North American ice sheet during the Wisconsin age: Geological Society of America Bulletin, v. 54, p. 325–362.

Gosse, J.C., and Phillips, F.M., 2001, Terrestrial in situ cosmogenic nuclides: Theory and application: Quaternary Science Reviews, v. 20, p. 1475–1560, doi: 10.1016/S0277-3791(00)00171-2.

Gosse, J.C., Grant, D.R., Klein, J., Klassen, R.A., Evenson, E.B., Lawn, B., and Middleton, R., 1993, Significance of altitudinal weathering zones in Atlantic Canada, inferred from in situ produced cosmogenic radionuclides: Geological Society of America Abstracts with Programs, v. 25, no. 6, p. A394.

Gosse, J.C., Evenson, E.B., Klein, J., Lawn, B., and Middleton, R., 1995a, Precise cosmogenic ^{10}Be measurements in western North America: Support for a global Younger Dryas cooling event: Geology, v. 23, p. 877–880, doi: 10.1130/0091-7613(1995)023<0877:PCBMIW>2.3.CO;2.

Gosse, J.C., Grant, D.R., Klein, J., and Lawn, B., 1995b, Cosmogenic ^{10}Be and ^{26}Al constraints on weathering zone genesis, ice cap basal conditions, and Long Range Mountains (Newfoundland) glacial history, in Proceedings of the Canadian Quaternary Association–Canadian Geomorphology Research Group (CANQUA-CGRC) Conference: Memorial University, Saint Johns, Newfoundland, p. CA19.

Gosse, J., Hecht, G., Mehring, N., Klein, J., Lawn, B., and Dyke, A., 1998, Comparison of radiocarbon and in situ cosmogenic nuclide-derived postglacial emergence curves for Prescott Island, central Canadian Arctic: Geological Society of America Abstracts with Programs, v. 30, no. 7, p. 298.

Gosse, J.C., Staiger, J.W., Fastook, J., Johnson, J., Utting, D., and Little, E., 2005, Tagging glacial erosion and till production for drift prospecting: Geological Society of America Abstracts with Programs, v. 37, no. 7, p. 398.

Gosse, J.C., Bell, T., Gray, J., Klein, J., Yang, G., and Finkel, R., 2006, Using cosmogenic isotopes to interpret the landscape record of glaciation, in Knight, P., ed., Nunataks in Newfoundland Glaciers and Earth's Changing Environment: Blackwell, p. 442–446.

Grant, D.R., 1977, Altitudinal weathering zones and glacial limits in western Newfoundland, with particular reference to Gros Morne National Park: Geological Survey of Canada Paper 77-1A, p. 455–463.

Grousset, F.E., Pujol, C., Labeyrie, L., Auffret, G., and Boelaert, A., 2000, Were the North Atlantic Heinrich events triggered by the behavior of the European ice sheets?: Geology, v. 28, p. 123–126, doi: 10.1130/0091-7613(2000)028<0123:WTNAHE>2.3.CO;2.

Hall, A.M., and Glasser, N.F., 2003, Reconstructing the basal thermal regime of an ice stream in a landscape of selective linear erosion: Glen Avon, Cairngorm Mountains, Scotland: Boreas, v. 32, p. 191–207, doi: 10.1080/03009480310001100.

Hall, A.M., and Sugden, D.E., 1987, Limited modification of midlatitude landscapes by ice sheets—The case of northeast Scotland: Earth Surface Processes and Landforms, v. 12, no. 5, p. 531–542.

Heinrich, H., 1988, Origin and consequences of cyclic ice rafting in the northeast Atlantic Ocean during the past 130,000 Years: Quaternary Research, v. 29, p. 142–152, doi: 10.1016/0033-5894(88)90057-9.

Hemming, S.R., and Hajdas, I., 2003, Ice-rafted detritus evidence from Ar-40/Ar-39 ages of individual hornblende grains for evolution of the eastern margin of the Laurentide ice sheet since 43 C-14 ky: Quaternary International, v. 99, p. 29–43, doi: 10.1016/S1040-6182(02)00110-6.

Hughes, T., Denton, G.H., and Grosswald, M.G., 1977, Was there a Late Würm arctic ice sheet?: Nature, v. 266, p. 596–602, doi: 10.1038/266596a0.

Hughes, T.J., Denton, G.H., and Fastook, J.L., 1985, The Antarctic Ice Sheet: An analog for Northern Hemisphere paleo-ice sheets?, in Woldenburk, M.J., ed., Models in Geomorphology: London, Allen and Unwin, p. 25–72.

Imbrie, J., and Imbrie, J.Z., 1980, Modeling the climatic response to orbital variations: Science, v. 207, p. 943–953.

Ives, J.D., 1957, Glaciation of the Torngat Mountains, northern Labrador: Journal of the Arctic Institute of North America, v. 10, p. 67–87.

Ives, J.D., 1978, Maximum extent of Laurentide Ice Sheet along east coast of North America during last glaciation: Arctic, v. 311, p. 24–53.

Ives, J.D., and Andrews, J.T., 1963, Studies in the physical geography of north-central Baffin Island, N.W.T.: Geographical Bulletin, v. 5, p. 5–48.

Jackson, L.E., 1980, Glacial history and stratigraphy of the Alberta portion of the Kananaskis Lakes map area: Canadian Journal of Earth Sciences, v. 17, p. 459–477.

Jackson, L.E., Phillips, F.M., Shimamura, K., and Little, E.C., 1997, Cosmogenic Cl-36 dating of the Foothills erratics train, Alberta, Canada: Geology, v. 25, p. 195–198, doi: 10.1130/0091-7613(1997)025<0195:CCDOTF>2.3.CO;2.

Jackson, L.E., Phillips, F.M., and Little, E.C., 1999, Cosmogenic Cl-36 dating of the maximum limit of the Laurentide Ice Sheet in southwestern Alberta: Canadian Journal of Earth Sciences, v. 36, p. 1347–1356, doi: 10.1139/cjes-36-8-1347.

Jansson, K.N., Stroeven, A.P., and Kleman, J., 2003, Configuration and timing of Ungava Bay ice streams, Labrador-Ungava, Canada: Boreas, v. 32, p. 256–262, doi: 10.1080/03009480310001146.

Jennings, A.E., 1993, The Quaternary history of Cumberland Sound, southeastern Baffin Island: The marine evidence: Geographie physique et Quaternaire, v. 47, p. 21–42.

Josenhans, H.W., Zevenhuizen, J., and Klassen, R.A., 1986, The Quaternary Geology of the Labrador Shelf: Canadian Journal of Earth Sciences, v. 23, p. 1190–1213.

Kaplan, M.R., and Miller, G.H., 2003, Early Holocene delevelling and deglaciation of the Cumberland Sound region, Baffin Island, Arctic Canada: Geological Society of America Bulletin, v. 115, p. 445–462, doi: 10.1130/0016-7606(2003)115<0445:EHDADO>2.0.CO;2.

Kaplan, M.R., Pfeffer, W.T., Sassolas, C., and Miller, G.H., 1999, Numerical modelling of the Laurentide Ice Sheet in the Baffin Island region: The role of a Cumberland Sound ice stream: Canadian Journal of Earth Sciences, v. 36, p. 1315–1326, doi: 10.1139/cjes-36-8-1315.

Kaplan, M.R., Miller, G.H., and Steig, E.J., 2001, Low-gradient outlet glaciers (ice streams) drained the Laurentide ice sheet: Geology, v. 29, p. 343–346, doi: 10.1130/0091-7613(2001)029<0343:LGOGIS>2.0.CO;2.

Klassen, R.W., 1994, Late Wisconsinan and Holocene history of southwestern Saskatchewan: Canadian Journal of Earth Sciences, v. 31, p. 1822–1837.

Kleman, J., 1994, Preservation of landforms under ice sheets and ice caps: Geomorphology, v. 9, p. 19–32, doi: 10.1016/0169-555X(94)90028-0.

Kleman, J., and Hättestrand, C., 1999, Frozen-bed Fennoscandian and Laurentide ice sheets during the Last Glacial Maximum: Nature, v. 402, p. 63–66, doi: 10.1038/47005.

Lal, D., 1991, Cosmic-ray labeling of erosion surfaces—In situ nuclide production rates and erosion models: Earth and Planetary Science Letters, v. 104, p. 424–439, doi: 10.1016/0012-821X(91)90220-C.

Lal, D., and Peters, B., 1967, Cosmic ray produced radioactivity on the earth, in Sitte, K., ed., Handbuch der Physik, v. XLVI/2: Berlin, Springer, p. 551–612.

Liverman, D.G.E., Catto, N.R., and Rutter, N.W., 1989, Laurentide glaciation in west-central Alberta: A single (Late Wisconsinan) event: Canadian Journal of Earth Sciences, v. 26, p. 266–274.

Løken, O.H., 1966, Baffin Island refugia older than 54,000 years: Science, v. 153, p. 1378–1380.

Marquette, G.C., Gray, J.T., Gosse, J.C., Courchesne, F., Stockli, L., Macpherson, G., and Finkel, R., 2004, Felsenmeer persistence under non-erosive ice in the Torngat and Kaumajet Mountains, Quebec and Labrador, as determined by soil weathering and cosmogenic nuclide exposure dating: Canadian Journal of Earth Sciences, v. 41, p. 19–38, doi: 10.1139/e03-072.

Marsella, K.A., Bierman, P.R., Davis, P.T., and Caffee, M.W., 2000, Cosmogenic Be-10 and Al-26 ages for the Last Glacial Maximum, eastern Baffin Island, Arctic Canada: Geological Society of America Bulletin, v. 112, p. 1296–1312, doi: 10.1130/0016-7606(2000)112<1296:CBAAAF>2.3.CO;2.

Martinson, D.G., Pisias, N.G., Hays, J.D., Imbrie, J., Moore, T.C., Jr., and Shackleton, N.J., 1987, Age dating and orbital theory of the ice ages: Development of a high-resolution 0 to 300,000-year chronostratigraphy: Quaternary Research, v. 27, p. 1–29, doi: 10.1016/0033-5894(87)90046-9.

McCuaig, S.J., Shilts, W.W., Evenson, E.B., and Klein, J., 1994, Use of cosmogenic ^{10}Be and ^{26}Al for determining glacial history of the South Bylot Island and Salmon River Lowlands, N.W.T., Canada: Geological Society of America Abstracts with Programs, v. 26, no. 7, p. A127.

Miller, G.H., and Dyke, A.S., 1974, Proposed extent of Late Wisconsin Laurentide ice on eastern Baffin Island: Geology, v. 2, p. 125–130, doi: 10.1130/0091-7613(1974)2<125:PEOLWL>2.0.CO;2.

Miller, G.H., Wolfe, A.P., Steig, E.J., Sauer, P.E., Kaplan, M.R., and Briner, J.P., 2002, The Goldilocks dilemma: Big ice, little ice, or "just-right" ice in the eastern Canadian Arctic: Quaternary Science Reviews, v. 21, p. 33–48, doi: 10.1016/S0277-3791(01)00085-3.

Nesje, A., and Dahl, S.O., 1990, Autochthonous block fields in southern Norway—Implications for the geometry, thickness, and isostatic loading of the Late Weichselian Scandinavian Ice Sheet: Journal of Quaternary Science, v. 5, p. 225–234.

Nishiizumi, K., Winterer, E.L., Kohl, C.P., Klein, J., Middleton, R., Lal, D., and Arnold, J.R., 1989, Cosmic-ray production rates of Be-10 and Al-26 in quartz from glacially polished rocks: Journal of Geophysical Research, B, Solid Earth and Planets, v. 94, p. 17,907–17,915.

Odell, N.E., 1933, The mountains of northern Labrador: Geographical Journal, v. 82, p. 193–210.

Patterson, C.J., 1998, Laurentide glacial landscapes: The role of ice streams: Geology, v. 26, p. 643–646, doi: 10.1130/0091-7613(1998)026<0643:LGLTRO>2.3.CO;2.

Peltier, W.R., 1996, Mantle viscosity and ice-age ice sheet topography: Science, v. 273, p. 1359–1364.

Pheasant, D.R., and Andrews, J.T., 1973, Wisconsin glacial chronology and relative sea level movements, Narpaing Fjord / Broughton Island area, eastern Baffin Island: Canadian Journal of Earth Sciences, v. 10, p. 1621–1641.

Phillips, F.M., Leavy, B.D., Jannik, N.O., Elmore, D., and Kubik, P.W., 1986, The accumulation of cosmogenic chlorine-36 in rocks: A method for surface exposure dating: Science, v. 231, p. 41–43.

Piper, D.J.W., and Macdonald, A., 2001, Timing and position of Late Wisconsinan ice margins on the upper slope seaward of Laurentian Channel: Geographie Physique et Quaternaire, v. 55, p. 131–140.

Prest, V.K., Grant, D.R., and Rampton, V.N., 1968, Glacial map of Canada: Geological Survey of Canada Map 1253A, scale 1:5,000,000.

Rae, A.C., Harrison, S., Mighall, T., and Dawson, A.G., 2004, Periglacial trimlines and nunataks of the Last Glacial Maximum: The Gap of Dunloe, southwest Ireland: Journal of Quaternary Science, v. 19, p. 87–97, doi: 10.1002/jqs.807.

Shilts, W.W., Cunningham, C.M., and Kaszycki, C.A., 1979, Keewatin Ice Sheet—Re-evaluation of the traditional concept of the Laurentide Ice Sheet: Geology, v. 7, p. 537–541, doi: 10.1130/0091-7613(1979)7<537:KISOTT>2.0.CO;2.

Staiger, J.K.W., Gosse, J.C., Johnson, J., Fastook, J., Gray, J., Little, E.C., Hilchey, A., and Finkel, R., 2004, Long-term (10^4-10^5 yr) differential glacial erosion beneath polythermal glacier ice in the Atlantic and Arctic Canadian highlands: Geological Society of America Abstracts with Programs, v. 36, no. 5, p. 70.

Staiger, J.K.W., Gosse, J.C., Johnson, J.H., Fastook, J., Gray, J.T., Stockli, D.F., Stockli, L., and Finkel, R., 2005, Quaternary relief generation by polythermal glacier ice: Earth Surface Processes and Landforms, v. 30, p. 1145–1159, doi: 10.1002/esp.1267.

Stea, R.R., Piper, D.J.W., Fader, G.B.J., and Boyd, R., 1998, Wisconsinan glacial and sea-level history of Maritime Canada and the adjacent continental shelf: A correlation of land and sea events: Geological Society of America Bulletin, v. 110, p. 821–845, doi: 10.1130/0016-7606(1998)110<0821:WGASLH>2.3.CO;2.

Steig, E.J., Wolfe, A.P., and Miller, G.H., 1998, Wisconsinan refugia and the glacial history of eastern Baffin Island, Arctic Canada: Coupled evidence from cosmogenic isotopes and lake sediments: Geology, v. 26, p. 835–838, doi: 10.1130/0091-7613(1998)026<0835:WRATGH>2.3.CO;2.

Stokes, C.R., and Clark, C.D., 2001, Palaeo-ice streams: Quaternary Science Reviews, v. 20, p. 1437–1457, doi: 10.1016/S0277-3791(01)00003-8.

Stokes, C.R., and Clark, C.D., 2002, Are long subglacial bedforms indicative of fast ice flow?: Boreas, v. 31, p. 239–249, doi: 10.1080/030094802760260355.

Sugden, D.E., 1977, Reconstruction of morphology, dynamics, and thermal characteristics of Laurentide Ice Sheet at its maximum: Arctic and Alpine Research, v. 9, p. 21–47, doi: 10.2307/1550407.

Sugden, D.E., and Watts, S.H., 1977, Tors, felsenmeer, and glaciation in northern Cumberland Peninsula, Baffin Island: Canadian Journal of Earth Sciences, v. 14, p. 2817–2823.

Tyrrell, J.B., 1898, The glaciation of north-central Canada: Journal of Geology, v. 6, p. 147–160.

Veillette, J.J., Dyke, A.S., and Roy, M., 1999, Ice-flow evolution of the Labrador Sector of the Laurentide Ice Sheet: A review with new data from northern Quebec: Quaternary Science Reviews, v. 18, p. 993–1019, doi: 10.1016/S0277-3791(98)00076-6.

Walcott, R.I., 1970, Isostatic response to loading of the crust in Canada: Canadian Journal of Earth Sciences, v. 7, p. 716–726.

Wolfe, A.P., Steig, E.J., and Kaplan, M.R., 2001, An alternative model for the geomorphic history of pre-Wisconsinan surfaces on eastern Baffin Island: A comment on Bierman et al. (Geomorphology 25 (1999) 25–39): Geomorphology, v. 39, p. 251–254.

Wolfe, A.P., Fréchette, B., Richard, P.J.H., Miller, G.H., and Forman, S.L., 2000, Paleoecology of a >90,000-year lacustrine sequence from Fog Lake, Baffin Island, Arctic Canada: Quaternary Science Reviews, v. 19, p. 1677–1699.

MANUSCRIPT ACCEPTED BY THE SOCIETY 11 APRIL 2006

The timing of glacier advances in the northern European Alps based on surface exposure dating with cosmogenic ^{10}Be, ^{26}Al, ^{36}Cl, and ^{21}Ne

Susan Ivy-Ochs[†]
Insitut für Teilchenphysik, ETH-Hönggerberg CH-8093 Zurich, Switzerland, and *Geographisches Institut, Universität Zurich-Irchel, CH-8057 Zurich, Switzerland*

Hanns Kerschner
Institut für Geographie, Universität Innsbruck, A-6020 Innsbruck, Austria

Anne Reuther
Insitut für Physische Geographie, Universität Regensburg, D-93040 Regensburg, Germany

Max Maisch
Geographisches Institut, Universität Zurich-Irchel, CH-8057 Zurich, Switzerland

Rudolf Sailer
Institut für Naturgefahren und Waldgrenzregionen, Bundesforschungs- und Ausbildungszentrum für Wald, A-6020 Innsbruck, Austria

Joerg Schaefer
Lamont Doherty Earth Institute, Palisades, New York 10964, USA

Peter W. Kubik
Hans-Arno Synal
Paul Scherrer Institut, c/o Institut für Teilchenphysik, ETH Hönggerberg, CH-8093 Zürich, Switzerland

Christian Schlüchter
Geologisches Institut, Universität Bern, CH-3012 Bern, Switzerland

ABSTRACT

Exposure dating of boulder and bedrock surfaces with ^{10}Be, ^{21}Ne, ^{26}Al, and ^{36}Cl allows us to constrain periods of glacier expansion in the European Alps. The age of 155 ka from a boulder of Alpine lithology located in the Jura Mountains (Switzerland) provides a minimum age for pre-LGM (Last Glacial Maximum), more extensive Alpine glaciations. During the LGM, glaciers expanded onto the foreland after 30 ka. By 21.1 ± 0.9 ka deglaciation had begun, and the Rhône Glacier abandoned the

[†]E-mail: ivy@phys.ethz.ch.

outer moraines. The age of 15.4 ± 1.4 ka provides a minimum age for formation of Gschnitz stadial moraines (Austria). They mark the first clear post-LGM readvance of mountain glaciers, when glacier termini were already situated well inside the mountains. Glacier advance at the onset of the Younger Dryas led to formation of Egesen I moraines, dated to 12.2 ± 1.0 ka at the Schönferwall site (Austria) and to 12.3 ± 1.5 ka at the outer moraine at Julier Pass (Switzerland). The age of 11.3 ± 0.9 ka for the inner moraine / rock glacier complex at Julier Pass corroborates the field evidence, which points to a marked increase in rock glacier activity and delayed moraine stabilization during the late Younger Dryas. An early Preboreal glacier advance, larger than the Little Ice Age advance(s) at 10.8 ± 1.0 ka, was recorded at Kartell cirque (Austria). A moraine doublet located a few hundred meters outside the A.D. 1850 moraines in Kromer Valley (Austria) was dated at 8.4 ± 0.7 ka. At least during termination 1, glacier volumes in the Alps varied in tune with climate oscillations, Heinrich event 1, the Younger Dryas cold phase, the Preboreal oscillation, and the 8.2 ka event.

Keywords: cosmogenic nuclides, LGM, Alpine Lateglacial, Holocene, glacier variations, moraine.

INTRODUCTION

At the beginning of the twentieth century, the classical fourfold system for the glaciations of the Alps was presented by Penck and Brückner (1901/1909). Distinct morphostratigraphic elements (for the most part terraces and moraines) located on the northern forelands of the Alps were the basis for setting up the following system: (1) Günz, as defined by older (or higher) "Deckenschotter" (cover gravel) deposits, (2) Mindel, as defined by younger (or lower) Deckenschotter deposits, (3) Riss, defined by moraine remnants and outwash related to the high terraces (Hochterrasse), and (4) Würm, delineated by moraines and outwash of the low terraces (Niederterrasse).

The Riss and Würm glacial periods were separated by the last interglacial. The system was subsequently expanded and repeatedly revised. Buoncristiani and Campy (2004), Fiebig et al. (2004), Schlüchter (2004), and van Husen (2004) summarize the present state of knowledge. Terminal moraines and associated deposits formed during the Last Glacial Maximum (late Würm) provide the most important morphological reference point in the forelands of the Alps (Fig. 1). The final disintegration of the foreland piedmont glaciers marks the start of the "Alpine Lateglacial" as defined by Penck and Brückner (1901/1909). During the Alpine Lateglacial, glaciers readvanced several times to successively smaller extents ("stadials"), building prominent moraine systems in the valleys and cirques (Table 1).

Temporal constraints for deciphering the timing of glacier variations have been provided by luminescence methods, radiocarbon, U/Th, and surface exposure dating with in situ–produced cosmogenic nuclides (cf. Preusser, 2004). Each dating method has its own niche and optimum age range, which is determined by both geological circumstances (such as the material that can be dated and its geological setting) and methodological constraints (e.g., the half-life of the nuclides). With surface exposure dating, the rock surfaces created during glacier variations can be dated directly (Gosse and Phillips, 2001), making surface exposure dating invaluable in formerly glaciated areas (Gosse, 2005; Reuther et al., 2006).

In this paper, we summarize the present state of knowledge on periods of glacier expansion in the Alps that have been temporally constrained using the cosmogenic nuclides ^{10}Be, ^{21}Ne, ^{26}Al, and ^{36}Cl. Reflecting the locations of the sites, the focus here is on the northern regions of the Alps. We discuss in detail eight sites: Montoz pre-LGM (Last Glacial maximum) boulders (Jura Mountains, Switzerland), the Wangen an der Aare LGM moraine (Solothurn, Switzerland), Grimsel Pass (Bern/Wallis, Switzerland), the Trins Gschnitz moraine (Tyrol, Austria), Daun and Egesen moraines at Julier Pass (Graubünden, Switzerland), the Egesen I Schönferwall moraine (Tyrol, Austria), and the Kartell (Tyrol, Austria) and Kromer (Vorarlberg, Austria) moraines. We discuss the exposure dating results in the context of the chronological framework of climate variations in the Alps and their vicinity. Specific site locations for the related radiocarbon, luminescence, and U/Th dates are found in the references given.

METHODS

Rock surfaces were sampled for exposure dating with hammer and chisel. Except where indicated, the flat tops of the largest boulders were sampled. Details on sample preparation can be found in Ivy-Ochs (1996) and Ochs and Ivy-Ochs (1997). Ratios of the radionuclide to stable nuclide were measured using accelerator mass spectrometry (AMS) at the PSI/ETH Zurich tandem accelerator facility (Synal et al., 1997). Neon isotopes were analyzed at the Institute of Isotope Geology and Mineral Resources of the ETH Zurich following the protocol given by Bruno et al. (1997).

The exposure ages presented here are based on half-lives of 1510 k.y. (^{10}Be), 716 k.y. (^{26}Al), and 301 k.y. (^{36}Cl). The production rates used for the age calculations are listed in Table 2. Major

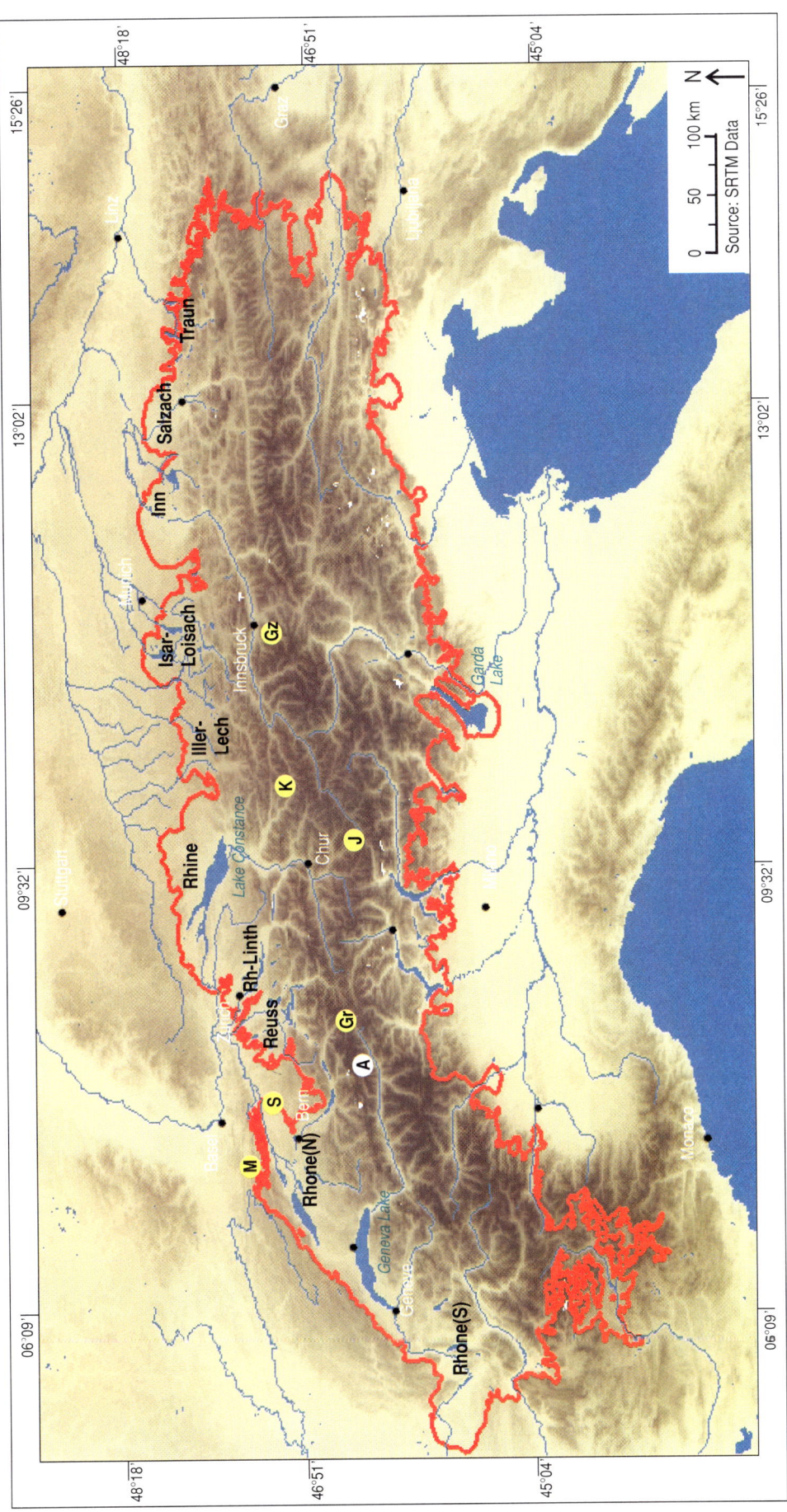

Figure 1. View of the Alps from Shuttle Radar Topographic Mission data. LGM ice extent based on Ehlers and Gibbard (2004). The main piedmont lobes of the northern forelands are labeled. Location of sites investigated with cosmogenic nuclides are indicated by the circles. The yellow circles stand for data discussed in detail in this paper. The white circles represent data from other studies. M—Montoz (this paper); S—Solothurn (Ivy-Ochs et al., 2004); Gr—Grimsel Pass (Ivy-Ochs, 1996; Kelly et al. 2006); A—Grosser Aletsch Glacier (Kelly et al., 2004b); J—Julier Pass (Ivy-Ochs et al. 1996, 1999); K—Kromer (Kerschner et al., 2006), Kartell and Schönferwall (this paper); Gz—Trins Gschnitz moraine (Ivy-Ochs et al. 2006).

TABLE 1. SUMMARY OF THE LATEGLACIAL STADIALS, EUROPEAN ALPS

Stadial	Moraine and glacier characteristics	Regional situation	ELA depression	Time-stratigraphic position
Kromer	Well-defined, blocky, multi-walled moraines, small rock glaciers. Type locality in Kromer Valley, Silvretta Mountains (7, 30)	Cirque and valley glaciers, clearly larger than LIA, but smaller than innermost Egesen phase in northwestern Austrian Alps, glaciers smaller than LIA in central Alps.	~75 m in western Austria	Possibly Misox cold phase (±8.2 ka event) (30)
Kartell	Well-defined, blocky, multi-walled moraines, small rock glaciers. Type localities in the Ferwall Group (Kartell cirque (10)).	Cirque and valley glaciers, clearly larger than LIA, but smaller than innermost Egesen phase.	−120 m at type locality (24)	Preboreal oscillation? (this paper)
Egesen	Sharp-crested, often blocky, multi-walled moraines (2) well documented in wide areas of the Alps. Three-phased readvance of valley glaciers and cirque glaciers; development of extensive rock glacier systems during later parts of the stadial (22, 24, 25). Type locality in the Stubai Valley SW of Innsbruck (2).	Cirque and valley glaciers, few dendritic glaciers (23).	−450 to −180 m for the maximum advance, depending on location (22)	Younger Dryas (e.g., 4, 17, 18)
Bølling-Allerød Interstadial	No field evidence for glacial advances, although various climatic fluctuations (colder phases) should have caused glaciers to advance. Deposits of the advances were likely overrun during Younger Dryas (20).	Cirque and valley glaciers (?).	Less than Egesen (20)	Bølling-Allerød
Daun	Well-defined but smoothed moraines, relatively few boulders, solifluction overprint during Younger Dryas (Egesen II); moraines usually missing in more oceanic areas of the Alps (overrun by Egesen?) (23). Smaller than Clavadel/Senders, perhaps "appendix" of Clavadel / Senders (indirectly 1). Type locality in the Stubai Valley SW of Innsbruck (3).	Glaciers slightly larger than local Egesen glaciers.	~ −400 to −250 m depending on location (7)	Before Bølling (11, 12)
Clavadel/Senders	Well defined, often sediment-rich moraines (11, 12, 13, 14). Clearly smaller than "Gschnitz". Type localities near Davos (Clavadel (11)) and Innsbruck (Senders (14)), probably equivalent to Zwischbergen stadial at Simplon Pass (15, 16).	Cirque and valley glaciers, some dendritic glaciers still intact.	~ −400 to −500 m depending on location (15, 16)	Before Bølling
Gschnitz	Steep-walled, somewhat blocky, large single moraines, no solifluction overprint below 1400 m. Widespread readvance of large valley glaciers on a timescale of several centuries (21). Glaciers advanced over ice-free terrain (8, 9, 14). Type locality in the Gschnitz Valley South of Innsbruck (3).	Many valley glaciers, some large dendritic glaciers still intact.	~ −650 to −700 m (7)	Before Bølling (5, 6, 8, 15, 16, 34)
Phase of early Lateglacial ice decay	General downwasting and recession of piedmont glaciers in the foreland with some oscillations of the glacier margins. Mainly ice marginal deposits; moraines indicating glacier advance in smaller catchments (26, 31, 32, 33). Glacial advances also due to ice-mechanical causes (26). Comprises the classical "Bühl" and "Steinach" stadials (1, 3, 8, 26).	Downwasting dendritic glaciers (5), increasing number of local glaciers.	Largely undefined, between ± Last Glacial Maximum and −800 m	Before Bølling, older than 15,400 ± 470 ^{14}C yr B.P. (8), ca. 18,020–19,010 cal yr B.P.
Last Glacial Maximum	Ice domes in the high Alps (27, 28, 29), outlet glaciers, piedmont glaciers on the foreland.	Piedmont lobes.	−1000 to >−1200 m	Final deglaciation 21,000 yr ago (19)

Note: References: 1—Senarclens-Grancy (1958); 2—Heuberger (1966); 3—Mayr and Heuberger (1968); 4—Patzelt (1972); 5—Patzelt (1975); 6—Patzelt (1995); 7—Gross et al. (1977); 8—van Husen (1977); 9—van Husen (1997); 10—Fraedrich (1979); 11—Maisch (1981); 12—Maisch (1982); 13—Maisch (1987); 14—Kerschner and Berktold (1982); 15—Müller (1982); 16—Müller (1984); 17—Kerschner (1986); 18—Ivy-Ochs et al. (1996); 19—Ivy-Ochs et al. (2004); 20—Ohlendorf (1998); 21—Kerschner et al. (1999); 22—Kerschner (2000); 23—Hertl (2001); 24—Sailer (2001); 25—Sailer et al. (1999); 26—Reitner (2005); 27—Florineth and Schlüchter (1998); 28—Florineth and Schlüchter (2000); 29—Kelly et al. (2004a); 30—Kerschner et al. (2006); 31—Keller and Krayss (1993); 32—Keller and Krayss (2005); 33—Keller (1988); 34—Ivy-Ochs et al. (2006). References before early 1950s: see Kerschner (1986). LIA—Little Ice Age; ELA—equilibrium line altitude.

and trace element data necessary for the ^{36}Cl calculations are found in Ivy-Ochs et al. (1996, 2004). Ages have been corrected for snow coverage or erosion on a site-by-site basis as shown in Table 3. No correction has been made for changes in magnetic field intensity over the exposure period. At the latitude of the Alps, this would be less than 1% (Masarik et al., 2001).

The errors given for each boulder age (Table 3) reflect analytical uncertainties (dominated by AMS measurement parameters). Where several nuclides have been measured in one boulder, an error-weighted mean has been calculated. Unless otherwise stated, moraine ages are averages. We consider the use of an error-weighted mean inappropriate as age results for certain boulders would be given more weight based largely on AMS uncertainties. Uncertainties for the mean ages of landforms are based on the 1σ confidence interval about the mean and include systematic uncertainties in the production rates as given in Table 2.

The radiocarbon dates discussed here have been calibrated using CALIB 5.0 (Stuiver et al., 1998, 2005). Uncalibrated radiocarbon dates are indicated by ^{14}C yr B.P., calibrated dates by cal. yr B.P. For dates over 21,000 ^{14}C yr B.P., we have used the data set of Hughen et al. (2004) to obtain an estimate of calendar dates. The latter should be considered estimates, as clear knowledge of the trend of the calibration curve beyond 25 ka is still under discussion (e.g., Bard et al., 2004). For the latter cases, the term "cal." will not be used, only "yr B.P."

The equilibrium line altitude (ELA) of a glacier is calculated with the accumulation area ratio (AAR) method with an AAR of 0.67, which gives the most reliable results for glaciers in the Alps. ELA depressions refer to the A.D. 1850 ("Little Ice Age") ELA, which is well known throughout the Alps (Gross et al., 1977).

DECKENSCHOTTER GLACIATIONS

The Deckenschotter units are poorly cemented, highly weathered, glaciofluvial gravel beds now found as isolated outcrops at elevations several tens up to hundreds of meters above the present-day drainage systems (Graf, 1993; Graf and Müller, 1999; Fiebig et al., 2004; van Husen, 2004). The precise age and the correlation of the various Deckenschotter deposits between the different foreland regions (Swiss Plateau, German and Austrian foreland) remains an open question (e.g., Fiebig et al., 2004). Parts of the Deckenschotter deposits may even date back to the late Pliocene (Bolliger et al., 1996). Preliminary results using cosmogenic nuclide burial methods (cf. Balco et al., 2005) to date Deckenschotter deposits have been presented (Fiebig and Häuselmann, 2005).

MOST EXTENSIVE GLACIATIONS

Located tens of kilometers beyond the LGM (late Würm) moraines, deposits of more extensive glaciations can be mapped. In the Swiss part of the northern Alpine foreland, these have been given the name "most extensive glaciations" (Schlüchter, 1988, 2004), and encompass the classical "Riss" glaciation(s). During this glaciation, the Rhône Glacier overtopped the southernmost chains of the Jura Mountains at several locations (e.g., Campy, 1992; Buoncristiani and Campy, 2004). This is documented by the finding of crystalline boulders of Alpine lithology at high elevation in the Jura Mountains (Penck and Brückner, 1901/1909; Hantke, 1978–1983), which themselves are dominated by calcareous bedrock. During the LGM, the glacier surface remained at lower altitudes (Fig. 2). Thus, the boulders that had been left by previous glaciations were not moved by glacier ice during younger glaciations. One of these boulders, located on a northwest-trending limestone ridge (Montoz, 1200 masl), has been dated with ^{10}Be and ^{21}Ne. The ^{10}Be exposure age is 106,000 ± 5000 yr, using no snow or erosion correction. The ^{21}Ne age is 128,000 ± 22,000 yr (Table 3). Using an erosion rate of 3 mm/k.y., a ^{10}Be exposure age of 155,000 yr is obtained, which we consider to be a minimum age for the most extensive glaciation(s). With an erosion rate of 4 mm/k.y., the calculated age is 195,000 yr. This result underlines the sensitivity of exposure ages in this age range to (assumed) erosion rates.

Another interpretation is that the ^{10}Be concentration in the boulder of the Montoz site has reached secular equilibrium, where erosion and radioactive decay balance production. If the boulder has been continuously exposed since deposition, one can calculate a maximum erosion rate of 5.3 mm/k.y.

LAST GLACIAL MAXIMUM (LATE WÜRM)

Reconstructions of the LGM ice topography are based on the location of ice-marginal features in the foreland, as well as trimlines and other glacial erosional features in the mountain valleys (e.g., Penck and Brückner, 1901/1909; Jäckli, 1970; van Husen, 1997; Florineth and Schlüchter, 1998, 2000; Kelly et

TABLE 2. PRODUCTION RATES USED FOR AGE CALCULATIONS

Cosmogenic isotope	Production rate	Reference
^{10}Be	atoms (g SiO$_2$)$^{-1}$ yr^{-1}	
Total (spallation, muon)	5.1 ± 0.3	Stone (2000)
muon reactions	2.6%	Stone (2000)
^{26}Al	atoms (g SiO$_2$)$^{-1}$ yr^{-1}	
Total (spallation, muon)	33.2 ± 2.9†	Stone (2000)
muon reactions	2.2%	Stone (2000)
^{36}Cl		
From Ca	atoms (g Ca)$^{-1}$ yr^{-1}	
spallation	48.8 ± 3.4	Stone et al. (1996)
muon reactions	5.2 ± 1.0	Stone et al. (1998)
From K	atoms (g K)$^{-1}$ yr^{-1}	
spallation	161 ± 9	Evans (2001)
muon reactions	10.2 ± 1.3	Evans (2001)
From Ti	13.5 atoms (g Ti)$^{-1}$ yr^{-1}	Masarik (2002)
From Fe	6.75 atoms (g Fe)$^{-1}$ yr^{-1}	Masarik (2002)
Thermal neutron flux	626 neutrons (g air)$^{-1}$yr^{-1}	Phillips et al. (2001)
^{21}Ne	atoms (g SiO$_2$)$^{-1}$ yr^{-1} 20.4 ± 3.9	Niedermann (2000)

†Based on the ^{10}Be production rate of Stone (2000) and the production ratio ^{26}Al:^{10}Be of 6.5 ± 0.4 (Kubik et al., 1998).

TABLE 3. COSMOGENIC NUCLIDE CONCENTRATIONS AND CALCULATED EXPOSURE AGES

Boulder/sample no.	Reference	Altitude (masl)	Thick (cm)	Shielding correction	Nuclide	10^5 atoms/g^{-1}	Exposure age (yr)	Snow corrected (yr)	Snow and erosion corrected† (yr)	Mean rock surface age (yr)
Jura Mountains, Montoz, 47.23°N										
MON	this paper	1200	3	0.986	^{10}Be	14.40 ± 0.63	106,000 ± 5000		155,000 ± 18,000	155,000 ± 18,000
					^{21}Ne	76.57 ± 13.09	128,000 ± 22,000			
Wangen an der Aare, 47.17°N										
ER1	4	580	6	1.000	^{10}Be	1.65 ± 0.10	19,680 ± 1240		20,750 ± 1310	21,050 ± 860
	4				^{26}Al	11.60 ± 0.74	21,400 ± 1370		21,510 ± 1380	
	4				^{36}Cl	3.02 ± 0.27	21,160 ± 1930		20,800 ± 1980	
ER2	4	585	5	0.921	^{10}Be	1.55 ± 0.68	19,770 ± 1050		20,920 ± 940	20,920 ± 940
ER7	4	610	4	1.000	^{10}Be	1.43 ± 0.10	16,400 ± 1130		17,090 ± 1180	17,070 ± 720
	4				^{26}Al	9.56 ± 0.51	16,890 ± 890		17,060 ± 900	
ER8	4	595	3	1.000	^{10}Be	1.62 ± 0.10	18,580 ± 1190		19,580 ± 1260	19,720 ± 720
	4				^{26}Al	10.7 ± 0.48	19,080 ± 860		19,560 ± 880	
Grimsel Pass, 46.57°N										
G103	1, 7	2660	2	0.989	^{10}Be	3.75 ± 0.28	9590 ± 720	11,720‡		11,720 ± 880
G106	1, 7	2660	2	0.987	^{10}Be	3.46 ± 0.25	8870 ± 650	10,830‡		10,760 ± 460
	1				^{26}Al	22.30 ± 1.18	8770 ± 460	10,720‡		
G110	1, 7	2680	4	0.910	^{10}Be	3.76 ± 0.27	10,580 ± 780		10,890	10,890 ± 800
G111	1, 7	2400	8	0.925	^{10}Be	3.43 ± 0.28	11,780 ± 970	12,970§		12,970 ± 1060
Trins Gschnitz moraine, 47.09°N										
Gamma 3	5	1215	4	0.878	^{10}Be	1.87 ± 0.10	14,550 ± 870		15,110 ± 920	15,110 ± 920
Gamma 5	5	1220	3	0.877	^{10}Be	1.88 ± 0.14	14,520 ± 1060	14,800#	15,380 ± 1130	15,380 ± 1130
Gamma 6	5	1225	2	0.961	^{10}Be	1.61 ± 0.16	11,370 ± 1120	11,570#	11,920 ± 1190	11,920 ± 1190
Gamma 7	5	1220	2	0.904	^{10}Be	1.76 ± 0.15	13,220 ± 1120	13,460#	13,940 ± 1200	13,940 ± 1200
Gamma 8a	5	1215	2	0.509	^{10}Be	1.21 ± 0.05	15,500 ± 990		16,130 ± 1040	16,130 ± 1040
Gamma 102a	5	1325	3	0.932	^{10}Be	1.94 ± 0.09	12,940 ± 610	13,180#	13,650 ± 650	13,650 ± 650
Gamma 103	5	1330	4	0.891	^{10}Be	2.03 ± 0.08	14,030 ± 590	14,290#	14,830 ± 720	14,830 ± 720

(continued)

TABLE 3. COSMOGENIC NUCLIDE CONCENTRATIONS AND CALCULATED EXPOSURE AGES (continued)

Boulder/sample no.	Reference	Altitude (masl)	Thick (cm)	Shielding correction	Nuclide	10^5 atoms/g^{-1}	Exposure age (yr)	Snow corrected (yr)	Snow and erosion corrected[†] (yr)	Mean rock surface age (yr)
Julier Pass moraines, 46.47°N										
II Dschember J202	this paper	2080	4	0.947	^{10}Be	3.12 ± 0.14	12,570 ± 580		13,210 ± 610	13,210 ± 610
JULIER 8	2	2175	6	0.960	^{10}Be	2.96 ± 0.15	11,210 ± 560		11,510 ± 580	11,510 ± 580
J10	2	2185	8	0.960	^{10}Be	2.87 ± 0.17	10,970 ± 660		11,260 ± 680	11,530 ± 530
	3				^{26}Al	20.40 ± 1.86	11,880 ± 1080		12,260 ± 1130	
	2				^{36}Cl	8.67 ± 0.79	12,200 ± 1110		11,560 ± 1300	
J12	2	2200	4	0.960	^{10}Be	3.60 ± 0.32	13,190 ± 1190		13,610 ± 1230	12,470 ± 630
	3				^{26}Al	20.90 ± 1.57	11,660 ± 870		12,030 ± 910	
	2				^{36}Cl	7.72 ± 0.66	12,810 ± 1100		12,130 ± 1270	
J14	this paper	2195	4	0.952	^{10}Be	2.87 ± 0.12	10,800 ± 440		11,080 ± 450	10,820 ± 370
	this paper				^{26}Al	17.40 ± 1.11	9980 ± 640		10,250 ± 660	
J15	2	2195	2	0.960	^{36}Cl	6.84 ± 0.68	11,680 ± 1170		11,040 ± 1330	11,040 ± 1330
J18	3	2210	4	0.960	^{10}Be	3.50 ± 0.21	12,740 ± 760		13,140 ± 800	13,130 ± 700
	3				^{26}Al	22.90 ± 2.52	12,660 ± 1390		13,100 ± 1460	
J104	3	2185	8	0.921	^{10}Be	2.86 ± 0.14	10,930 ± 550		11,220 ± 570	11,220 ± 570
J200	this paper	2200	3	0.960	^{10}Be	3.05 ± 0.13	11,090 ± 470		11,390 ± 450	11,610 ± 400
	this paper				^{26}Al	21.50 ± 1.42	11,960 ± 790		12,350 ± 820	
J201b	this paper	2160	4	0.945	^{10}Be	2.65 ± 0.11	10,120 ± 410		10,370 ± 420	10,370 ± 420
Ferwall moraines										
Schönferwall, F1, 47.07°N	this paper	1740	3	0.932	^{10}Be	2.26 ± 0.16	11,580 ± 800	11,800[#]	12,170 ± 850	12,170 ± 850
F2	this paper	1725	2	0.706	^{10}Be	1.79 ± 0.12	12,130 ± 790		12,520 ± 820	12,520 ± 820
F3	this paper	1720	2	0.927	^{10}Be	2.17 ± 0.10	11,250 ± 530	11,460[#]	11,810 ± 560	11,810 ± 560
F4b	this paper	1715	2	0.960	^{10}Be	2.30 ± 0.12	11,560 ± 620	11,780[#]	12,140 ± 660	12,140 ± 660
Kartell, K1, 47.05°N	this paper	2190	2	0.840	^{10}Be	2.61 ± 0.19	10,660 ± 780	10,870[#]	11,180 ± 820	11,180 ± 820
K2	this paper	2205	3	0.920	^{10}Be	2.69 ± 0.13	9950 ± 470	10,140[#]	10,410 ± 500	10,410 ± 500
K3a	this paper	2210	2	0.642	^{10}Be	2.02 ± 0.16	10,590 ± 830		10,880 ± 860	10,880 ± 860
Kromer, Kr1, 46.91°N	6	2220	1	0.983	^{10}Be	2.46 ± 0.12	8410 ± 400			8410 ± 400
Kr2	6	2220	1	0.966	^{10}Be	2.51 ± 0.12	8690 ± 410			8690 ± 410
Kr3	6	2175	1	0.961	^{10}Be	2.32 ± 0.12	8380 ± 420			8380 ± 420
Kr4	6	2185	2	0.982	^{10}Be	2.28 ± 0.10	8010 ± 360			8010 ± 360
Kr5	6	2135	1	0.958	^{10}Be	2.30 ± 0.09	8550 ± 340			8550 ± 340

Note: References: 1—Ivy-Ochs (1996); 2—Ivy-Ochs et al. (1996); 3—Ivy-Ochs et al. (1999); 4—Ivy-Ochs et al. (2004); 5—Ivy-Ochs et al. (2006); 6—Kerschner et al. (2006); 7—Kelly et al. (2006).
[†]Erosion rate of 3.0 ± 0.5 mm/k.y.
[‡]Snow correction of 2 m for 6 months of year (density of 0.3 g/cm^3) (for details see Kelly et al., 2006).
[§]Snow correction of 1 m for 6 months of year (for details, see Kelly et al., 2006).
[#]Snow correction of 30 cm for 4 months of year (for details, see Ivy-Ochs et al., 2006).

Figure 2. Digital elevation model based on DHM25 data set (used with permission of Swisstopo©) for the region near the terminal moraines of the Solothurn lobe of the Rhône Glacier (vertical exaggeration is three times). Generalized extent of the Rhône Glacier during the LGM at Wangen an der Aare and Solothurn LGM stadials based on Nussbaum (1910) and Ledermann (1978). Scale varies with perspective.

al., 2004a; Keller and Krayss, 1993, 2005). On the foreland, glaciers coalesced and spread out into broad piedmont lobes (Fig. 1). The ELA depression associated with the LGM glaciers was ~1200–1500 m (Haeberli, 1991; van Husen, 1997; Keller and Krayss, 2005).

The northern arm of the Rhône Glacier formed a broad nose-shaped lobe in the region of Solothurn, with the glacier terminus near the town of Wangen an der Aare (600 masl). There, the former ice margin is delineated by steep-walled, subparallel moraine sets, as well as broad, less distinct moraines. Four boulders, located on a broad ridge of the outermost right lateral moraines just inside a major meltwater drainage system (Fig. 2), were dated using ^{10}Be, ^{26}Al, and ^{36}Cl (Ivy-Ochs et al., 2004). The ages range from 17,070 ± 720 to 21,050 ± 860 yr. The oldest age (21,050 ± 860 yr) was obtained from the top of the largest boulder (Fig. 3). Our interpretation is that it represents a minimum age for the start of deglaciation. Due to its size, postdepositional exhumation of that particular boulder can be excluded.

DEGLACIATION AT HIGH ALTITUDES IN THE ALPS

Ice domes formed above the heads of the main valleys during the LGM (Florineth and Schlüchter, 1998, 2000; Kelly et al., 2004a). From there, ice flowed outward into the main valleys and thence onto the foreland. Due to the geometry of the ice surface, transfluences over many of the high passes existed. Glacial erosional features indicate that ice of the upper Rhône Glacier flowed northward over Grimsel Pass (2164 masl) into the catchment of the Aare Glacier (Florineth and Schlüchter, 1998). Samples were taken on a transect from Nägelisgrätli to Grimsel Pass (Fig. 4) to determine when ice flow over the pass ceased (Ivy-Ochs, 1996; Kelly et al., 2006). The ^{10}Be and ^{26}Al ages range from 10,760 ± 460 to 12,970 ± 1060 yr (Table 3), with the younger ages stemming from the higher-elevation sites (Nägelisgrätli). This is the opposite of what one would expect for a simple scenario of glacier surface lowering during deglaciation. The oldest age (G111) is from the site closest to the pass. This provides a minimum age of 12,970 ± 1060 yr for breakdown of the northward transfluence of the Rhône Glacier.

On the Nägelisgrätli ridgeline, Florineth and Schlüchter (1998) recognized a second generation of glacial erosional features above 2500 masl. These features indicate ice flow to the south and southeast, i.e., coming out of the cirques. The younger

Figure 3. Erratic boulder ER1 located on the outermost moraine at Wangen an der Aare (Solothurn, Switzerland). The oil drum near the tree on the right side is 1 m high (photo: M. Ochs). Boulder location shown in Figure 2.

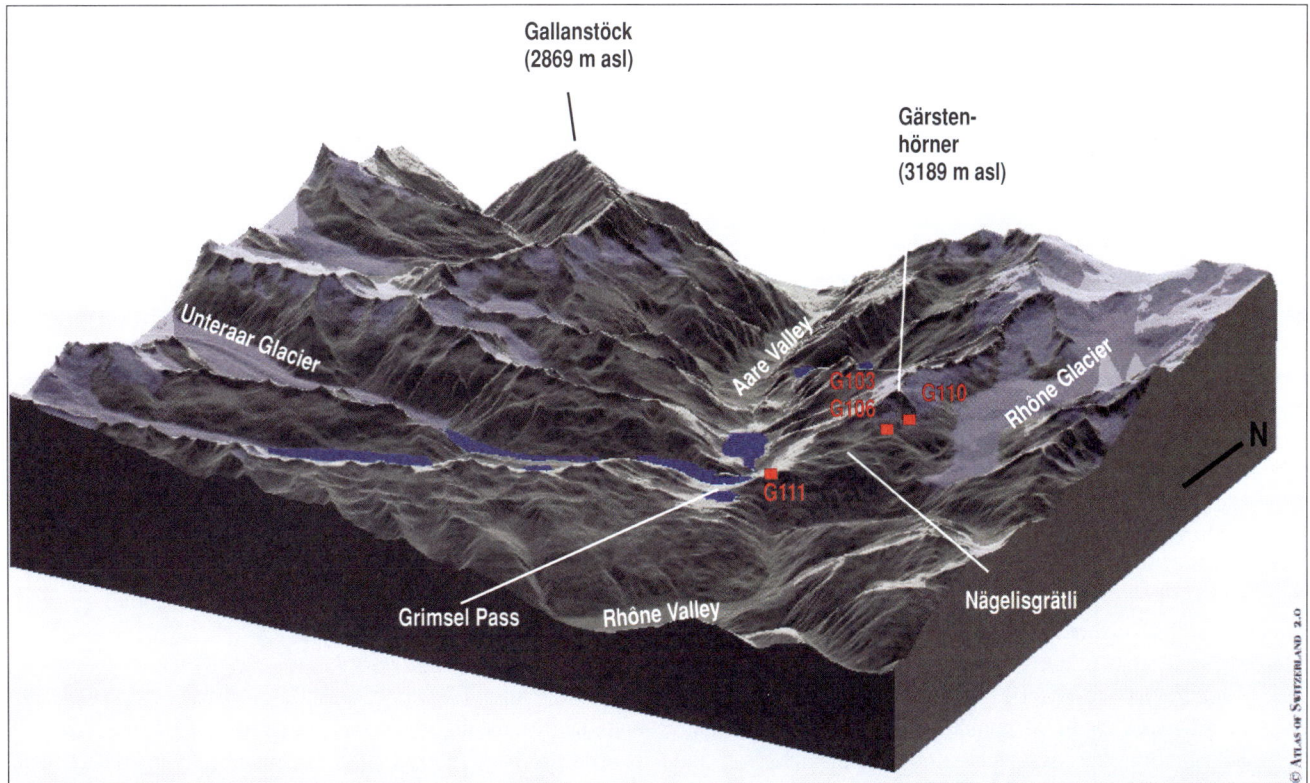

Figure 4. Three-dimensional model of Grimsel Pass site. Extract from Atlas der Schweiz (permission of reprint from Swisstopo©). Red boxes mark the sampling sites. Scale varies with perspective.

features crosscut the older north-indicating features between 2500 and 2600 masl. Exposure ages (^{10}Be and ^{26}Al) for polished quartz veins of the surface with only the younger features are 10,760 ± 460 and 11,720 ± 880 yr. They provide constraints for the final retreat of cirque glaciers south of Gärstenhörner (cf. Florineth and Schlüchter, 1998). The ^{10}Be age of 10,890 ± 800 yr (G110) from a steep sidewall overhanging the present-day Rhône Glacier provides additional support for the presence of a local ice body until the early Holocene. These dates indicate that small cirque glaciers continued to actively undercut their headwalls into Holocene time, modifying and further emphasizing the trough shoulders of the main valleys.

THE ALPINE LATEGLACIAL STADIALS

Following the peak of the LGM, the final disintegration of the foreland piedmont lobes (Schlüchter, 2004; van Husen, 2004) marks the onset of the "Alpine Lateglacial" (referred to as "Lateglacial" below) (Penck and Brückner, 1901/1909). Moraine systems in the mountain valleys indicate repeated glacier advances ("stadials") during the Lateglacial (Table 1). In most cases, a moraine can be assigned to a stadial by evaluating the following diagnostic features (e.g., Penck and Brückner, 1901/1909; Heuberger, 1966; Gross et al., 1977; Maisch, 1981, 1982, 1987): (1) the relative position of the moraines, (2) the moraine morphology and related periglacial features, and (3) the depression of the ELA (ΔELA) of the glacier. The hypothesis is that glacier positions with similar ELA depressions and similar morphologic characteristics located in comparable climate regions occurred at the same time (Gross et al., 1977; Maisch, 1981, 1987).

The absolute ages for most of the Lateglacial stadials are poorly constrained. Numerous minimum radiocarbon ages suggest that all of them, except the Egesen (Younger Dryas equivalent), Kromer, and Kartell, occurred during the Oldest Dryas (Maisch, 1981, 1982, 1987).

Early Lateglacial Ice Decay

For the most part, no clear end moraines, which would indicate stillstand or readvance, have been recognized in the lower parts of the main valleys (Reitner, 2005), but can be found in smaller catchments near the northern fringe of the Alps (Weissbad stadial; Keller, 1988; Keller and Krayss, 2005). This time period corresponds to the classical "Bühl" stadial (Penck and Brückner, 1901/1909; Mayr and Heuberger, 1968; van Husen, 1977) (Table 1). A subsequent minor readvance was recorded by lateral moraines and ice-marginal features located at Steinach am Brenner near Innsbruck (Tyrol, Austria) (Senarclens-Grancy, 1958; Mayr and Heuberger, 1968). This readvance is difficult to identify at other locations (cf. Reitner, 2005).

Gschnitz Stadial

The Gschnitz stadial shows for the first time a clear and widespread readvance of glaciers in the Alps. In the central Alps, the ELA depression was on the order of 650–700 m (Gross et al., 1977; Maisch, 1987; Kerschner et al., 1999). The moraines are often several tens of meters high and sharp-crested; in many places they contain abundant coarse debris.

At the type locality in Trins (Gschnitz Valley, Tyrol, Austria; Fig. 1), the moraine complex consists of an arcuate end moraine with lateral moraines and related ice-marginal features. Kettle holes in the moraine complex suggest delayed moraine stabilization. Seven boulders from the moraine at Trins have been dated with ^{10}Be (sample locations shown in Ivy-Ochs et al., 2006). The measured exposure ages range from 11,920 ± 1190 to 16,130 ± 1040 yr (Table 3). Pollen and radiocarbon data from several sites across the Alps indicate that the Gschnitz stadial occurred before the onset of the Bølling interstadial (before 14,700 yr ago). Therefore, our interpretation is that the Gschnitz moraine at Trins stabilized no later than 15,400 ± 1400 yr ago. This is the mean of the exposure ages of the four oldest boulders. The younger ages are interpreted to reflect exhumation, boulder toppling, or even tree growth on top of the boulders (see also Ivy-Ochs et al., 2006). We note that the mean age is indistinguishable from the oldest boulder age (16,130 ± 1040 yr).

Clavadel/Senders Stadial and Daun Stadial

The Gschnitz advance was followed by a marked period of marked glacier downwasting. The next readvance is called the Clavadel stadial and is based on deposits in Sertigtal (near Davos) in eastern Switzerland (Maisch, 1981, 1982). It is probably equivalent to an advance recognized in Senderstal (northern Stubaier Alpen, Tyrol, Austria) (Kerschner and Berktold, 1982; Kerschner, 1986) and in Zwischbergental (Simplon Pass area, Switzerland) (Müller, 1982). In suitable locations, the moraines are tens of meters high and multiwalled. Nevertheless, in most cases they lack large boulders suitable for surface exposure dating. The ELA depression during the Clavadel/Senders stadial was on the order of 400–500 m (Maisch, 1981, 1982; Kerschner and Berktold, 1982).

The Daun stadial (in the sense of Heuberger, 1966) followed the Clavadel/Senders stadial. ELA depressions were 300–400 m. Daun moraines are often subdued and broad-crested; boulders are infrequent and often appear to be sunken into or weathering out of the deposit. Their overall position suggests that this group of moraines may possibly be nothing else than the final advances during the Clavadel/Senders stadial.

Remnants of a left lateral moraine on the eastern side of Julier Pass (Graubünden, Switzerland) were attributed to the Daun stadial (Il Dschember site) (Suter, 1981). The moraine is ~10 m high and wide with a flat crest and ~20 m long. From a boulder partially embedded in the moraine, a ^{10}Be age of 13,210 ± 610 yr was obtained. This is a minimum age for the Daun stadial as the boulder appears to be weathering out of the till.

Egesen Stadial

Moraines formed during the Egesen stadial are found in numerous valleys throughout the Alps (e.g., Heuberger, 1968; Maisch, 1981, 1982; Kerschner et al., 2000, and references therein). The moraine complexes can often be subdivided into Egesen I (Egesen maximum), Egesen II (Bocktentälli), and Egesen III (Maisch, 1981, 1987). Egesen I and even more notably the Egesen II (Bocktentälli) moraines are steep-walled, sharp-crested and often quite rich in large boulders. The ELA depression varies from less than 200 m in the inner-Alpine dry region to more than 400 m along the maritime northern and northwestern fringe of the Alps (Kerschner et al., 2000).

Surface exposure dating results from two sites will be discussed in detail here. These are the Julier Pass Lagrev Egesen moraine complex (Graubünden, Switzerland) (Ivy-Ochs et al., 1996) and the Egesen I moraine at Schönferwall (Tyrol, Austria). Similar exposure ages have been obtained from the Grosser Aletsch Glacier (Bern/Wallis, Switzerland) (Kelly et al., 2004b) (Fig. 1).

Egesen Stadial: Julier Pass Lagrev Site

At the Julier Pass site (Suter, 1981; Maisch, 1981), landforms and glacial deposits stemming from advances of the Lagrev Glacier during both the Egesen I maximum and Egesen II (Bocktentälli) phases are present (Fig. 5). Nine granodiorite and diorite boulders were dated with ^{10}Be, ^{26}Al, and ^{36}Cl (boulder locations shown in Ivy-Ochs et al., 1996, 1999). Data from all nuclides are plotted in Figure 6. Individual nuclide boulder ages and error-weighted mean of ^{10}Be, ^{26}Al, and ^{36}Cl ages are given in Table 3. Overall, we see excellent agreement between exposure ages from the three different nuclides.

The outer moraine consists of well-developed left and right lateral moraines, each 400–500 m long. Ages from the outer moraine ridge range from 10,820 ± 370 to 13,130 ± 700 yr. The mean age for the outer ridge is 12,300 ± 1500 yr based on ^{10}Be concentrations, and 11,700 ± 1500 yr based on ^{10}Be, ^{26}Al, and ^{36}Cl (Fig. 6).

During a second, slightly smaller advance, debris of the earlier end moraine was entrained in the advancing glacier. The inner moraine is a broad arcuate deposit (~20 m high). The terminal moraine developed into a small rock glacier, indicating the presence of permafrost at that altitude (2200 m) after the deposition of the moraine. Three boulders were analyzed for ^{10}Be, ^{26}Al, and ^{36}Cl. The boulder ages range from 11,220 ± 570 to 11,530 ± 530. The mean ^{10}Be age of the moraine/rock glacier complex is 11,300 ± 900 yr, and 11,300 ± 1300 yr based on all three nuclides.

A relict rock glacier is present in the former glacier tongue area just inside the right lateral moraine (Fig. 5) (cf. Frauenfelder et al., 2001). The last phase of rock glacier activity may have ended 10,370 ± 420 yr ago, based on ^{10}Be from a large boulder perched on one of the inner ridges of this rock glacier.

Egesen Stadial: Schönferwall Site

In the Schönferwall and Ochsental Valley (southwest of St. Anton am Arlberg, Austria; Fig. 1), a complete moraine series

Figure 5. Photograph of Julier Pass Lagrev site (Graubünden, Switzerland). View is up into the Lagrev Glacier catchment. Julier Pass is to the right. To the left, the pass road heads down toward the Inn Valley (Upper Engadine) (photo: M. Maisch).

of the Egesen stadial and younger readvances up to Little Ice Age (LIA) moraines is preserved. A detailed map of the Egesen I, II, and III moraines, which are separated by ~1–2 km from each other, is found in Sailer et al. (1999). A steep-walled ridge extending from an elevation of 1820 masl down to 1680 masl forms the right lateral part of the moraine complex. Numerous boulders several meters in diameter are embedded in coarse till. Four boulders located along the frontal part of the ridge were sampled (boulder locations shown in Fig. 7). The four ^{10}Be ages range from 11,810 ± 560 to 12,520 ± 820 yr, giving a very tight age distribution and a mean moraine age of 12,200 ± 1000 yr.

Kartell and Kromer Stadials

In many valleys, a glacier advance between the innermost Egesen stadial moraines and the LIA moraines can be distinguished. In the 1970s, it was considered to be early Preboreal in age, because the consensus was that glaciers had already melted back and readvanced to their LIA extent during the Preboreal (Patzelt and Bortenschlager, 1973). As these moraines were particularly well developed in the Kromer Valley (western Silvretta Mountains, Vorarlberg, Austria), they were assigned to a "Kromer stadial" (Gross et al., 1977). Later, similar moraines were found in the neighboring Ferwall Group and named "Kartell stadial" (Fraedrich, 1979). As the ELA depression was 75 m in the Kromer Valley and 120 m in the Kartell cirque, which was considered to be comparable, the stadial was later called the "Kromer/Kartell" stadial.

The moraine in the Kartell cirque (south of St. Anton am Arlberg, Tyrol, Austria) is a well-developed arc located less than one kilometer in front of the LIA moraine. The Kartell moraine is a single, several-meter-high ridge that is composed of clast-supported boulders 1–3 m in diameter with little to no fine- or medium-grained sediment. Two boulders are located directly on the crest (K1 and K2); boulder K3 lies within the former tongue

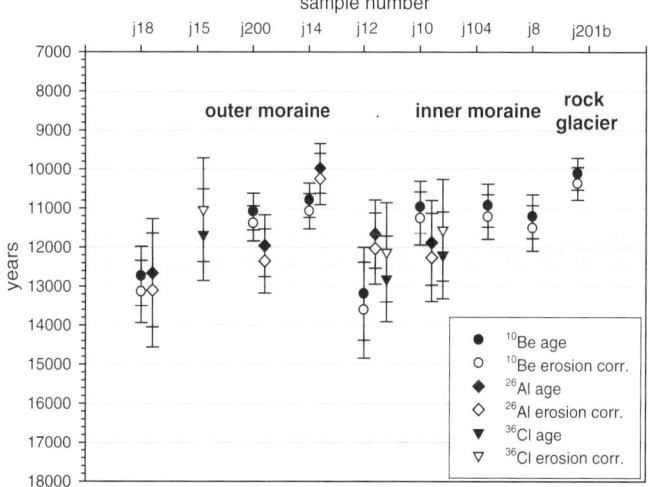

Figure 6. Plot of all surface exposure ages from the Julier Pass Lagrev Egesen moraine complex. The shaded band indicates the boundaries of the Younger Dryas period (12.7–11.6 ka) (Schwander et al., 2000). Boulders J18, J15, J200, J14, and J12 are from the outer ridge, J10, J104, and J8 are from the inner ridge, and J201b is from the relict rock glacier located just inside the right lateral ridge.

region (Fig. 8). The three ^{10}Be ages range between 10,410 ± 500 and 11,180 ± 820 yr with a mean of 10,800 ± 1000 yr.

In a second step, the moraines at the type locality of the Kromer stadial were sampled with the aim to verify the dating of the Kartell moraines. The Kromer moraines are two closely spaced arcs of blocky debris (Hertl, 2001; Kerschner et al., 2006). Inside the moraine, the area of the former glacier tongue is densely covered with angular to subangular boulders resting either on bedrock or on a thin veneer of till (Fig. 9). Sampling sites and details on moraine morphology are given in Kerschner et al. (2006). Five boulders were ^{10}Be dated with ages ranging from 8010 ± 360 to 8690 ± 410 yr. The mean of the five ages is 8400 ± 700 yr. Kromer moraines are not of the same age as the Kartell moraines, but are ~2000 yr younger.

DISCUSSION

More than a century of detailed mapping both in the Alps and on the foreland provides a unique opportunity for exposure dating in a well-constrained field situation. Nevertheless, several challenges have been faced.

Pre-LGM deposits of the Alps are difficult to date with cosmogenic nuclides. The earlier glaciations are represented by till and/or outwash layers that are often exposed in gravel pits. Such deposits are unsuitable for rock surface exposure dating as described here. Where older moraines beyond the LGM limits are present, intense weathering during warm periods as well as frost shattering and periglacial processes active during subsequent cold periods has led to degradation of the moraines and boulders (cf. Zreda et al., 1994; Putkonen and Swanson, 2003).

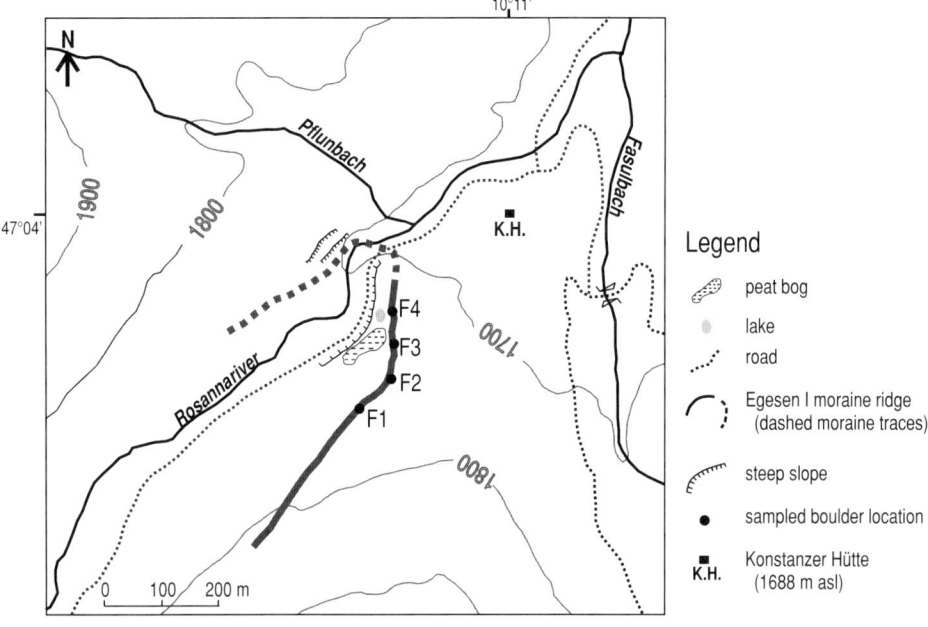

Figure 7. Map of the Schönferwall Egesen I moraine (Tyrol, Austria) (modified from Sailer, 2001) with the location of the sampling sites.

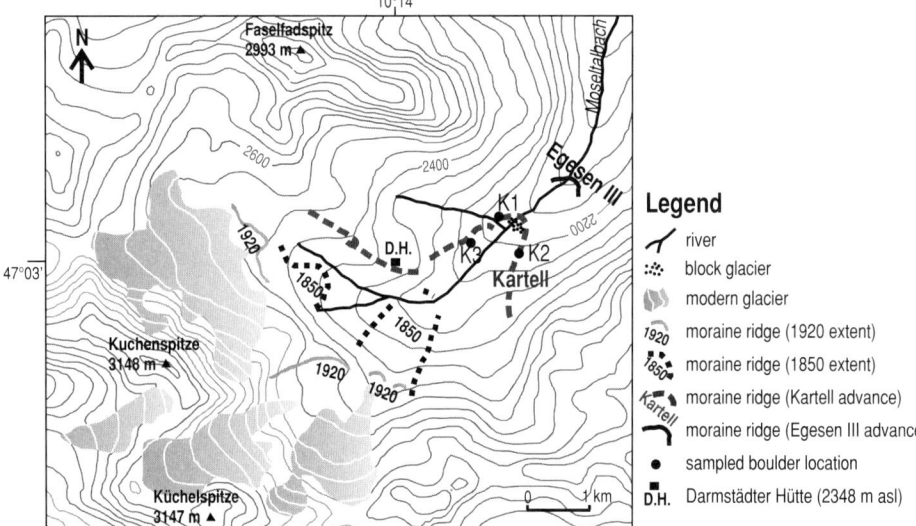

Figure 8. Map of the Kartell moraine (Tyrol, Austria) (modified from Sailer, 2001) with the location of the sampling sites. The present as well as the A.D. 1920 and A.D. 1850 ice extents are shown.

Overall, exposure age data from the younger moraines (Egesen, Kartell, Kromer) cluster much better than data from the older moraines (LGM, Gschnitz). For the younger moraines, the mean moraine ages are indistinguishable from the oldest age of the data set. We note that the younger moraines have been subjected to degradational processes for shorter periods of time. In addition, the two youngest moraines dated (Kartell and Kromer) are made up of wedged-together blocks with little to no sediment matrix. Exhumation or toppling is impossible in such a case.

By looking at the data critically on a site-by-site basis, we were able to glean important insight into the timing of climate change in the northern regions of the Alps based on surface exposure dating. Indeed, many of the landforms were undatable prior to the advent of cosmogenic nuclide methods. Below we discuss the timing of the various phases of glacier advance in the Alps as based on exposure dating (Fig. 10). We compare these results to previously published dating results (radiocarbon, luminescence, and U/Th) and information on climate variations derived from paleobotanical and paleoglaciological studies.

The oldest exposure-dated boulder related to Alpine glaciations comes from the Montoz site in the Swiss Jura Mountains. The ^{10}Be age, 155,000 yr, supports the concept that pre-LGM, more extensive glaciation(s) occurred before the last interglacial (cf. Penck and Brückner, 1901/1909; Fiebig et al., 2004; Schlüchter, 2004). Dates from several sites in and around the Alps place the last interglacial between 135 ka and 115 ka (Preusser, 2004). Data from Spannagel Cave in Austria indicate temperatures much like today's 135,000 yr ago (Spötl et al., 2002).

In the Swiss part of the northern Alpine foreland, early Würm ice expansions onto the lowlands have been recognized, although none were as extensive as that of the late Würm (LGM) (Schlüchter et al., 1987; Preusser et al., 2003; Preusser and Schlüchter, 2004).

Figure 9. Photograph of the Kromer moraine, which extends from the center foreground to the left. Little Ice Age moraine is in the center of the photograph. Farther up-valley is the Schweizer Glacier. Peaks along the crest are 2800–2850 masl (photo: H. Kerschner).

A number of radiocarbon dates from the northern foreland show that, at the beginning of the LGM, glaciers expanded onto the foreland after approximately 30,000 years ago. A mammoth tusk deposited in front of the advancing Rhône Glacier yielded a date of 25,370 ± 190 ^{14}C yr B.P. (Schlüchter, 2004), providing a maximum age limit of 28,500–30,200 yr B.P. for deposition of the outermost terminal moraines at Wangen an der Aare. In the complex Rhine Glacier area farther to the east (Fig. 1), radiocarbon dates show that the Rhine/Linth Glacier reached its maximum extent ("Killwangen maximum") sometime between 28,060 ± 340 ^{14}C yr B.P. (ca. 31,000–32,300 yr B.P.) and 19,820 ± 190 ^{14}C yr B.P. (23,450–24,010 cal. yr B.P.) (Schlüchter and Röthlisberger, 1995; Schlüchter, 2004). A date of 24,910 ± 215 ^{14}C yr B.P.

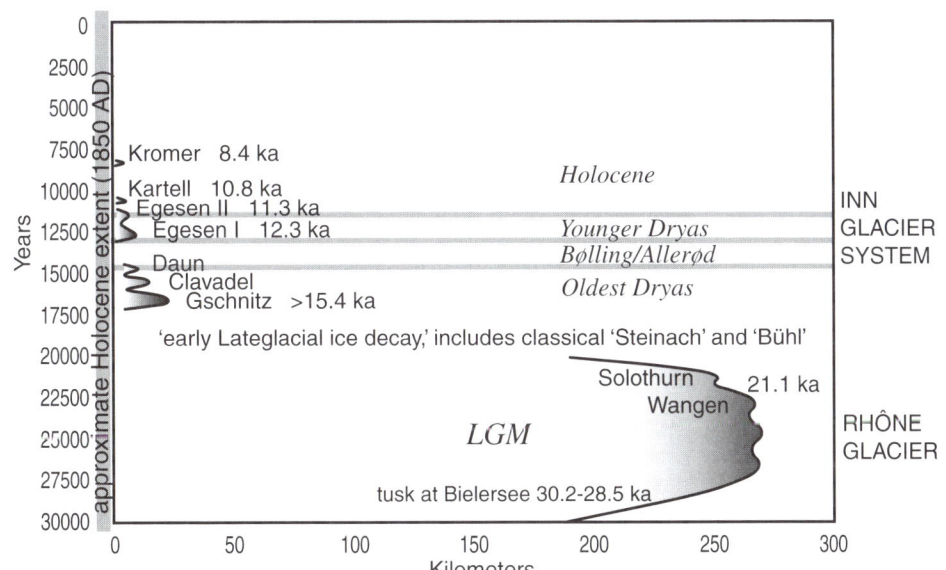

Figure 10. Time-distance diagram for the LGM and younger advances of Alpine glaciers. Site names and ^{10}Be exposure dating results are shown. Note that the LGM glacier shown is the Rhône Glacier, while the Gschnitz, Egesen, and Kartell sites all lie in the catchment of the Inn Glacier. The Kromer site is situated in the Rhine (Ill) Glacier catchment. Finsterhennen is the only site where a radiocarbon date directly related to an exposure-dated site is found.

(ca. 28,300–29,500 yr B.P.) was obtained from a mammoth tusk under the lowermost till of the late Würm advance of the Rhine Glacier in the Lake Constance region (Schreiner, 1992). Still farther to the east, in the catchment of the Inn Glacier (Fig. 1), the Baumkirchen site (Tyrol, Austria) is situated ~120 km upstream from the terminal moraines. There, the stratigraphically highest wood fragment out of 13 samples underlying basal till gave a date of 26,800 ± 1300 ^{14}C yr B.P. (Fliri et al., 1970; Fliri, 1973; Patzelt and Resch, 1986), corresponding to ca. 29,400–30,600 yr B.P. The onset of deglaciation at 21,100 ± 900 yr ago is recorded by the direct dating of the largest boulder at the Wangen an der Aare site. For the northern foreland, the peak of the LGM in the Alps was broadly synchronous with the global ice-volume maximum (e.g., Clark and Mix, 2000). In comparison to the sluggishly responding ice sheets, the collapse of the Alpine piedmont lobes occurred early during the last termination, right after the peak in ice volume had been reached.

At the end of the LGM, the piedmont lobes separated into glaciers filling the main valleys. Van Husen (2000) estimated that it took several hundred to perhaps a thousand years for downwasting of the foreland glaciers, which is equivalent to a retreat of glacier tongues of several tens of meters per year (Keller and Krayss, 2005). The rate of downwasting and retreat was likely increased by calving into the lakes (Poscher, 1993; van Husen, 2000) that formed at that time (e.g., Müller, 1995; Schindler, 2004). Wide areas of the Alps including the lower parts of the main valleys were already free of glacier cover early in the Lateglacial. A minimum age is given by a radiocarbon date of 15,400 ± 470 ^{14}C yr B.P. (18,020–19,100 cal. yr B.P.) (van Husen, 1977) from the Rödschitz peat bog. This site is located inside the moraines of the Steinach stadial of the Traun Glacier system (Fig. 1), some 80 km up-valley from the LGM terminal moraines at Gmunden (Oberösterreich, Austria).

In the high Alps, after the peak of the LGM, ice domes developed into systems of large dendritic glaciers that extended into the major drainages. Many of these may have remained intact until some time between the Gschnitz stadial and the onset of the Bølling warm phase. The ^{10}Be data indicate that the transfluence of the Rhône Glacier northward over Grimsel Pass into the Aare watershed broke down no later than 13,000 ± 1100 yr ago (cf. Kelly et al., 2006). This result is supported by a number of pollen analytical studies and radiocarbon dates from the high mountain passes, which are summarized in Ivy-Ochs et al. (2006). These studies show that many of the inner-Alpine valleys were already free of ice by the last phase of the Oldest Dryas (e.g., Welten, 1982; Ammann et al., 1994; Burga and Perret, 1998).

Field evidence indicates that the Gschnitz stadial glaciers advanced over ice-free terrain (van Husen, 1977; Kerschner and Berktold, 1982); thus a distinct period of warming preceded the readvance. With an ELA depression of ~700 m, glaciers were already confined to the inner valleys of the Alps during the Gschnitz stadial. At Trins, the glacier end was ~160 km upstream of the LGM end moraines of the Inn Glacier. The glacier was 18–20 km long. The overall morphological situation and the long reaction time of the Gschnitz Glacier suggest that the glacier advance and the related cold event lasted ~500 yr. ^{10}Be data indicate that stabilization of the terminal moraine occurred no later than 15,400 ± 1400 yr ago. Radiocarbon dates from correlative sites (cf. Ivy-Ochs et al., 2006) imply that the Gschnitz stadial is older than 13,250 ± 250 ^{14}C yr B.P. (Patzelt, 1975, 1995) (15,600–16,260 cal. yr B.P.). A paleoclimatic interpretation of the Gschnitz Glacier at the type locality shows that precipitation was reduced by 50%–70% of modern values, and summer temperature was ~8.5–10 K lower than during the middle of the twentieth century (Kerschner et al., 1999; Ivy-Ochs et al., 2006). Based on the age of the moraine, and the cold and dry climate at the time of its formation, we suggest that the Gschnitz stadial was the response of Alpine glaciers to cooling associated with Heinrich event 1 in the North Atlantic Ocean (Hemming, 2004), possibly its second phase (Sarnthein et al., 2001).

Pollen data from bogs located proximal to Daun moraines (C. Burga in Maisch, 1981) strongly support a pre-Bølling age of the Daun and thus the Clavadel/Senders stadial. Similarly, morphological relationships between the Daun and Egesen moraines indicate a considerable warming of the climate between the two (Heuberger, 1966; Kerschner, 1978b). This point of view is supported by the ^{10}Be age of 13,200 ± 600 yr from the Julier Pass Daun site.

During the Bølling/Allerød warm period, several brief cold events were recorded by changes in δ^{18}O in lake sediments and in biological proxies in and around the Alps (von Grafenstein et al., 1999, 2000; Ammann et al., 2000; Schwander et al., 2000). Each of these cold pulses was likely severe enough to cause glaciers to advance, but glaciers probably remained smaller than during the later Egesen advances (Ohlendorf, 1998). If glaciers did advance, the moraines were overrun by the subsequent Egesen glaciers.

Egesen stadial moraines have long been associated with the Younger Dryas cold period (Patzelt, 1972; Kerschner, 1978a, 1978b; Maisch, 1981). In suitable locations, many minor moraines are preserved in between the three main moraine systems (e.g., Heuberger, 1966). This indicates a markedly unstable climate, characterized by a succession of glacier-friendly periods (i.e., either colder or wetter or both) (Sailer et al., 1999). The ^{10}Be results from the Schönferwall site (12,200 ± 1000 yr) and the outer moraine at Julier Pass (12,300 ± 1500 yr) provide a good estimate of the time of stabilization of Egesen I moraines. Rock glacier development from moraines of the second advance (Egesen II; cf. Sailer and Kerschner, 1999) is documented by the inner moraine at Julier Pass. It dates to 11,300 ± 900 yr, pointing to landform stabilization around the Younger Dryas / Preboreal boundary. Similarly, the ^{10}Be age of 10,400 ± 400 yr for the relict rock glacier at Julier Pass suggests a prolonged period of cold and dry conditions lasting into the early Holocene. As a consequence, permafrost remained active and led to the conservation of ice in the inner moraine complex.

The ^{10}Be data show that the Egesen maximum is equivalent to the first phase of the Younger Dryas cold period that began 12,700 yr ago (Dansgaard et al., 1993; Schwander et al., 2000;

Johnsen et al., 2001). A data set of ~160 Egesen maximum (early Younger Dryas) ELA depressions provides a good basis for reconstructing spatial fields of precipitation and precipitation change (for details see Kerschner et al., 2000; Kerschner, 2005). Under the assumption of a summer temperature depression of –3.5 to –4 K (Burga and Perret, 1998), precipitation during the early Younger Dryas was similar to modern values or even somewhat higher (~10%) along the maritime northern and northwestern fringe of the Alps and up to 30% lower in the central, sheltered parts of the Alps. Field data suggest that precipitation was also reduced by 20%–30% south of the main Alpine divide. In the later part of the Younger Dryas (Egesen II / Bocktentälli), rock glaciers developed as temperatures remained rather cold but moisture delivery to the Alps decreased (Sailer and Kerschner, 1999). In addition, the ages of glacially abraded bedrock surfaces at the base of the nunataks in the Grimsel Pass area (Nägelisgrätli site) show that glacial erosion by small local glaciers continued at higher altitudes until the early Holocene, as the higher parts of the cirques remained above the ELA at least until the end of the Younger Dryas.

Based on pollen and lithologic changes, a brief cooling event during the first half of the Preboreal, often termed "Palü", had been identified (e.g., Burga, 1987). However, due to a prolonged radiocarbon plateau in that time range, dating of the event (events?) remained ambiguous (e.g., Küttel, 1977; Zoller et al., 1998). In Swiss lake sediments, a brief cold snap that began several hundred years after the end of the Younger Dryas was pinpointed with $\delta^{18}O$ (Schwander et al., 2000). It was correlated with the Preboreal oscillation (11,363–11,100 yr) as recognized in Greenland ice cores (O'Brien et al., 1995; Johnsen et al., 2001). This may be consistent with the glacier advance that led to formation of the Kartell moraine, which stabilized 10,800 ± 1000 yr ago. Similarly, small rock glaciers, for example at Julier Pass, remained active until 10,400 ± 400 yr ago. This evidence implies that climate in the Alps at the end of the Younger Dryas was not characterized by a sudden and prolonged temperature increase. In contrast, rock glacier activity and even advances of small glaciers during the early Preboreal were the response to continued cool and possibly also rather dry conditions that persisted for hundreds of years after the end of the Younger Dryas.

Based on ^{10}Be, the moraine doublet in Kromer Valley stabilized 8400 ± 700 yr ago (Kerschner et al., 2006). This is contemporaneous with the Misox cold phase registered in pollen records in the central Alps (e.g., Zoller et al., 1998) and the CE (Central European) cold phase 3 (Haas et al., 1998). The ages suggest that the moraines were deposited during a phase of glacier-friendly climate into which the 8.2 ka event was embedded (Alley and Ágústsdóttir, 2005; Rohling and Pälike, 2005). Apparently, small glaciers at the maritime northwestern slope of the Alps reacted to the colder and primarily wetter conditions characteristic of the early phase of the 8.2 ka climatic downturn (Rohling and Pälike, 2005) and advanced. A paleoclimatic interpretation of the glacier advance showed that it was favored by increased moisture transport from the west and northwest and to a lesser extent by a drop in summer temperature. In contrast, large glaciers in the central Alps, which were more or less constantly as small or even smaller than today during the early Holocene (Nicolussi and Patzelt, 2000, 2001; Hormes et al., 2001), showed only minor advances, which peaked ca. 8400 yr ago (Nicolussi and Patzelt, 2001). On the other hand, the fact that rock glaciers developed from the Kromer moraines at altitudes 200–300 m lower than today's active rock glaciers in the region indicates a depression of mean annual temperature on the order of –1.4 to –2.1 K (Kerschner et al., 2006). This agrees well with the temperature values for that time period given by von Grafenstein et al. (1998, 1999) based on $\delta^{18}O$ analysis in lake sediments from southern Germany.

CONCLUSIONS

1. Initial results of ^{10}Be and ^{21}Ne from one boulder at the Montoz site bear out that concept that the most extensive glaciation(s) of the Alps occurred before the last interglacial. A reasonable minimum age estimate is 155 ka.

2. During the late Würm, the piedmont lobes on the foreland reached their maximum extent broadly synchronously with the global LGM. Deglaciation began no later than 21.1 ± 0.9 ka. Notably, ice melted back out of the foreland regions fast and early in the last termination.

3. Data from the Nägelisgrätli site near Grimsel Pass indicate that the northward transfluence of the LGM Rhône Glacier had already broken down by 13.0 ± 1.1 ka. The slopes just below the former nunataks, which were above the local ELA until the end of the Younger Dryas, were quasi-continuously covered by ice, firn, or snowfields at least until the early Holocene.

4. The first clear readvance of independent glaciers in the Alps is represented by moraines of the Gschnitz stadial. The Gschnitz cold period likely began at or before 16.0 ka and lasted until no later than 15.4 ± 1.4 ka. During Gschnitz time, precipitation was two-thirds less than today, and summer temperatures were lower by 8.5–10 K. Based on the age and the cold, dry climate, we suggest that the Gschnitz Glacier advance was the response to Heinrich event 1 in the North Atlantic Ocean.

5. Egesen I moraines were formed as glaciers advanced in response to the Younger Dryas climatic cooling at 12.7 ka. The outer moraine at Julier Pass and the Egesen I moraine at Schönferwall stabilized right around or perhaps just before 12.3 ± 1.5 ka and 12.2 ± 1.0 ka, respectively. Toward the end of the Younger Dryas, glaciers waned and rock glaciers formed. This indicates that conditions remained cold yet became progressively drier. This is especially well exemplified by the inner moraine deposits at the Julier Pass Lagrev site, which date to 11.3 ± 0.9 ka.

6. Field data and surface exposure dates show that climate instability continued well past the Younger Dryas / Preboreal boundary and likely encompassed the Preboreal oscillation. The date of 10.8 ± 1.0 ka from moraines located less than a kilometer in front of LIA moraines in the Kartell cirque shows that in certain catchments small glaciers readvanced during the earliest Holocene.

7. Before 8.4 ± 0.7 ka, in Kromer Valley small cirque glaciers advanced to between the LIA position and the innermost Egesen moraines. Glaciers at the northern to northwestern fringe of the Alps advanced as a reaction to the cooler and primarily wetter conditions during the first phase of the 8.2 ka event, while larger glaciers in the drier parts of the Alps remained behind their LIA limits.

ACKNOWLEDGMENTS

We thank A. Hertl, G. Patzelt, and M. Schuh for help during sampling. Insightful discussions with M. Fiebig, H.-R. Graf, G. Patzelt, F. Preusser, and J. Reitner improved this manuscript. We thank M. Kaplan and an anonymous reviewer for critical review. This research was partially funded by ETH grant 0-20-624-92 and Swiss National Foundation grant 21-043469.95/1 to C. Schlüchter, and by Austrian Science Foundation (FWF) grants P-12600 GEO and P-15108 to H. Kerschner. We sincerely acknowledge the dedication of our tandem accelerator crew. The Zurich AMS facility is jointly operated by the Swiss Federal Institute of Technology, Zurich, and by Paul Scherrer Institute, Villigen, Switzerland.

REFERENCES CITED

Alley, R.B., and Ágústsdóttir, A.M., 2005, The 8k event and consequences of a major Holocene abrupt climate change: Quaternary Science Reviews, v. 24, p. 1123–1149, doi: 10.1016/j.quascirev.2004.12.004.

Ammann, B., Lotter, A.F., Eicher, U., Gaillard, M.-J., Wohlfarth, B., Häberli, W., Lister, G., Maisch, M., Niessen, F., and Schlüchter, C., 1994, The Würmian Late-glacial in lowland Switzerland: Journal of Quaternary Science, v. 9, p. 119–125.

Ammann, B., Birks, H.J.B., Brooks, S.J., Eicher, U., von Grafenstein, U., Hofmann, W., Lemdahl, G.V., Schwander, J., Tobolski, K., and Wick, L., 2000, Quantification of biotic responses to rapid climatic changes around the Younger Dryas—A synthesis: Palaeogeography, Palaeoclimatology, Palaeoecology, v. 159, p. 313–347, doi: 10.1016/S0031-0182(00)00092-4.

Balco, G., Stone, J.O.H., and Mason, J.A., 2005, Numerical ages for Plio-Pleistocene glacial sediment sequences by ^{26}Al/^{10}Be dating of quartz in buried paleosols: Earth and Planetary Science Letters, v. 232, p. 179–191, doi: 10.1016/j.epsl.2004.12.013.

Bard, E., Rostek, F., and Menot-Combes, G., 2004, A better radiocarbon clock: Science, v. 303, no. 5655, p. 178–179, doi: 10.1126/science.1091964.

Bolliger, T., Fejfar, O., Graf, H.-R., and Kälin, D.W., 1996, Vorläufige Mitteilung über Funde von pliozänen Kleinsäugern aus den Höheren Deckenschottern des Irchels (Kt. Zürich): Eclogae Geologicae Helvetiae, v. 89, p. 1043–1048.

Bruno, L.A., Baur, H., Graf, T., Schlüchter, C., Signer, P., and Wieler, R., 1997, Dating of Sirius Group tillites in the Antarctic Dry Valleys with cosmogenic ^{3}He and ^{21}Ne: Earth and Planetary Science Letters, v. 147, p. 37–54, doi: 10.1016/S0012-821X(97)00003-4.

Buoncristiani, J.-F., and Campy, M., 2004, The paleogeography of the last two glacial episodes in France: The Alps and Jura, in Ehlers, J., and Gibbard, P.L., eds., Quaternary glaciations—Extent and chronology, Part I: Europe: London, Elsevier, p. 101–110.

Burga, C.A., 1987, Gletscher- und Vegetationsgeschichte der südrätischen Alpen seit der Späteiszeit (Puschlav, Livigno, Bormiese): Denkschriften der Schweizerischen Naturforschenden Gesellschaft, v. 101, 161 p.

Burga, C.A., and Perret, R., 1998, Vegetation und Klima der Schweiz seit dem jüngeren Eiszeitalter: Thun, Ott Verlag, 805 p.

Campy, M., 1992, Paleogeographical relationships between alpine and Jura glaciers during the two last Pleistocene glaciations: Palaeogeography, Palaeoclimatography, Palaeoecology, v. 93, p. 1–12, doi: 10.1016/0031-0182(92)90180-D.

Clark, P.U., and Mix, A.C., 2000, Ice sheets by volume: Nature, v. 406, p. 689–690, doi: 10.1038/35021176.

Dansgaard, W., Johnsen, S.J., Clausen, H.B., Dahl-Jensen, D., Gundestrup, N.S., Hammer, C.U., Hvidberg, C.S., Steffensen, J.P., Sweinbjörnsdottir, A.E., Jouzel, J., and Bond, G., 1993, Evidence of general instability of past climate from a 250-kyr ice-core record: Nature, v. 364, p. 218–220, doi: 10.1038/364218a0.

Ehlers, J., and Gibbard, P.L., editors, 2004, Quaternary glaciations—Extent and chronology, Part I: Europe: London, Elsevier, 475 p.

Evans, J., 2001, Calibration of the production rates of cosmogenic ^{36}Cl from Potassium [Ph.D. thesis]: Canberra, Australian National University, 142 p.

Fiebig, M., and Häuselmann, P., 2005, A first attempt to date "Deckenschotter" with cosmogenic nuclides: Bern, Switzerland, INQUA–SEQS, Subcommission on European Quaternary Stratigraphy meeting, p. 11.

Fiebig, M., Buiter, S.J.H., and Ellwanger, D., 2004, Pleistocene glaciations of South Germany, in Ehlers, J., and Gibbard, P.L., eds., Quaternary glaciations—Extent and chronology, Part I: Europe: London, Elsevier, p. 147–154.

Fliri, F., 1973, Beiträge zur Geschichte der alpinen Würmvereisung-Forschungen am Bänderton von Baumkirchen (Inntal, Nordtirol): Zeitschrift für Geomorphologie N.F.: Supplementband, v. 16, p. 1–14.

Fliri, F., Bortenschlager, S., Felber, H., Heissel, W., Hilscher, H., and Resch, W., 1970, Der Bänderton von Baumkirchen (Tirol): Zeitschrift für Gletscherkunde und Glazialgeologie, v. 6, p. 5–35.

Florineth, D., and Schlüchter, C., 1998, Reconstructing the Last Glacial Maximum (LGM) ice surface geometry and flowlines in the central Swiss Alps: Eclogae Geologicae Helvetiae, v. 91, p. 391–407.

Florineth, D., and Schlüchter, C., 2000, Alpine evidence for atmospheric circulation patterns in Europe during the Last Glacial Maximum: Quaternary Research, v. 54, p. 295–308, doi: 10.1006/qres.2000.2169.

Fraedrich, R., 1979, Spät- und postglaziale Gletscherschwankungen in der Ferwallgruppe (Tirol/Vorarlberg): Dusseldorfer Geographische Schriften, v. 12, p. 1–161.

Frauenfelder, R., Haeberli, W., Hoelzle, M., and Maisch, M., 2001, Using relict rockglaciers in GIS-based modelling to reconstruct Younger Dryas permafrost distribution patterns in the Err-Julier area, Swiss Alps: Norsk Geografisk Tidsskrift, Norwegian Journal of Geography, v. 55, p. 195–202, doi: 10.1080/00291950152746522.

Gosse, J.C., 2005, The contribution of cosmogenic nuclides to unraveling alpine paleo-glacier histories, in Huber, U., et al., eds., Global change and mountain regions: A state of knowledge overview: Dordrecht, Springer, p. 39–50.

Gosse, J.C., and Phillips, F.M., 2001, Terrestrial in situ cosmogenic nuclides: Theory and application: Quaternary Science Reviews, v. 20, p. 1475–1560, doi: 10.1016/S0277-3791(00)00171-2.

Graf, H.R., 1993, Die Deckenschotter der zentralen Nordschweiz [Ph.D. thesis]: ETH Zürich, No. 10205, 151 p.

Graf, H.-R., and Müller, B., 1999, Das Quartär: Die Epoche der Eiszeiten, in Bolliger, T., ed., Geologie des Kantons Zürich: Thun, Ott, p. 71–95.

Gross, G., Kerschner, H., and Patzelt, G., 1977, Methodische Untersuchungen über die Schneegrenze in alpinen Gletschergebieten: Zeitschrift für Gletscherkunde und Glazialgeologie, v. 12, p. 223–251.

Haas, J.N., Richoz, I., Tinner, W., and Wick, L., 1998, Synchronous Holocene climatic oscillations recorded on the Swiss Plateau and at timberline in the Alps: The Holocene, v. 8, p. 301–309, doi: 10.1191/095968398675491173.

Haeberli, W., 1991, Zur Glaziologie der letzteiszeitlichen Alpenvergletscherung, in Frenzel, B., ed., Klimageschichtliche Probleme der letzten 130000 Jahre: Paleoklimaforschung, v. 1, p. 409–419.

Hantke, R., 1978–1983, Eiszeitalter, Die jüngste Erdgeschichte der Schweiz und ihrer Nachbargebiete: Thun, Ott, v. 1-3, 1901 p.

Hemming, S., 2004, Heinrich events: Massive late Pleistocene detritus layers of the North Atlantic and their global climate imprint: Reviews of Geophysics, v. 42, p. 1–43, doi: 10.1029/2003RG000128.

Hertl, A., 2001, Untersuchungen zur spätglazialen Gletscher- und Klimageschichte der österreichischen Silvrettagruppe [Ph.D. thesis]: Innsbruck, Universität Innsbruck, 313 p.

Heuberger, H., 1966, Gletschergeschichtliche Untersuchungen in den Zentralalpen zwischen Sellrain und Ötztal: Wissenschaftliche Alpenvereinshefte, v. 20, p. 1–126.

Heuberger, H., 1968, Die Alpengletscher im Spät- und Postglazial: Eiszeitalter und Gegenwart, v. 19, p. 270–275.

Hormes, A., Müller, B.U., and Schlüchter, C., 2001, The Alps with little ice: Evidence for eight Holocene phases of reduced glacier extent in the

central Swiss Alps: The Holocene, v. 11, p. 255–265, doi: 10.1191/095968301675275728.
Hughen, K., Lehman, S., Southon, J., Overpeck, J., Marchal, O., Herring, C., and Turnbull, J., 2004, ^{14}C activity and global carbon cycle changes over the past 50,000 years: Science, v. 303, p. 202–207, doi: 10.1126/science.1090300.
Ivy-Ochs, S., 1996, The dating of rock surfaces using in situ produced ^{10}Be, ^{26}Al and ^{36}Cl, with examples from Antarctica and the Swiss Alps [Ph.D. thesis]: ETH Zürich, No. 11763, 196 p.
Ivy-Ochs, S., Schlüchter, C., Kubik, P.W., Synal, H.-A., Beer, J., and Kerschner, H., 1996, The exposure age of an Egesen moraine at Julier Pass, Switzerland, measured with the cosmogenic radionuclides ^{10}Be, ^{26}Al and ^{36}Cl: Eclogae Geologicae Helvetiae, v. 89, p. 1049–1063.
Ivy-Ochs, S., Schlüchter, C., Kubik, P., and Denton, G., 1999, Moraine exposure dates imply synchronous Younger Dryas glacier advances in the European Alps and in the southern Alps of New Zealand: Geografiska Annaler, v. 81A, p. 313–323, doi: 10.1111/j.0435-3676.1999.00060.x.
Ivy-Ochs, S., Schäfer, J., Kubik, P.W., Synal, H.-A., and Schlüchter, C., 2004, The timing of deglaciation on the northern Alpine foreland (Switzerland): Eclogae Geologicae Helvetiae, v. 97, p. 47–55.
Ivy-Ochs, S., Kerschner, H., Kubik, P.W., and Schlüchter, C., 2006, Glacier response in the European Alps to Heinrich event 1 cooling: The Gschnitz stadial: Journal of Quaternary Science, v. 21, p. 115–130, doi: 10.1002/jqs.955.
Jäckli, A., 1970, Die Schweiz zur letzten Eiszeit: Atlas der Schweiz, map 6: Wabern-Bern, Eidg. Landestopographie.
Johnsen, S.J., Dahl-Jensen, D., Gundestrup, N., Steffensen, J.P., Clausen, H.B., Miller, H., Masson-Delmotte, V., Sveinbjörnsdottir, A.E., and White, J., 2001, Oxygen isotope and palaeotemperature records from six Greenland ice-core stations: Camp Century, Dye-3, GRIP, GISP2: Renland and NorthGRIP: Journal of Quaternary Science, v. 16, p. 299–307, doi: 10.1002/jqs.622.
Keller, O., 1988, Ältere spätwürmzeitliche Gletschervorstösse und Zerfall des Eisstromnetzes in den nördlichen Rhein-Alpen (Weissbad-Stadium/Bühl-Stadium): Physische Geographie, v. 27A, 241 p., and v. 27B, 291 p.
Keller, O., and Krayss, E., 1993, The Rhine-Linth Glacier in the upper Würm: A model of the last Alpine glaciation: Quaternary International, v. 18, p. 15–27, doi: 10.1016/1040-6182(93)90049-L.
Keller, O., and Krayss, E., 2005, Der Rhein-Linth-Gletscher im letzten Hochglazial: Vierteljahresschrift der Naturforschenden Gesellschaft in Zürich, v. 150, p. 19–32 and p. 69–85.
Kelly, M.A., Buoncristiani, J.-F., and Schlüchter, C., 2004a, A reconstruction of the Last Glacial Maximum (LGM) ice-surface geometry in the western Swiss Alps and contiguous Alpine regions in Italy and France: Eclogae Geologicae Helvetiae, v. 97, p. 57–75.
Kelly, M.A., Kubik, P.W., Blanckenburg von, F., and Schlüchter, C., 2004b, Surface exposure dating of the Great Aletsch Glacier Egesen moraine system, western Swiss Alps, using the cosmogenic nuclide ^{10}Be: Journal of Quaternary Science, v. 19, p. 431–441.
Kelly, M.A., Ivy-Ochs, S., Kubik, P.W., von Blanckenburg, F., and Schlüchter, C., 2006, Exposure ages of glacial erosional features in the Grimsel Pass region, central Swiss Alps: Boreas (in press).
Kerschner, H., 1978a, Paleoclimatic inferences from Late Würm rock glaciers, eastern central Alps, western Tyrol, Austria: Arctic and Alpine Research, v. 10, p. 635–644, doi: 10.2307/1550684.
Kerschner, H., 1978b, Untersuchungen zum Daun- und Egesenstadium in Nordtirol und Graubünden (methodische Überlegungen): Geographischer Jahresbericht aus Österreich, v. 36, p. 26–49.
Kerschner, H., 1986, Zum Sendersstadium im Spätglazial der nördlichen Stubaier Alpen, Tirol: Zeitschrift für Geomorphologie N.F.: Supplementband, v. 61, p. 65–76.
Kerschner, H., 2005, Glacier-climate models as palaeoclimatic information sources: Examples from the Alpine Younger Dryas period, in Huber, U.M., et al., eds., Global change and mountain regions: A state of knowledge overview: Dordrecht, Springer, p. 73–82.
Kerschner, H., and Berktold, E., 1982, Spätglaziale Gletscherstände und Schuttformen im Senderstal, nördliche Stubaier Alpen, Tirol: Zeitschrift für Gletscherkunde und Glazialgeologie, v. 17, p. 125–134.
Kerschner, H., Ivy-Ochs, S., and Schlüchter, C., 1999, Paleoclimatic interpretation of the early late-glacial glacier in the Gschnitz Valley, central Alps, Austria: Annals of Glaciology, v. 28, p. 135–140.
Kerschner, H., Kaser, G., and Sailer, R., 2000, Alpine Younger Dryas glaciers as paleo-precipitation gauges: Annals of Glaciology, v. 31, p. 80–84.
Kerschner, H., Hertl, A., Gross, G., Ivy-Ochs, S., and Kubik, P.W., 2006, Surface exposure dating of moraines in the Kromer Valley (Silvretta Mountains, Austria)—Evidence for glacial response to the 8.2 ka event in the eastern Alps: The Holocene, v. 16, p. 7–15.
Kubik, P.W., Ivy-Ochs, S., Masarik, J., Frank, M., and Schlüchter, C., 1998, ^{10}Be and ^{26}Al production rates deduced from an instantaneous event within the dendron-calibration curve, the landslide of Köfels, Ötz Valley Austria: Earth and Planetary Science Letters, v. 161, p. 231–241, doi: 10.1016/S0012-821X(98)00153-8.
Küttel, M., 1977, Pollenanalytische und geochronologische Untersuchungen zur Piottino-Schwankung (Jüngere Dryas): Boreas, v. 6, p. 259–274.
Ledermann, H., 1978, Geologischer Atlas der Schweiz, Atlasblatt 72 (1127 Solothurn), mit Erläuterungen: Bern, Schweizerische Geologische Kommission, 36 p.
Maisch, M., 1981, Glazialmorphologische und gletschergeschichtliche Untersuchungen im Gebiet zwischen Landwasser- und Albulatal (Kt. Graubünden, Schweiz): Physische Geographie, v. 3, 215 p.
Maisch, M., 1982, Zur Gletscher- und Klimageschichte des alpinen Spätglazials: Geographica Helvetica, v. 37, p. 93–104.
Maisch, M., 1987, Zur Gletschergeschichte des alpinen Spätglazials: Analyse und Interpretation von Schneegrenzdaten: Geographica Helvetica, v. 42, p. 63–71.
Masarik, J., 2002, Numerical simulation of in situ production of cosmogenic nuclides: Geochimica et Cosmochimica Acta, v. 66, p. A491.
Masarik, J., Frank, M., Schäfer, J.M., and Wieler, R., 2001, Correction of in situ cosmogenic nuclide production rates for geomagnetic field intensity variations during the past 800,000 years: Geochimica et Cosmochimica Acta, v. 65, p. 2995–3003, doi: 10.1016/S0016-7037(01)00652-4.
Mayr, F., and Heuberger, H., 1968, Type areas of Late Glacial and Post Glacial deposits in Tyrol, eastern Alps, in Richmond, G.M., ed., Glaciation of the Alps: University of Colorado Studies, Series in Earth Sciences, v. 7, p. 143–165.
Müller, B.U., 1995, Das Walensee-/Seeztal - eine Typusregion alpiner Talgenese, Vom Entstehen und Vergehen des grossen Rheintal-/Zürichsees: Sargans, Sarganserländer Druck AG, 219 p.
Müller, H.N., 1982, Zum alpinen Spätglazial: Das Zwischenbergstadium: Zeitschrift für Gletscherkunde und Glazialgeologie, v. 17, p. 135–142.
Müller, H.N., 1984, Spätglaziale Gletscherschwankungen in den westlichen Schweizer Alpen (Simplon-Süd und Val de Nendaz, Wallis) und im nordisländischen Tröllaskagi-Gebirge (Skidadalur): Küng, Näfels, 205 p.
Nicolussi, K., and Patzelt, G., 2000, Discovery of early Holocene wood and peat on the forefield of the Pasterze Glacier, eastern Alps, Austria: The Holocene, v. 10, p. 191–199, doi: 10.1191/095968300666855842.
Nicolussi, K., and Patzelt, G., 2001, Untersuchungen zur holozänen Gletscherentwicklung von Pasterze und Gepatschferner (Ostalpen): Zeitschrift für Gletscherkunde und Glazialgeologie, v. 36, p. 1–87.
Niedermann, S., 2000, The ^{21}Ne production rate in quartz revisited: Earth and Planetary Science Letters, v. 183, p. 361–364, doi: 10.1016/S0012-821X(00)00302-2.
Nussbaum, F., 1910, Das Endmoränengebiet des Rhonegletschers von Wangen a. A.: Mitteilungen der Naturforschenden Gesellschaft Bern, v. 1761, p. 141–168.
O'Brien, S.R., Mayewski, P.A., Meeker, L.D., Meese, D.A., Twickler, M.S., and Whitlow, S.I., 1995, Complexity of Holocene climate as reconstructed from a Greenland ice core: Science, v. 270, p. 1962–1964.
Ochs, M., and Ivy-Ochs, S., 1997, The chemical behavior of Be, Al, Fe, Ca, and Mg during AMS target preparation modeled with chemical speciation calculations: Nuclear Instruments and Methods in Physics Research, v. B123, p. 235–240.
Ohlendorf, C., 1998, High alpine lake sediments as chronicles for regional glacier and climate history in the Upper Engadine, southeastern Switzerland [Ph.D. thesis]: ETH Zürich, No. 12705, 203 p.
Patzelt, G., 1972, Die spätglazialen Stadien und postglazialen Schwankungen von Ostalpengletschern: Berichte der Deutschen Botanischen Gesellschaft, v. 85, p. 47.
Patzelt, G., 1975, Unterinntal - Zillertal - Pinzgau – Kitzbühel, Spät- und postglaziale Landschaftsentwicklung, in Fliri, F., and Leidlmair, A., eds., Tirol—ein geographischer Exkursionsführer: Innsbrucker Geographische Studien, v. 2, p. 309–329.
Patzelt, G., 1995, The pollen profile of the peat bog at the Lans Lake, 12, Alpine Traverse, in Schirmer, W., ed., Quaternary field trips in central Europe: München, Pfeil, p. 670–671.

Patzelt, G., and Bortenschlager, S., 1973, Die postglazialen Gletscher- und Klimaschwankungen in der Venedigergruppe (Hohe Tauern, Ostalpen): Zeitschrift für Geomorphologie N.F.: Supplementband, v. 16, p. 25–72.

Patzelt, G., and Resch, W., 1986, Quartärgeologie des mittleren Tiroler Inntales zwischen Innsbruck und Baumkirchen (Exkursion C am 3. April 1986): Jahresbericht und Mitteilungen des oberrheinischen geologischen Vereins N.F., v. 68, p. 43–66.

Penck, A., and Brückner, E., 1901/1909, Die Alpen im Eiszeitalter: Leipzig, Tauchnitz, v. 1–3, 1199 p.

Phillips, F.M., Stone, W.D., and Fabryka-Martin, J.T., 2001, An improved approach to calculating low-energy cosmic-ray neutron fluxes near the land/atmosphere interface: Chemical Geology, v. 175, p. 689–701, doi: 10.1016/S0009-2541(00)00329-6.

Poscher, G., 1993, Neuergebnisse der Quartärforschung in Tirol, in Geologie des Oberinntaler Raumes - Schwerpunkt Blatt 144 Landeck: Arbeitstagung der Geologischen Bundesanstalt 1993: Wien, Geologische Bundesanstalt, p. 7–27.

Preusser, F., 2004, Towards a chronology of the late Pleistocene in the northern Alpine foreland: Boreas, v. 33, p. 195–210, doi: 10.1080/030094804 10001271.

Preusser, F., and Schlüchter, C., 2004, Dates from an important early Late Pleistocene ice advance in the Aare Valley, Switzerland: Eclogae Geologicae Helvetiae, v. 97, p. 245–253.

Preusser, F., Geyh, M.A., and Schlüchter, C., 2003, Timing of Late Pleistocene climate change in lowland Switzerland: Quaternary Science Reviews, v. 22, p. 1435–1445, doi: 10.1016/S0277-3791(03)00127-6.

Putkonen, J., and Swanson, T., 2003, Accuracy of cosmogenic ages for moraines: Quaternary Research, v. 59, p. 255–261, doi: 10.1016/S0033-5894(03)00006-1.

Reitner, J., 2005, Quartärgeologie und Landschaftsentwicklung im Raum Kitzbühel - St. Johann i.T. - Hopfgarten (Nordtirol) vom Riss bis in das Würm-Spätglazial (MIS 6-2) [Ph.D. thesis]: Wien, Universität Wien, 190 p.

Reuther, A.U., Ivy-Ochs, S., and Heine, K., 2006, Application of surface exposure dating in glacial geomorphology and the interpretation of moraine ages: Zeitschrift für Geomorphologie Supplementband (in press).

Rohling, E.J., and Pälike, H., 2005, Centennial-scale climate cooling with a sudden cold event around 8,200 years ago: Nature, v. 434, p. 975–979, doi: 10.1038/nature03421.

Sailer, R., 2001, Spätseizeitliche Gletscherstände in der Ferwallgruppe [Ph.D. thesis]: Innsbruck, Universität Innsbruck, 205 p.

Sailer, R., and Kerschner, H., 1999, Equilibrium line altitudes and rock glaciers in the Ferwall Group (western Tyrol, Austria) during the Younger Dryas cooling event: Annals of Glaciology, v. 28, p. 141–145.

Sailer, R., Kerschner, H., and Heller, A., 1999, Three-dimensional reconstruction of Younger Dryas glaciers with a raster-based GIS: Glacial Geology and Geomorphology, 1999/rp01: http://ggg.qub.ac.uk/ggg/.

Sarnthein, M., Stattegger, K., Dreger, D., Erlenkeuser, H., Grootes, P., Haupt, B., Jung, S., Kiefer, T., Kuhnt, W., Pflaumann, U., Schäfer-Neth, Ch., Schulz, H., Schulz, M., Seidov, D., Simstich, J., van Kreveld, S., Vogelsang, E., Völker, A., and Weinelt, M., 2001, Fundamental modes and abrupt changes in North Atlantic circulation and climate over the last 60 ky—Concepts, reconstruction and numerical modeling, in Schäfer, P., et al., eds., The northern North Atlantic: A changing environment: Berlin, Springer, p. 365–410.

Schindler, C., 2004, Zum Quartär des Linthgebiets zwischen Luchsingen, dem Walensee und dem Zürcher Obersee, in Landesgeologie BWG, ed., Beiträge zur Geologischen Karte der Schweiz: Bern, Landesgeologie, 158 p.

Schlüchter, C., 1988, The deglaciation of the Swiss Alps: A paleoclimatic event with chronological problems: Bulletin de l'Association française pour l'étude du Quarternaire, 1988-2/3, p. 141–145.

Schlüchter, C., 2004, The Swiss glacial record—A schematic summary, in Ehlers, J., and Gibbard, P.L., eds., Quaternary glaciations—Extent and chronology, Part I: Europe: London, Elsevier, p. 413–418.

Schlüchter, C., and Röthlisberger, C., 1995, 100,000 Jahre Gletschergeschichte, in Schweizerische Gletscherkommission, ed., Gletscher im ständigen Wandel: Zürich, vdf Hochschulverlag AG an der ETH Zürich, p. 47–63.

Schlüchter, C., Maisch, M., Suter, J., Fitze, P., Keller, W.A., Burga, C.A., and Wynistorf, E., 1987, Das Schieferkohlen-Profil von Gossau (Kanton Zürich) und seine stratigraphische Stellung innerhalb der letzten Eiszeit: Vierteljahrsschrift der Naturforschenden Gesellschaft in Zürich, v. 132, no. 3, p. 135–174.

Schreiner, A., 1992, Einführung in die Quartärgeologie: Stuttgart, Schweizerbart, 256 p.

Schwander, J., Eicher, U., and Ammann, B., 2000, Oxygen isotopes of lake marl at Gerzensee and Leysin (Switzerland), covering the Younger Dryas and two minor oscillations, and their correlation to the GRIP ice core: Palaeogeography, Palaeoclimatology, Palaeoecology, v. 159, p. 203–214, doi: 10.1016/S0031-0182(00)00085-7.

Senarclens-Grancy, W., 1958, Zur Glazialgeologie des Ötztales und seiner Umgebung: Mitteilungen der Geologischen Gesellschaft Wien, v. 49, p. 257–313.

Spötl, C., Mangini, A., Frank, N., Eichstädter, R., and Burns, S.J., 2002, Start of the last interglacial period at 135 ka: Evidence from a high Alpine speleothem: Geology, v. 30, p. 815–818, doi: 10.1130/0091-7613(2002)030<0815:SOTLIP>2.0.CO;2.

Stone, J.O., 2000, Air pressure and cosmogenic isotope production: Journal of Geophysical Research, v. 105, no. B10, p. 23,753–23,760, doi: 10.1029/2000JB900181.

Stone, J.O.H., Allan, G.L., Fifield, L.K., and Cresswell, R.G., 1996, Cosmogenic chlorine-36 from calcium spallation: Geochimica et Cosmochimica Acta, v. 60, p. 679–692, doi: 10.1016/0016-7037(95)00429-7.

Stone, J.O.H., Evans, J.M., Fifield, L.K., Allan, G.L., and Cresswell, R.G., 1998, Cosmogenic chlorine-36 production in calcite by muons: Geochimica et Cosmochimica Acta, v. 62, p. 433–454, doi: 10.1016/S0016-7037(97)00369-4.

Stuiver, M., Reimer, P.J., Bard, E., Beck, J.W., Burr, G.S., Hughen, K.A., Kromer, B., McCormac, G., van der Pflicht, J., and Spurk, M., 1998, INTCAL98 radiocarbon age calibration, 24,000–0 cal BP: Radiocarbon, v. 40, p. 1041–1083.

Stuiver, M., Reimer, P.J., and Reimer, R.W., 2005, CALIB 5.0.1 [WWW program and documentation]: Belfast, ^{14}CHRONO Centre, Queen's University Belfast: http://www.calib.org (June 2006).

Suter, J., 1981, Gletschergeschichte des Oberengadins: Untersuchung von Gletscherschwankungen in der Err-Julier-Gruppe: Physische Geographie, v. 2, 170 p.

Synal, H.-A., Bonani, G., Döbeli, M., Ender, R.M., Gartenmann, P., Kubik, P.W., Schnabel, C., and Suter, M., 1997, Status report of the PSI/ETH AMS facility: Nuclear Instruments and Methods in Physics Research, v. B123, p. 62–68.

van Husen, D., 1977, Zur Fazies und Stratigraphie der jungpleistozänen Ablagerungen im Trauntal (mit quartärgeologischer Karte): Jahrbuch der Geologischen Bundesanstalt, v. 120, p. 1–130.

van Husen, D., 1997, LGM and Late-glacial fluctuations in the eastern Alps: Quaternary International, v. 38-39, p. 109–118, doi: 10.1016/S1040-6182(96)00017-1.

van Husen, D., 1999, 2000, Geological processes during the Quaternary: Mitteilungen der Österreichischen Geologischen Gesellschaft, v. 92, p. 135–156.

van Husen, D., 2004, Quaternary glaciations in Austria, in Ehlers, J., and Gibbard, P.L., eds., Quaternary glaciations—Extent and chronology, Part I: Europe: London, Elsevier, p. 1–13.

von Grafenstein, U., Erlenkeuser, H., Müller, J., Jouzel, J., and Johnsen, S., 1998, The cold event 8200 years ago documented in oxygen isotope records of precipitation in Europe and Greenland: Climate Dynamics, v. 14, p. 73–81, doi: 10.1007/s003820050210.

von Grafenstein, U., Erlenkeuser, H., Brauer, A., Jouzel, J., and Johnson, S.J., 1999, A mid-European decadal isotope-climate record from 15,500 to 5,000 years BP: Science, v. 284, p. 1654–1657, doi: 10.1126/science.284.5420.1654.

von Grafenstein, U., Eicher, U., Erlenkeuser, H., Ruch, P., Schwander, J., and Ammann, B., 2000, Isotope signature of the Younger Dryas and two minor oscillations at Gerzensee (Switzerland): Palaeoclimatic and palaeolimnologic interpretation based on bulk and biogenic carbonates: Palaeogeography, Palaeoclimatology, Palaeoecology, v. 159, p. 215–229, doi: 10.1016/S0031-0182(00)00086-9.

Welten, M., 1982, Vegetationsgeschichtliche Untersuchungen in den westlichen Schweizer Alpen: Bern-Wallis: Denkschriften der Schweizerischen Naturforschenden Gesellschaft, v. 95, p. 1–105.

Zoller, H., Athanasiadis, N., and Heitz-Weniger, A., 1998, Late-glacial and Holocene vegetation and climate change at the Palü Glacier, Bernina Pass, Grisons Canton, Switzerland: Vegetation History and Archaebotany, v. 7, p. 241–249, doi: 10.1007/BF01146197.

Zreda, M., Phillips, F.M., and Elmore, D., 1994, Cosmogenic ^{36}Cl accumulation in unstable landforms, 2: Simulations and measurements on eroding moraines: Water Resources Research, v. 30, p. 3127–3136, doi: 10.1029/94WR00760.

MANUSCRIPT ACCEPTED BY THE SOCIETY 11 APRIL 2006

Applications of morphochronology to the active tectonics of Tibet

Frederick J. Ryerson
Institute of Geophysics and Planetary Physics, Lawrence Livermore National Laboratory, Livermore, 94550 California, USA

Paul Tapponnier
Laboratoire de Tectonique, Mécanique de la Lithosphère, UMR7578, CNRS, Institut de Physique du Globe de Paris, 75252 Paris cedex 05, France

Robert C. Finkel
Center for Accelerator Mass Spectrometry, Lawrence Livermore National Laboratory, Livermore, 94550 California, USA

Anne-Sophie Mériaux
Institute of Geophysics and Planetary Physics, Lawrence Livermore National Laboratory, Livermore, 94550 California, USA

Jérôme Van der Woerd
Institut de Physique du Globe de Strasbourg, CNRS-UMR 7516, Strasbourg, France

Cécile Lasserre
Département Terre Atmosphère Océan, École Normale Supérieure, 24, rue Lhomond, 75231 Paris cedex 05, France

Marie-Luce Chevalier
Laboratoire de Tectonique, Mécanique de la Lithosphère, UMR7578, CNRS, Institut de Physique du Globe de Paris, 75252 Paris cedex 05, France, and *Institute of Geophysics and Planetary Physics, Lawrence Livermore National Laboratory, Livermore, 94550 California, USA*

Xi-wei Xu
Institute of Geology, China Earthquake Administration, Beijing 100029, People's Republic of China

Hai-bing Li
Laboratory of Continental Dynamics, Institute of Geology, Chinese Academy of Geological Sciences, Beijing 100037, People's Republic of China

Geoffrey C.P. King
Laboratoire de Tectonique, Mécanique de la Lithosphère, UMR7578, CNRS, Institut de Physique du Globe de Paris, 75252 Paris cedex 05, France

ABSTRACT

The Himalayas and the Tibetan Plateau were formed as a result of the collision of India and Asia, and provide an excellent opportunity to study the mechanical response of the continental lithosphere to tectonic stress. Geophysicists are divided in their views on the nature of this response, advocating either (1) homogeneously distributed deformation with the lithosphere deforming as a fluid continuum or (2) highly localized deformation with the lithosphere deforming as a system of blocks. The resolution of this issue has broad implications for understanding the tectonic response of continental lithosphere in general. Homogeneous deformation is supported by relatively low decadal, geodetic slip-rate estimates for the Altyn Tagh and Karakorum faults. Localized deformation is supported by high millennial, geomorphic slip rates constrained by both cosmogenic and radiocarbon dating on these faults. Based upon the agreement of rates determined by radiocarbon and cosmogenic dating, the overall linearity of offset versus age correlations, and the plateau-wide correlation of landscape evolution and climate history, the disparity between geomorphic and geodetic slip-rate determinations is unlikely to be due to the effects of surface erosion on the cosmogenic age determinations. Similarly, based upon the consistency of slip rates over various observation intervals, secular variations in slip rate appear to persist no longer than 2000 yr and are unlikely to provide reconciliation. Conversely, geodetic and geomorphic slip-rate estimates on the Kunlun fault, which does not have significant splays or associated thrust faults, are in good agreement, indicating that there is no fundamental reason why these complementary geodetic and geomorphic methods should disagree. Similarly, the geodetic and geomorphic estimates of shortening rates across the northeastern edge of the plateau are in reasonable agreement, and the geomorphic rates on individual thrust faults demonstrate a significant eastward decrease in the shortening rate. This rate decrease is consistent with the transfer of slip from the Altyn Tagh fault to genetically related thrust mountain building at its terminus. Rates on the Altyn Tagh fault suggest a similar decrease in rate, but the current data set is too small to be definitive. Overall, the high, late Pleistocene–Holocene geomorphic slip velocities on the major strike-slip faults of Tibet suggest that these faults absorb as much of India's convergence relative to Siberia as the Himalayan Main Frontal Thrust on the southern edge of the plateau.

Keywords: deformation, active faults, Tibetan Plateau, continental deformation, cosmogenic nuclides.

INTRODUCTION

With the exception of the ocean basins, the Tibetan Plateau is among the most prominent physiographic features on the surface of Earth, with an area equivalent to ~2/3 that of the conterminous United States and an average elevation of ~5000 m. Much as the study of the ocean basins helped to provide the fundamental constraints on global tectonics and the rheology of the oceanic lithosphere, the evolution of Tibet provides a similar opportunity to constrain the mechanical behavior of the continental lithosphere and the effects of plate tectonics within the continental regime. The ongoing Indo-Asian collision provides an opportunity to investigate the evolution of tectonic features within and around the collision zone over a period from ca. 50–55 Ma to the present. As the current manifestation of Indo-Asian convergence, active tectonic processes form an important component of the long-term tectonic evolution of Asia. The rates of such processes, such as slip on major, active strike-slip faults or uplift along thrust systems, are critical to assessing how ongoing convergence is accommodated. If indeed "the present is the key to the past," the rates of active processes and the inferred deformation mechanisms should inform our understanding of the long-term tectonic evolution of central Asia.

One of the central debates involving the evolution of the Tibetan Plateau concerns whether the deformation of the continental lithosphere is homogeneously distributed or localized along the boundaries of blocks undergoing little internal deformation. The latter is consistent with the basic tenets of plate tectonics, e.g., rigid lithospheric plates, while the former implies that the rheology of the continental and oceanic lithosphere may be fundamentally different, with the continents deforming more like fluids (Avouac and Tapponnier, 1993; England and Houseman, 1986; England and Molnar, 1997a, 1997b; Peltzer and Saucier, 1996; Peltzer and Tapponnier, 1988). These end-member models make very different predictions regarding the rates of slip on the large strike-slip faults, such as the Altyn Tagh, Karakorum, Kunlun, and Haiyuan faults that define much of the boundary of the Tibetan Plateau (Fig. 1). Viewed as a by-product of distributed deformation in the lower crust and continental lithosphere, relatively low rates of slip (<1 cm/yr) on these bounding strike-slip

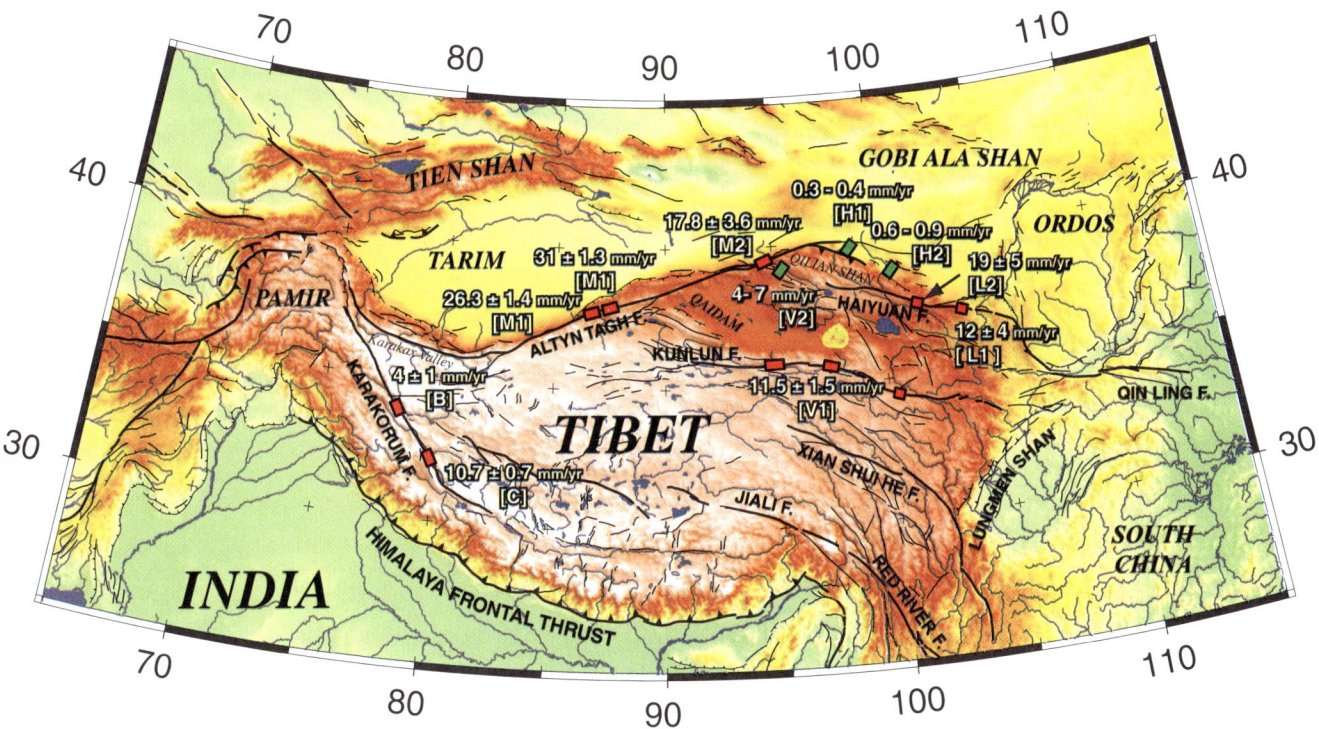

Figure 1. Kinematic map of active faults in Tibet displaying the geomorphic slip-rate determinations constrained by cosmogenic, radiocarbon, and OSL dating. Red boxes are locations of slip-rate determinations on strike-slip faults. M1—Mériaux et al., 2004; M2—Mériaux et al., 2005; V1—Van der Woerd et al., 2002b; L1—Lasserre et al., 1999; L2—Lasserre et al., 2002; C—Chevalier et al., 2005a; B—Brown et al., 2002a. The green boxes are sites where shortening rates have been determined by dating uplifted terraces. V2—Van der Woerd et al., 2001; H1—Hetzel et al., 2002; H2—Hetzel et al., 2004.

faults suggest widely distributed, homogeneous deformation (Bendick et al., 2000; Chen et al., 2000; Shen et al., 2001a; Washburn et al., 2001). At the other extreme, if these represent localized lithospheric shear zones, e.g., "lithospheric faults" equivalent to intracontinental plate boundaries, then slip rates from 10 to 30 mm/yr are expected (Avouac and Tapponnier, 1992; Peltzer and Saucier, 1996; Tapponnier et al., 2001). The resolution of this issue, which has implications beyond the tectonics of Tibet, requires quantitative information, not only on the deep structure of these faults (Herquel et al., 1999; Wittlinger et al., 1998), but also on their slip rates.

The ability to measure slip rate on active faults has been revolutionized by a number of recent technical developments. The global positioning system, GPS, is in wide use, and there have been a number of GPS surveys focused on the determination of decadal slip rates in central Asia (Banerjee and Burgmann, 2002; Bendick et al., 2000; Bilham et al., 1997; Chen et al., 2000, 2004; Larson et al., 1999; Shen et al., 2001b; Wallace et al., 2004; Wang et al., 2002; Zhang et al., 2004). More recently, interferometric synthetic aperture radar (InSAR), previously applied in imaging relatively large coseismic strains (Massonnet et al., 1993; Peltzer et al., 1994, 2001b; Peltzer and Rosen, 1995), has been applied to the interseismic regime, generating decadal slip-rate determinations (Peltzer et al., 2001a; Wright et al., 2004). Both GPS and InSAR have been applied in attempts to constrain the velocity field associated with the Indo-Asian collision, and in determining slip rates on individual faults. The third element of this revolution is the extension of rate estimates obtained by dating offset landscape features, "morphochronology," due to the more general availability of cosmic-ray exposure dating by accelerator mass spectrometry (AMS). Surface exposure dating using in situ ^{14}C, ^{10}Be, ^{26}Al, ^{21}Ne, and ^{36}Cl extends the temporal range of such determinations beyond that accessible to radiocarbon dating, as well as enabling the dating of surfaces that do not contain radiocarbon-datable materials. Geodetic and geomorphic measurements should be complementary, recording fault motion over vastly different time windows. Geomorphic measurements have the advantage of integrating over many earthquake cycles, while geodetic methods are more sensitive to interseismic strain and differential motion perpendicular to the fault trace. We know of no a priori reason why far-field, decadal geodetic measurements and near-field, millennial geomorphic measurements should not agree.

The application of morphochronology to active tectonics requires the preservation of tectonically offset landscape features and datable surfaces that can be used to constrain their age. In many respects, the Tibetan Plateau is an ideal location for the application of cosmic-ray exposure dating. Much of the Tibetan landscape has been influenced by glacial action and late

Pleistocene climate change, and locations can be found where glacial features such as moraines and glacial valleys have been tectonically offset. In other areas, climate change has modulated fluvial processes resulting in the deposition of alluvial fans and in the deposition and incision of fluvial terraces. These features serve as temporal markers whose offset can be used to indicate the rate of tectonic deformation. The relatively arid climate of the plateau has allowed many of these features to be preserved, recording offsets as old as ca. 250,000 yr B.P. (Chevalier et al., 2005a). In addition, due to its high elevation, Tibet is an ideal location for the application of cosmogenic dating, as the production rates here are as high as any on the planet, allowing both smaller and younger samples to be dated with usable precision.

In this paper, we will review the assumptions and methods employed in recent efforts to constrain the rates of motion along some of the major faults in Asia by cosmogenic dating, as well as address some of the criticisms that have been leveled at this method. We will focus on the large-scale spatial variation in millennial slip rates across the plateau and along its boundaries, comparing these with various kinematic models of Indo-Asian deformation. Of particular interest is the variation in slip rate along the northeastern boundary of the plateau where uplift on active thrusts has been linked to slip along the major strike-slip faults (Meyer et al., 1998). The genetic link between the eastward extrusion of central Tibet and uplift along the northeastern boundary prescribes and eastward decrease in strike-slip velocity that should be quantitatively reflected in uplift rates. Where possible, we will compare these rate estimates obtained by cosmogenic dating with those from (1) radiocarbon dating, (2) geodetic methods, and (3) Cenozoic geologic rates.

DEVELOPMENT OF GEOMORPHIC OFFSETS AND COSMOGENIC DATING OF DEPOSITIONAL SURFACES

Using geomorphic offsets to determine slip rates on active faults requires the creation of datable landscape features that are then passively preserved (Sieh and Jahns, 1984; Weldon and Sieh, 1985). The interpretational framework for determining slip rates on strike-slip faults based on the lateral displacement of terrace risers and fluvial channels is illustrated in Figure 2. Here an alluvial surface, T3, deposited atop an active strike-slip fault is subsequently incised, and a younger terrace, T2, is emplaced along the active stream. Ideally, the resulting T3/T2 riser would be continually refreshed by fluvial activity on T2. The T3/T2 riser becomes a passive offset marker only when T2 is abandoned due to a new episode of incision, which in turn leads to the formation of the underlying T1 terrace and the T2/T1 riser. Renewed incision indicates that the threshold between aggradation and degradation has been crossed, a condition that may be driven by climate-related variations in sediment load and/or precipitation rate. While the time at which the aggradation/degradation threshold is reached at a given point along a river may vary (Bull, 1991), terrace abandonment at a particular point should represent a discrete temporal event. The displacements recorded by terrace risers are necessarily minimum displacements, as they may be reworked by subsequent fluvial erosion. For instance, flooding by the stream occupying T1 could cause it to temporarily reoccupy T2. This could result in lateral erosion of the T3/T2 riser, reducing the observed offset.

The abandonment age of a terrace is typically defined by its youngest surface exposure age and/or the radiocarbon dates obtained from the shallowest subsurface sample (cf. Mériaux et al., 2004). Coupling the abandonment of T2 with the overlying T3/T2 offset yields a maximum estimate of the slip rate, the "strath abandonment model" age (Mériaux et al., 2005), and is relevant to a scenario in which occupation of a terrace level is of short duration and no riser offset is accumulated prior to abandonment. In the case where a strath terrace has been occupied by a stream system for a period that is long relative to earthquake recurrence, and some permanent riser offset is accumulated prior to abandonment, the offset should be linked with the onset of emplacement of the surface underlying the riser. This may be the case for surfaces constructed by high-energy, ephemeral braided streams where deposition is likely diachronous. In such instances, the emplacement age can be approximated by the oldest exposure age in the surface age distribution. This "strath emplacement model" yields a minimum estimate of the slip rate.

Both "strath" models link the age of the riser to the age of the underlying terrace, but use different extremes in the age distribution to estimate the slip rate. An absolute lower bound on the slip rate is obtained by combining an offset with the abandonment age of the overlying terrace. This "fill-channel model" is appropriate when the strath terrace underlying the riser of interest is flooded after abandonment and buried by new "fill" deposits that yield an erroneously young age for the riser offset. However, as the riser cannot be older than the terrace into which it is incised, the abandonment age of the upper terrace yields an absolute minimum approximation of the slip rate. The appropriateness of "strath" as opposed to "fill" approximations is based upon the geomorphic characterization of the terraces in the field. The "fill" approximation is useful in determining an absolute minimum rate regardless of the nature of the terraces, and is more generally useful in determining rates from stream channel offsets, as a stream channel cannot predate the surface it incises.

Fluvial terraces can also be used to determine vertical uplift rates when a normal or thrust fault displaces active stream terraces (Fig. 3) (for normal faults see Avouac and Peltzer, 1993). Here tectonic activity uplifts the terraces in the hanging wall of the thrust, producing a disequilibrium stream gradient that the fluvial system seeks to eliminate by renewed incision. Continued uplift and incision can produce a series of uplifted terraces perched above an active river such as those found in the hanging walls of the Tanghenan Shan thrusts near Yanchiwan (Fig. 4). Again, dating of the perched surfaces can be combined with vertical offsets and fault geometry to obtain uplift and convergence rates (cf. Hetzel et al., 2002, 2004; Van der Woerd et al., 2001). In this case, the distinction between strath or fill terraces is of

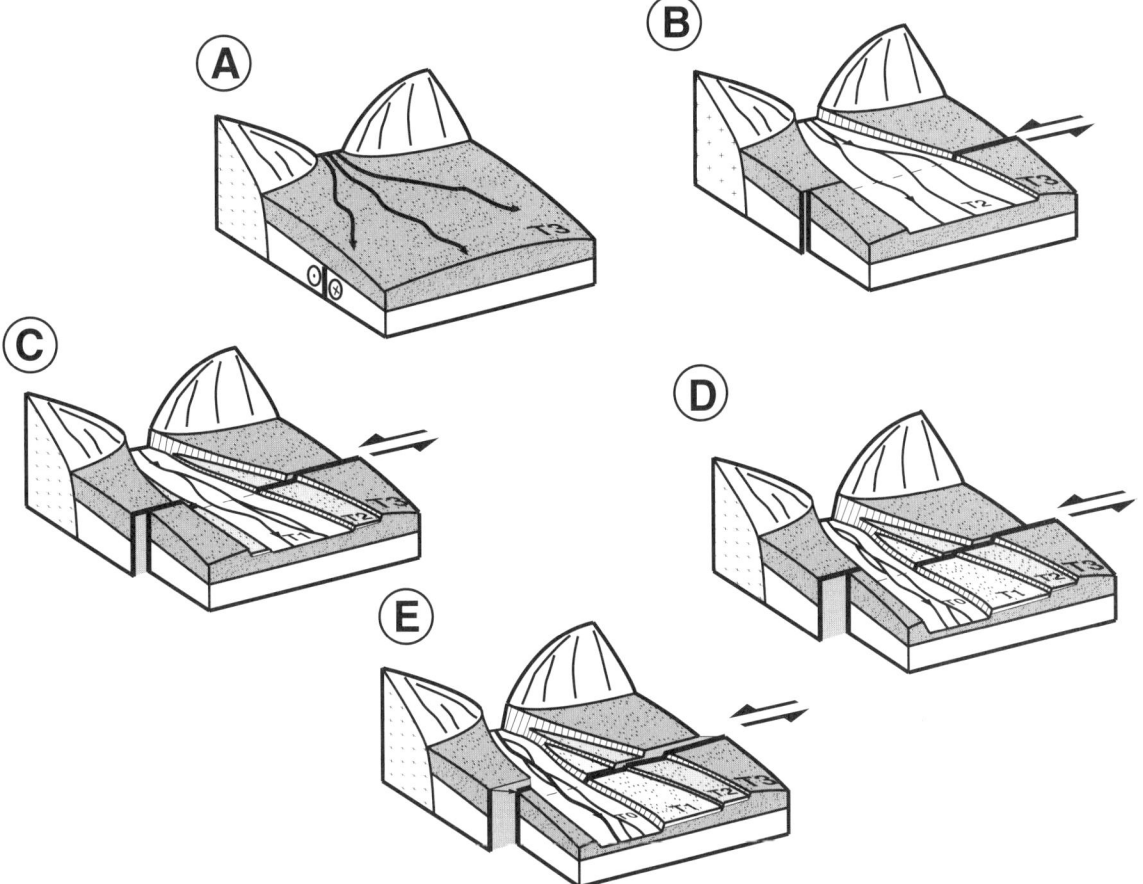

Figure 2. Sequential development of offset terraces associated with strike-slip movement along an active fault. Block diagrams showing plausible sequence of terrace emplacement and stream entrenchment disrupted by strike-slip faulting across bajada in range-front piedmont basin. (A) Emplacement of large fan T3 (fill) at time of large sedimentary discharge. Fault trace is buried. (B) After period of energetic discharge, stream incises channel T2. T3 surface is abandoned and begins to record faulting, but the riser is constantly refreshed by lateral cutting. (C) During new episode of entrenchment, T2 (strath) is abandoned and riser T3/T2, now passive marker, begins to record lateral displacement. Age of T2 abandonment dates riser offset. (D) Successive episodes of terrace beveling and entrenchment of stream due to climatic variation and stream profile evolution lead to formation of several terraces whose risers are offset differently. All episodes may not be recorded by all rivers, and older terraces can be eroded by further incision (see right bank of stream, for example). (E) Similar situation with small vertical slip component. Vertical offset accumulates when terrace is abandoned by stream. Hence, vertical offset of T1 (or T2) is correlated to horizontal T2/T1 (or T3/T2) riser offset, and correlated offsets have ages of T1, or T2, respectively.

little consequence, and the greater vertical separation of terraces makes flooding of abandoned surfaces less likely.

Dating Depositional Surfaces

Quaternary slip rates are almost exclusively determined by dating depositional as opposed to bedrock surfaces. Cosmic-ray exposure ages of cobbles from a single terrace often show scatter that is attributed to a number of factors including bio- and cryoturbation of the surface, postabandonment surface contamination, and predepositional exposure or inheritance. Diachronous emplacement, characterized by a discrete time interval between the initiation of emplacement and abandonment, adds to this scatter. Interpretation of the dispersion in cosmic-ray exposure ages is difficult, and only a limited number of explicit constraints can be applied. For instance, comparison with radiocarbon dates from the same location has been used to help constrain the age of abandonment (Mériaux et al., 2004), though there is a priori no reason that radiocarbon dates should not also display diachroneity.

Subsurface sampling of amalgamated samples has been used to constrain predepositional exposure (Anderson et al., 1996; Hancock et al., 1999; Repka et al., 1997). The method makes use of the fact that the total concentration of a cosmogenic nuclide in a statistically significant sample of cobbles will be the sum of (1) the average predepositional exposure inventory and (2) that

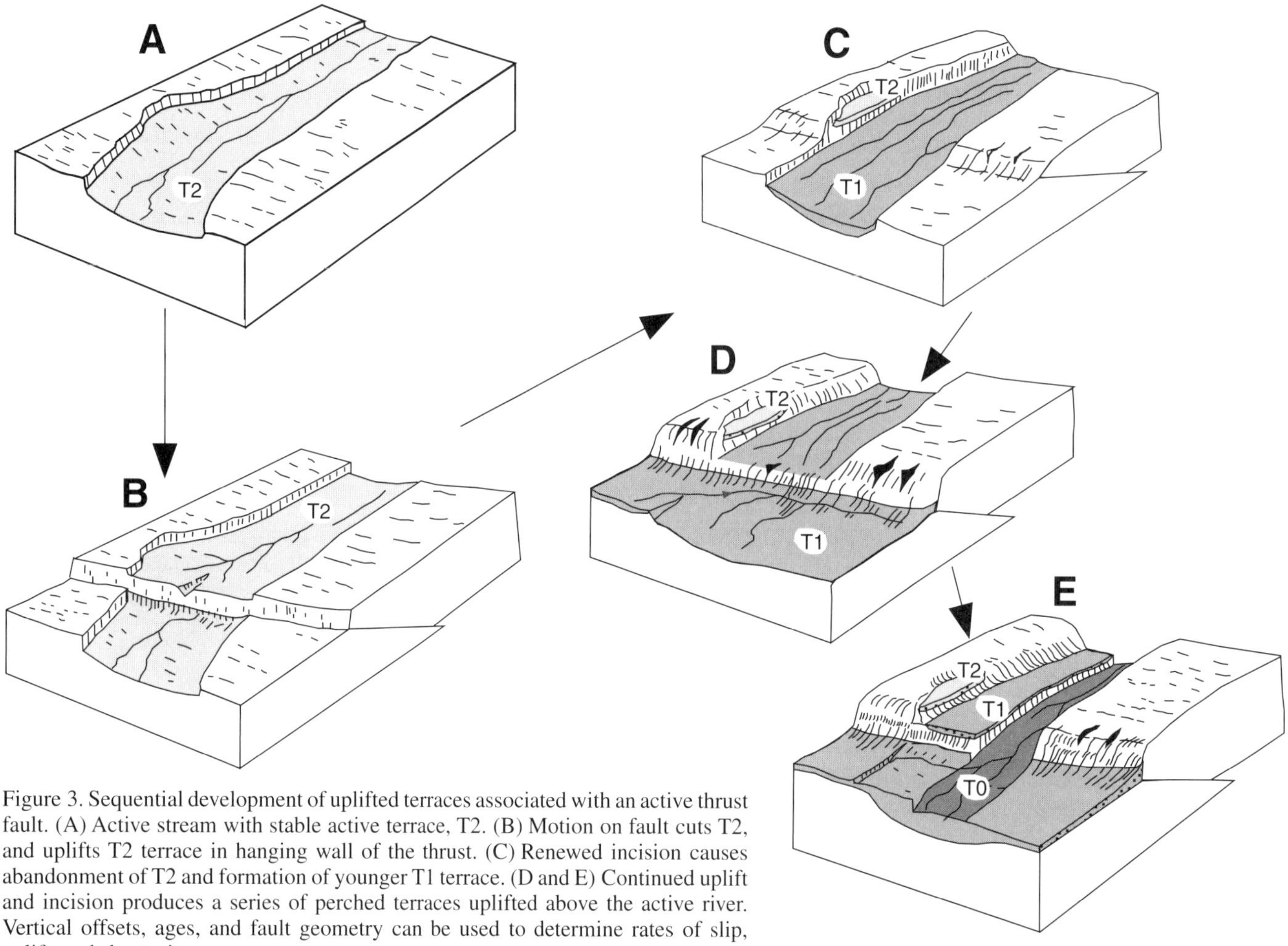

Figure 3. Sequential development of uplifted terraces associated with an active thrust fault. (A) Active stream with stable active terrace, T2. (B) Motion on fault cuts T2, and uplifts T2 terrace in hanging wall of the thrust. (C) Renewed incision causes abandonment of T2 and formation of younger T1 terrace. (D and E) Continued uplift and incision produces a series of perched terraces uplifted above the active river. Vertical offsets, ages, and fault geometry can be used to determine rates of slip, uplift, and shortening.

Figure 4. Uplifted fluvial terraces in the hanging wall of the Tanghenan Shan thrust near Yanchiwan (~39°N, 96°E).

which accumulated locally by exposure in the surface of interest (Fig. 5). The concentration of the latter component decreases exponentially with depth as production is attenuated due to shielding. Cosmogenic nuclide concentrations in totally shielded samples should reflect only the predepositional exposure inventory. This inherited component can then be subtracted to yield the concentration of the cosmogenic nuclide produced during exposure at the site of deposition. The concentration of a cosmogenic nuclide accumulated in a single cobble prior to deposition cannot be quantitatively constrained. However, depth-corrected ages from the analysis of individual subsurface cobbles does provide an upper bound on the apparent age of the surface (Fig. 5C) (cf. Mériaux et al., 2005). Analysis of subsurface samples cannot distinguish predepositional exposure inherited during storage and transport from that which may have accumulated locally due to diachronous deposition of the surface. Such information must be gleaned from the distribution of single surface cobble ages in concert with subsurface sampling and radiocarbon dating where possible. The distribution of cosmogenic ages from individual surface cobbles may be used to define upper and lower

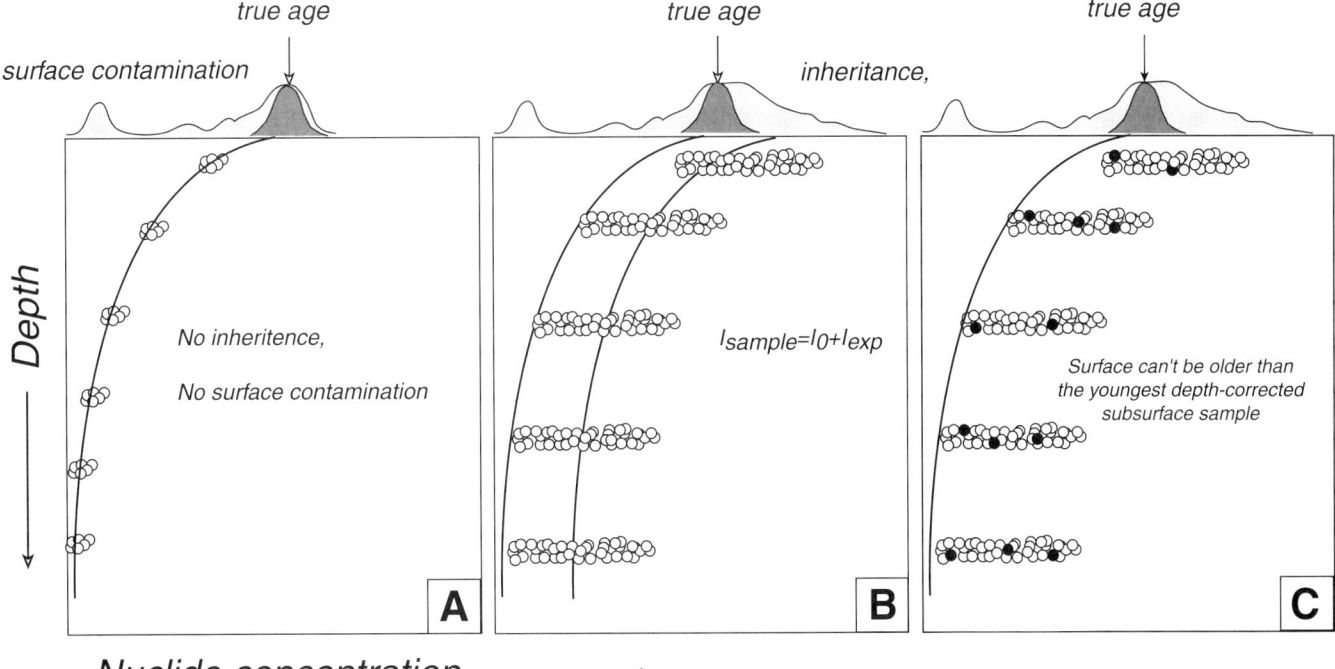

Figure 5. Relationship of surface and subsurface cosmogenic isotope compositions. (A) Surface age frequency distribution in which the "abandonment age" (dark gray) has been mixed with a younger population introduced by surface contamination, but contains no inherited component. Subsurface samples are free of surface contamination, and as there is no inheritance, follow an exponential decrease with depth, reaching zero concentration at the shielding depth. (B) Surface age frequency distribution in which the "abandonment age" (dark gray) has been mixed with a younger population introduced by surface contamination, and an older population associated with inheritance. Subsurface samples are surface contamination free, but inherited component varies from sample to sample. Amalgamation of multiple samples allows the average inheritance, I_0, and average exposure concentrations, I_{exp}, to be determined (Anderson et al., 1996; Hancock et al., 1999; Perg et al., 2002; Repka et al., 1997). (C) Similar to B, but with sampling of individual cobbles (filled symbols) instead of amalgamated samples. Since subsurface samples can only be perturbed by inheritance, a limiting isochron can be established, providing a maximum age for the surface (Mériaux et al., 2004).

brackets on the slip rate. A depth profile, or the depth-corrected ages of single subsurface cobbles, may be used to more explicitly constrain this estimate. If we assume that all the age dispersion is due to diachronous deposition—not inheritance—then the cobble ages, regardless of the interpretational model, provide the maximum age estimate or a minimum model-dependent slip rate. In Tibet, the issue is not that inheritance-affected surface ages yield slip rates that are too low, but rather that the geomorphic rates are high relative to geodetic measurements (Chevalier et al., 2005a; Wright et al., 2004).

GEOMORPHIC SLIP-RATE DETERMINATIONS IN CENTRAL ASIA

Quaternary faulting and seismicity suggest that a major portion of the active deformation in central Asia is partitioned between strike-slip faults that rival or surpass the San Andreas fault in length, and fold-and-thrust mountain belts (Fig. 1). Over a distance of more than 2000 km, the Altyn Tagh fault forms the northern edge of the plateau, and the association of this fault with the break in topography implies a genetic link with uplift of the plateau. Along its central and eastern sections, the Altyn Tagh fault splays into the subparallel Kunlun and Haiyuan faults. The Altyn Tagh, Kunlun, and Haiyuan faults are all left-lateral and are thought to guide the eastward extrusion of Tibet (Tapponnier and Molnar, 1977, 1979). Near its eastern terminus, the Altyn Tagh fault merges with the young, fold-and-thrust ranges of the Tanghenan Shan, Taxueh Shan, and Qilian Shan (Fig. 1). If these active tectonic features are related, a systematic decrease in slip rates should be observed as slip on the Altyn Tagh fault is partitioned onto the Kunlun fault, Haiyuan fault, and subperpendicular thrusts (cf. Meyer et al., 1998; Xu et al., 2005). Geomorphic rates using cosmogenic isotopes have been determined on the Kunlun, Altyn Tagh, and Haiyuan faults of northern Tibet, as well as the thrust systems in the Qilian Shan and Tanghenan Shan. Here we will present a synopsis of the recent geomorphic rate determinations on these features, and evaluate the spatial variation in slip rates with respect to existing velocity models.

If active, sinistral strike-slip faults along the northern edge of the plateau guide the eastward extrusion of Tibet, then similar dextral features must exist along the southern boundary of the plateau (Avouac and Tapponnier, 1993). Convergence along the Himalaya and eastward-directed strike-slip motion along the Altyn Tagh fault is kinematically inconsistent with the existence of a single plate of continental lithosphere composing this entire region, and eastward motion must be accommodated along other features that decouple motion between India–southern Tibet and northern Tibet. The Karakorum fault, a major dextral, strike-slip fault in southwestern Tibet (Fig. 1), forms the western part of this "Tibetan plate." East of Mount Kailas, dextral motion cannot be assigned to a single well-defined tectonic feature. Rather, dextral motion has been proposed along an en echelon right-lateral shear zone, the Karakorum-Jiali fracture zone, that lies along a chord connecting the eastern and western syntaxes and transfers motion from the Karakorum fault to other faults, such as the Jiali fault, in southeastern Asia (Armijo et al., 1989). Block rheological models require that the rates of motion along the faults in southern Tibet must be comparable to those that delineate its northern boundary (cf. Avouac and Tapponnier, 1993). Quantitative rates of motion on the Karakorum fault and Karakorum-Jiali fracture zone are scarce relative to those in the north, and only a preliminary assessment of these models is possible or warranted at this time.

Rates of Strike-Slip Movement in Tibet

The Kunlun Fault

The Kunlun fault system is ~1600 km long and extends between 87°E and 105°E, from the high Qiangtang platform of northern Tibet, a dry permafrost plateau over 5000 masl, down to the humid, forested and deeply incised (2000 m a.s.l. on average) mountainous rim of southern China's Sichuan basin (Fig. 1). On a large scale, the Kunlun fault system follows the northeastern border of the highest part of the Tibetan Plateau and is composed of twelve principal, linear segments (Fig. 6) (Van der Woerd et al., 1998, 2000, 2002b). It has been one of the more historically active strike-slip faults in central Asia, and the largest earthquakes recorded in the twentieth century in northern Tibet (Gu et al., 1989) ruptured segments of these sinistral faults or nearby thrusts. In particular, three large earthquakes—January 7, 1937, April 19, 1963, and March 24, 1971—ruptured the central stretch of the Kunlun fault, with magnitudes of 7.5, 7.1, and 6.4, respectively (e.g., Cui and Yang, 1979; Gu et al., 1989; Tapponnier and Molnar, 1977). Together, the 1963 and 1937 surface breaks were reported to reach a cumulative length of ~300 km, with maximum horizontal displacements of 7–8 m (e.g., Li and Jia, 1981; Molnar and Lyon Caen, 1989). On November 8, 1997, the Manyi earthquake, the largest earthquake ever recorded instrumentally in northern Tibet (Ms = 7.9), ruptured one of the westernmost, N80°E-striking segments of the Kunlun fault, between 86°E and 90°E (Peltzer et al., 1999), a feature previously mapped on Landsat images by Tapponnier and Molnar (1977). The rupture length was in excess of 150 km, and the maximum horizontal slip, ~7 m (Peltzer et al., 1999). Then on November 14, 2001, the M 7.8 Kokoxili earthquake ruptured the Kusai Hu segment of the Kunlun fault over a length of ~300 km (Van der Woerd et al., 2002a).

Prior to the availability of cosmogenic dating, the average Quaternary slip rate at the Kunlun Pass was estimated to be between 10 and 20 mm/yr (Kidd and Molnar, 1988), and GPS results suggested a rate as small as 6 mm/yr (Burchfiel et al., 1998) or ~8–11 mm/yr (Zhang et al., 2004). The millennial geomorphic slip rate has been determined at 13 locations along the Xidatan-Dongdatan, Dongxi Co, and Maqen segments of the fault (Fig. 6), a stretch spanning ~600 km of the Kunlun fault system (Van der Woerd et al., 1998, 2000, 2002b). In the west, at three sites along the Xidatan-Dongdatan segment of the fault, near 94°E, terrace riser offsets ranging from 24 to 110 m, with cosmogenic ages ranging from ca. 1800 to ca. 8200 yr B.P., yield a mean left-lateral slip rate of 11.7 ± 1.5 mm/yr. At two sites along the Dongxi-Anyemaqin segment of the fault, near 99°E, terrace riser offsets ranging from 57 to 400 m with ^{14}C ages ranging from 5400 to 37,000 yr B.P. yield a minimum slip rate of ~10 mm/yr, comparable to that in the west. Farther east, near 100.5°E, along the Maqen segment of the fault, the 180 m offset of a lateral moraine emplaced between the Last Glacial Maximum (20 ka) and 11,100 yr B.P. yields a mean slip rate of 12.5 ± 2.5 mm/yr. The slip rate thus appears to be constant along the 600 km of the Kunlun fault studied. The average rate is 11.5 ± 2.0 mm/yr over the last 37,000 yr (Fig. 7) and is in agreement with the higher, most recent GPS estimates (Zhang et al., 2004).

A number of important conclusions can be drawn from the late Quaternary distribution of slip along the Kunlun fault system. From a methodological standpoint, the same rate is obtained using either ^{10}Be or ^{26}Al cosmogenic dating of quartz or radiocarbon dating of organic material. This indicates that (1) the production rates at these elevations and latitudes are reasonably accurate and (2) that the cosmogenic ages are not significantly influenced by predepositional inheritance or erosion. At the largest scales, the spatial constancy of the rate is consistent with the simple geometry of the fault system, which at any longitude is characterized by a single strand. Also, unlike the Altyn Tagh fault, the Kunlun fault does not appear to have any clear-cut spatial relationship to high-angle shortening features—large thrusts and growing mountain ranges—that characterize the northeastern edge of the plateau. Hence, the potential for slip partitioning involving other genetically related faults is minimized. The temporal constancy of the slip rate demonstrates that the Kunlun fault does not manifest any secular velocity variations for time periods greater than ~2000 yr, the age of the youngest offset measured. Recently, based on three-dimensional paleoseismology at Wrightwood, California, Weldon et al. (2004) show that the slip rate on the San Andreas fault varied between 8.9 and 2.4 cm/yr over the last 2000 yr, spanning 14 seismic cycles. The timescale over which this variation is observed is essentially equivalent to that averaged by the youngest dated feature on the Kunlun fault. The similarity of the cosmogenically derived slip rate on the Kunlun fault to that derived from the essentially instantaneous geodetic results suggest that secular variations in slip rate do not persist over ~2000 yr.

The Haiyuan Fault

Together with the Altyn Tagh and Kunlun faults, the Haiyuan fault is one of the main left-lateral strike-slip faults defining

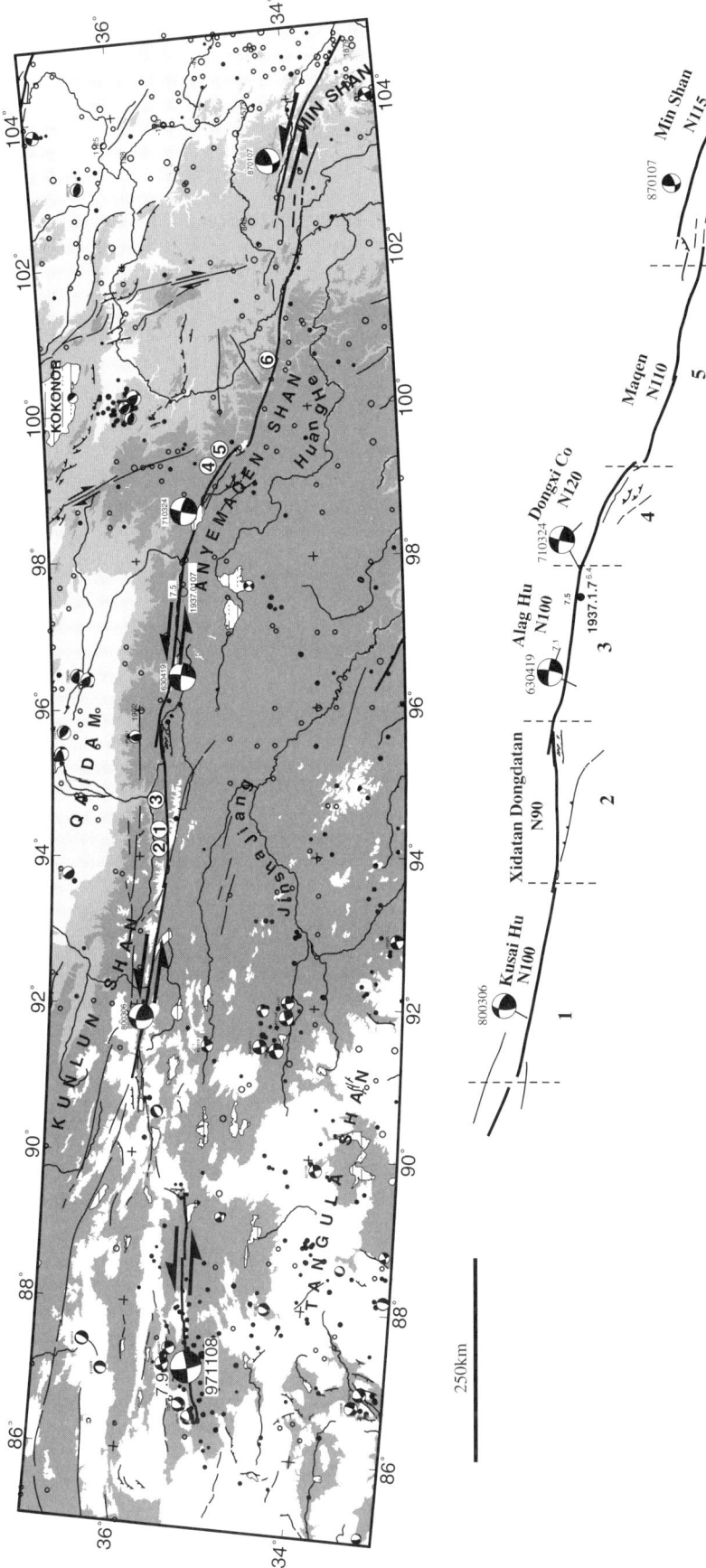

Figure 6. Large-scale segmentation of Kunlun fault (after Van der Woerd et al., 2002b). From 91°E to 105°E, six principal segments are identified. The Min Shan segment (103°E to 105°E) steps ~50 km to the north. West of 91°E, several segments, with strikes between 70° and 120°, linked with normal faults form broad western horsetail termination. Instrumental (USGS 1977–1998) and historical (Gu et al., 1989) seismicity, filled and open circles, respectively, for magnitude Ms > 4. Focal mechanisms from Molnar and Lyon-Caen (1989) and USGS.

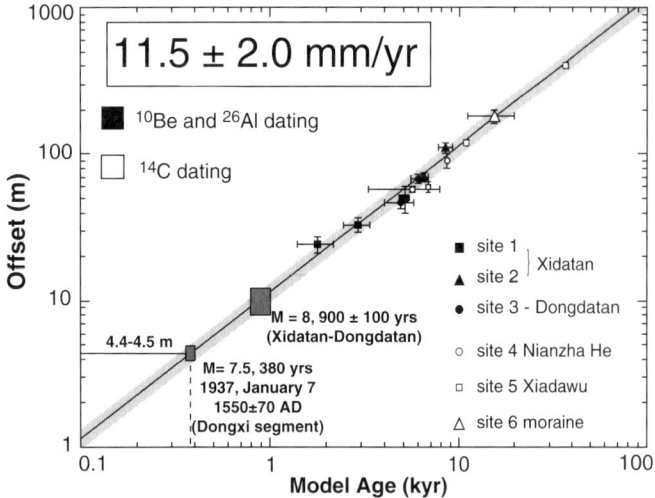

Figure 7. Summary of late Pleistocene–Holocene sinistral slip rates deduced from cosmogenic ^{10}Be-^{26}Al and ^{14}C dating of alluvial terraces at six sites along the Kunlun fault. Consistency between independent values obtained with different dating techniques implies uniform average slip rate of 11.5 ± 2.0 mm/yr along 600 km of fault (Van der Woerd et al., 1998, 2000, 2002b).

the northeastern edge of Tibet. This 1000 km long fault accommodates the eastward component of movement of Tibet relative to the Gobi-Ala Shan platform to the north (Tapponnier and Molnar, 1977; Zhang et al., 1988). The large 1920 Haiyuan (M = 8.7) and 1927 Gulang (M = 8–8.3) earthquakes occurred on and near the Haiyuan fault (Fig. 8), and attest to its capacity to produce large earthquakes (Deng, 1986; Gaudemer et al., 1995; Zhang et al., 1987). In this connection, an ~220 km long seismic gap of great potential hazard has been identified along the western stretch of the fault near Tianzhu (Gaudemer et al., 1995). The strike of the Haiyuan fault is roughly parallel to that of the Kunlun fault (Fig. 1), but unlike the Kunlun, which is defined by a single strand at most latitudes, the main Leng Long Ling segment of the Haiyuan fault splays into the Gulang and Maomao Shan faults at ~102.5°E (Fig. 8). The partitioning of slip along these splays has been established by dating geomorphic offsets on the Maomao Shan (Lasserre et al., 1999) and Leng Long Ling (Lasserre et al., 2002) segments of the fault.

Meyer (1991) and Gaudemer et al. (1995) identified moraine and glacial valley edge offsets, on the order of 200–270 m, at several sites in the Leng Long Ling using SPOT images. In the absence of absolute chronological data, they inferred the offsets to have accrued in the period since the LGM or the onset of the Holocene, yielding rates between 10 and 26 mm/yr. More recently, Lasserre et al. (2002) mapped a 200 ± 40 m offset of the Xiying He moraine on the Leng Long Ling section of the fault. In situ ^{10}Be and ^{26}Al ages of samples collected from the ridge of the moraines cluster around 10,300 ± 339 yr. This age is interpreted as the time of last reshaping of the moraine before the valley glacier withdrew south of the fault around the end of the Younger Dryas (ca. 11,000 yr B.P.). Assuming that the 200 ± 40 m moraine offset was recorded after glacial retreat across the fault constrains the late Pleistocene slip rate on the Leng Long Ling segment of the Haiyuan fault to be 19 ± 5 mm/yr. The cosmic-ray exposure dating of the Xiying He moraine offset thus confirms and refines the earlier estimates of Meyer (1991) and Gaudemer et al. (1995).

West of the junction between the Gulang and Maomao Shan segments of the Haiyuan fault, ^{14}C dating of fluvial terrace risers near Songshan, on the Maomao Shan segment (Fig. 8), yields a slip rate of 12 ± 4 mm/yr averaged over ~14,000 yr (Lasserre et al., 1999). The eastward decrease in rate reflects the partitioning of slip between the Gulang and Maomao Shan splays of the fault. In this connection, the slip contribution of the Gulang fault was previously estimated to be 4.3 ± 2.1 mm/yr (Gaudemer et al., 1995) and is in agreement with the velocity decrease on the main southern segments of the Haiyuan fault, east of where the Gulang and Maomao Shan splays merge. A long-term sinistral slip component as high as ~2 cm/yr between northeastern Tibet and the Gobi-Ala Shan platform east of 100°E has important implications for our understanding of the deformation of central Asia (Peltzer and Saucier, 1996). The geodetic rate of 8 ± 2 mm/yr inferred from regional GPS campaigns, which does not resolve the rates on the individual segments (Chen et al., 2000), is in agreement with the geomorphic rate from the Maomao Shan segment, but somewhat lower than the geomorphic rate obtained from the Leng Long Ling segment. This may reflect either the still small number of measurement epochs and stations used in such studies, the large error on the geomorphic determination, or more fundamental differences in short- and long-term crustal fault mechanics.

The Altyn Tagh Fault

The 2500 km long, left-lateral Altyn Tagh fault system defines the northern edge of the Tibetan Plateau from Muztagh Ata Tagh, near 75°E, to Yabraishan, east of 102°E (Fig. 1), and has been interpreted as the main continental plate boundary guiding the northeastward extrusion of Tibet relative to the Tarim basin (Peltzer and Tapponnier, 1988; Tapponnier and Molnar,

Figure 8. (A) Tectonic map of Indo-Asian collision zone (after Lasserre et al., 2002). Shaded region shows location of SPOT mosaic (B). Faults are from fieldwork and SPOT and Landsat image analysis. The portion of the fault defined as the "Tianzhu seismic gap" (Gaudemer et al., 1995) is drawn in white. Locations of 1920 and 1927 (white dots) M8 earthquakes are from Gaudemer et al. (1995). Arrows point to study sites where slip rates on Haiyuan fault were determined from measurements and dating of geomorphic features offset by the fault. [L1]—Maomao Shan segment, 12 ± 4 mm/yr (Lasserre et al., 1999); [L2]—Leng Long Ling segment, 19 ± 5 mm/yr (Lasserre et al., 2002). The rate on the Gulang fault [G1], 4.3 ± 2.1 mm/yr, is an estimate from Gaudemer et al. (1995).

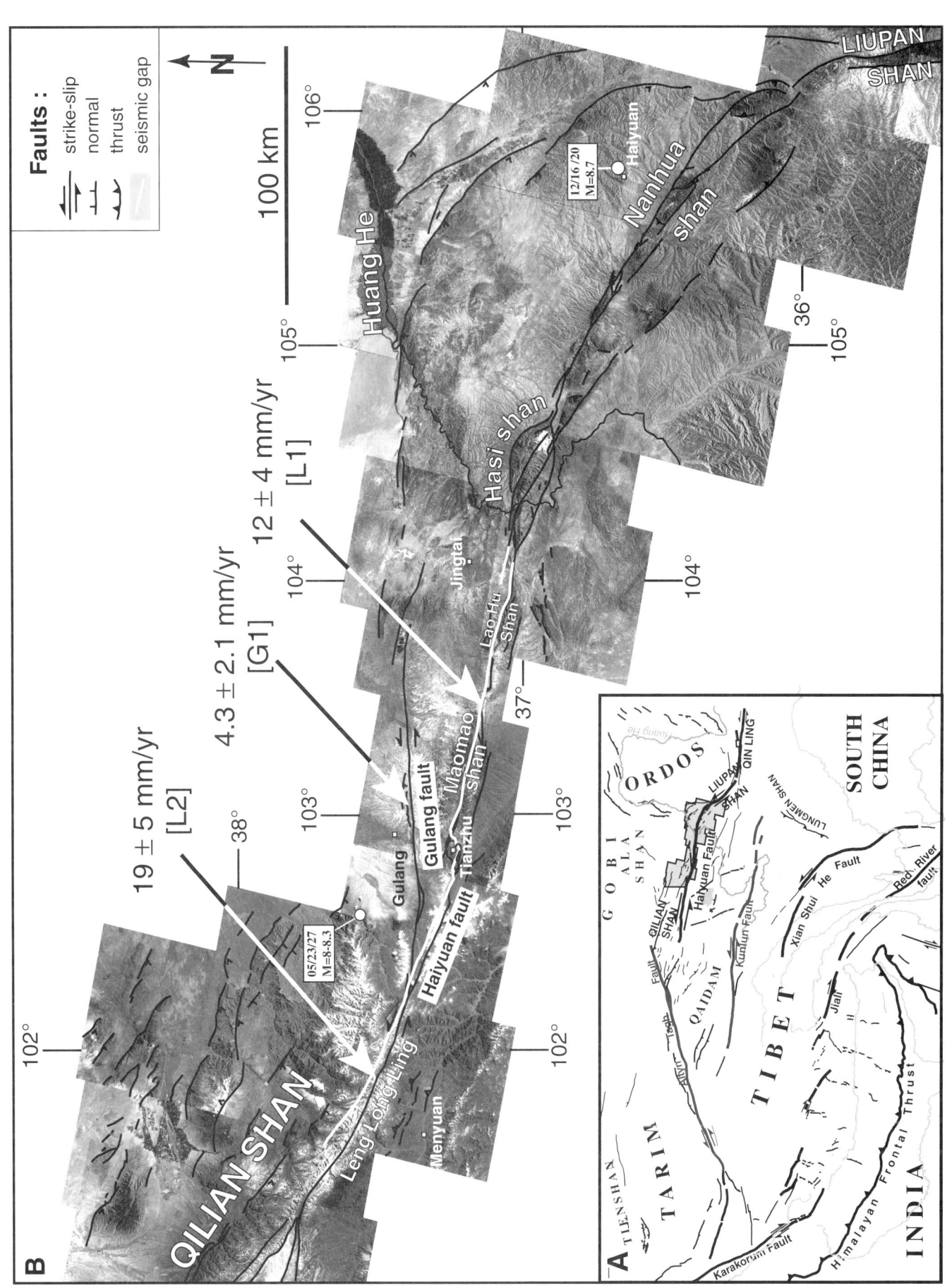

1977). Seismic tomography suggests that a deep shear zone extend beneath the fault to the base of the lithosphere (Wittlinger et al., 1998, 2004), but the magnitude of the millennial and Cenozoic slip rates and total offsets along the fault remain controversial, with estimates of the sinistral slip rate ranging from 2 mm/yr to 40 mm/yr (Bendick et al., 2000; Ge et al., 1992; Mériaux et al., 2004; Peltzer et al., 2000; Shen et al., 2001b; Wang et al., 2001; Washburn et al., 2001; Zhang et al., 2004).

A number of total-offset markers along the Altyn Tagh fault have now been defined and the long-term Cenozoic, geologic slip rates inferred. The majority of the pre-Tertiary piercing points are found along the central and eastern portions of the fault and yield offsets ranging from 260 to 500 km (Gehrels et al., 2003; Ritts and Biffi, 2000; Sobel et al., 2001; Yang et al., 2001; Yin et al., 2002; Yue et al., 2001). Near the western termination of the fault, Peltzer and Tapponnier (1988) noted that Paleozoic plutons in the western Kunlun have been offset by >500 km across the Altyn Tagh fault. A more recent reevaluation of this offset using additional age constraints yields a similar estimate of the total offset, 475 ± 75 km (Cowgill et al., 2003). Based upon the history of sedimentation for the southwestern Tarim basin, Qaidam basin, and Hexi corridor, Yin et al. (2002) infer that thrusting in the western Kunlun and along the Qiman Tagh and northern Qaidam thrust systems began prior to 46 Ma and ca. 49 Ma, respectively. Interpreting these thrust systems as termination or branching faults of the Altyn Tagh fault, they conclude that the fault must have been active since ca. 49 Ma, and that, combined with the 475 ± 75 km offset of Paleozoic plutons in the western Kunlun, the long-term slip rate is ~9 mm/yr. However, the temporal constraints must be viewed as somewhat controversial, as the proposed initiation of slip on the Altyn Tagh fault would predate that initiation of slip on the Red River shear zone, ca. 36 Ma (Gilley et al., 2003), and slip on the Gangdese thrust system, ca. 27 Ma (Yin et al., 1994). Both features lie significantly closer to the Indo-Asian boundary, and the proposed chronology would imply an "out-of-sequence" onset of continental deformation. In this connection others, e.g., Yue et al. (2001), have proposed an Oligo-Miocene initiation of slip on the Altyn Tagh fault. As such, the 9 mm/yr slip rate is best considered as a lower bound on the integrated Cenozoic slip rate. Similarly, Yue et al. (2001) have reconstructed the offsets for the Xorkol basin (91°–92°E), north of the Altyn Tagh fault, and its inferred Oligocene and post–Early Miocene source regions to the south, obtaining offsets of ~380 km and 300 km, respectively, yielding a long-term slip rate of 12–16 mm/yr. The Qilian Shan lies at the eastern termination of the Altyn Tagh fault, a region of active shortening. Any post-Miocene shortening would result in an underestimate of the observed offset, and this observation may also constitute a lower bound.

Mériaux et al. (2004) have determined the ages of fluvial and glacial geomorphic markers left-laterally displaced along the central segment of the Altyn Tagh fault using radiocarbon and ^{10}Be-^{26}Al cosmic-ray exposure dating (Fig. 1). Two sites near Tura (~37.6°N, 86.6°E) were investigated: Cherchen He and Sulamu Tagh. Here, the Altyn Tagh fault is characterized by a single strand, and geomorphic slip-rate determinations, therefore, capture the full rate on the fault. The sites are geomorphologically distinct, with Cherchen He dominated by fluvial processes recording offsets between 166 and 420 m, and Sulamu Tagh by glacial action with offsets between 470 and 3660 m. Nine offsets with ages between 6 and 113 ka yield a constant average slip rate of 26.9 ± 6.9 mm/yr (Fig. 9), well in excess of the Cenozoic and geodetic slip-rate estimates (Bendick et al., 2000; Wallace et al., 2004). The lack of secular variation of the rate is consistent with and extends the results obtained on the Kunlun fault (Van der Woerd et al., 1998, 2000, 2002b).

With respect to the use and consistency of cosmogenic dating as applied to active tectonics, the Cherchen He site is of particular interest. At the Cherchen River site west of Tura, the Altyn Tagh fault cuts fluvial terraces, offsetting their risers (T2/T1) and an abandoned channel ("PC" in Fig. 10). The timing of terrace abandonment was determined by ^{10}Be and ^{26}Al cosmic-ray surface exposure dating of quartz cobbles from the surfaces of T1 and T2 and subsurface samples from T1. The average ages from cosmogenic dating of surface samples are 14.2 ± 1.3 ka for T2 and 6.5 ± 0.7 ka for T1 (Fig. 11). The youngest depth-corrected ages for subsurface samples in the T1 depth profile yield a maximum age for the T1 surface (younger than 6 ka), in good agreement with the youngest surface ages. The radiocarbon ages of three charcoal samples found in T1 on the east side of the site

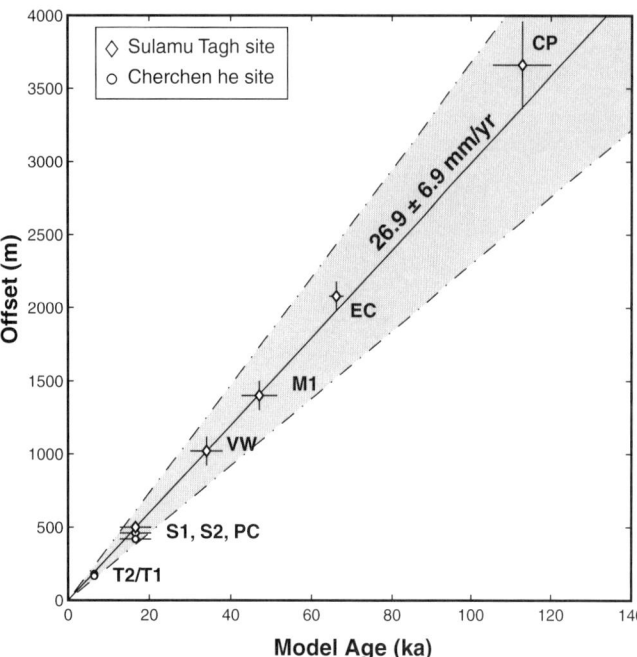

Figure 9. Slip-rate summary on the central Altyn Tagh fault during the last glacial cycle (Mériaux et al., 2004). Open symbols are the preferred slip-rate estimates yielding a bracket of 25–35 mm/yr; the lower bound is largely controlled by the Cherchen He data. The linear regression of the nine data points that determine the upper slip-rate estimates yields a rate of 26.9 ± 6.9 mm/yr.

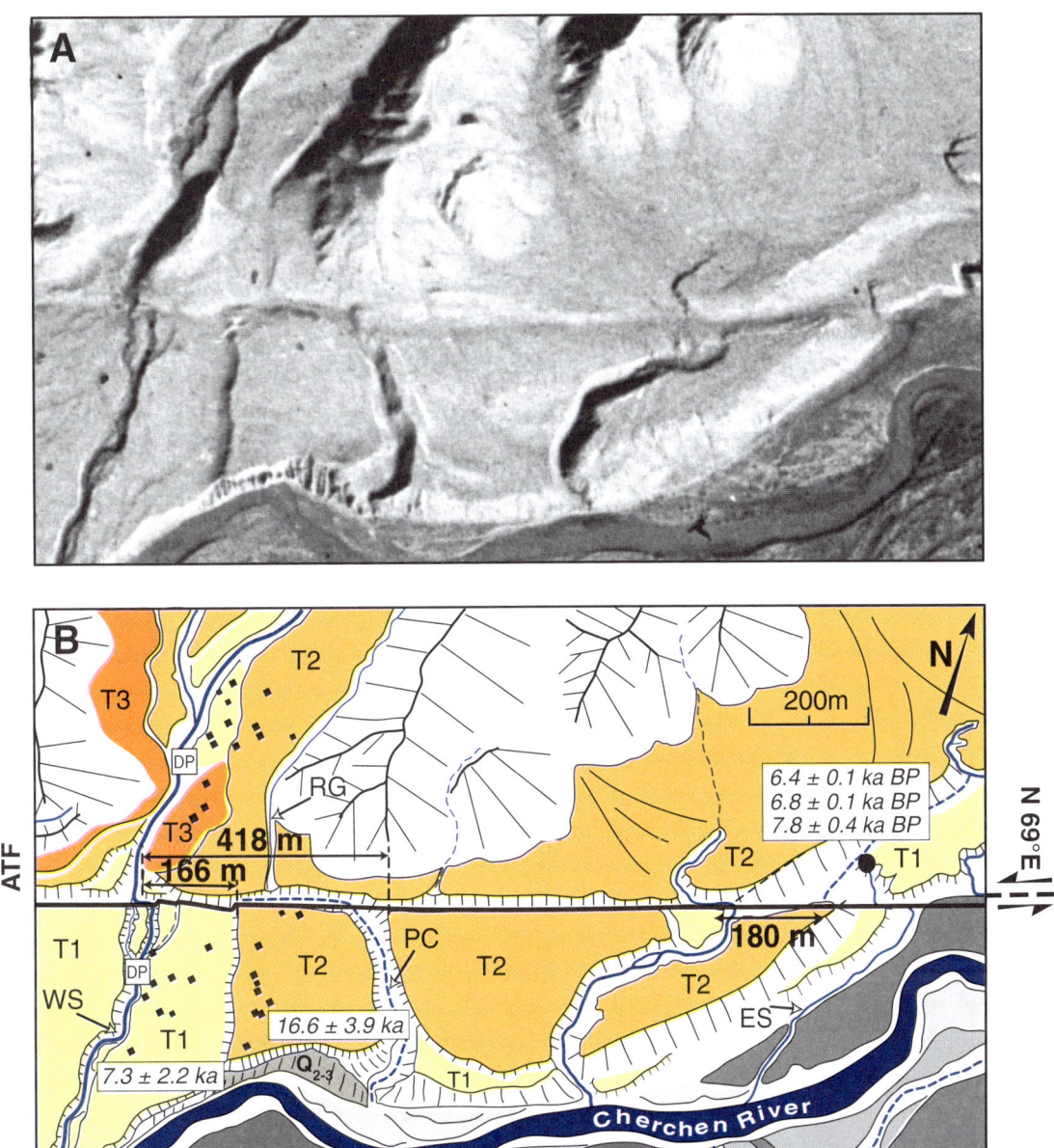

Figure 10. Offset terrace risers and channels of tributaries of Cherchen He west of Tura (37.6°N, 86.4°E; ~3000 masl). (A) Corona image (DS 1025-2118DA031) (Mériaux et al., 2004). (B) Results of field mapping. Locations of surface samples are designated as ◆. DP—locations of subsurface sampling for cosmogenic dating. Terrace surfaces are numbered and shaded as a function of increasing age and elevation. WS—western stream; ES—eastern stream; PC and RG—paleochannel and regressive channel in incised T2 terrace, respectively. Measured riser and stream offsets, and average terrace ages are shown.

corroborate that this terrace was abandoned after 6.4 ± 0.1 ka. The agreement between the youngest ^{10}Be surface and subsurface cosmogenic ages and the radiocarbon data confirms that the older cosmogenic ages are "outliers" in the surface population likely due to predepositional exposure. The 166 ± 10 and 180 ± 10 m offsets of the T2/T1 risers postdate the abandonment of T1, an incisional strath terrace, at ca. 6 ka, and the 418 ± 10 m offset of the abandoned fluvial channel must postdate abandonment of the T2 surface it incises. The age-offset pairs yield compatible rates of 25.7 ± 3.3 mm/yr, 28.1 ± 3.2 mm/yr, and 29.5 ± 2.9 mm/yr. The 100,000 yr rate derived from offset glacial features at Sulamu Tagh, ~50 km to the east, is 31.0 ± 1.3 mm/yr, in agreement at the 2σ level.

The concordance of age determinations based on surface and subsurface cosmogenic ages and those based on "stratigraphically correct" radiocarbon dating preclude any discernible effects of erosion and/or inheritance on the ~15,000 yr slip rate at Cherchen He. That the 100,000 yr rate at Sulamu Tagh is consistent

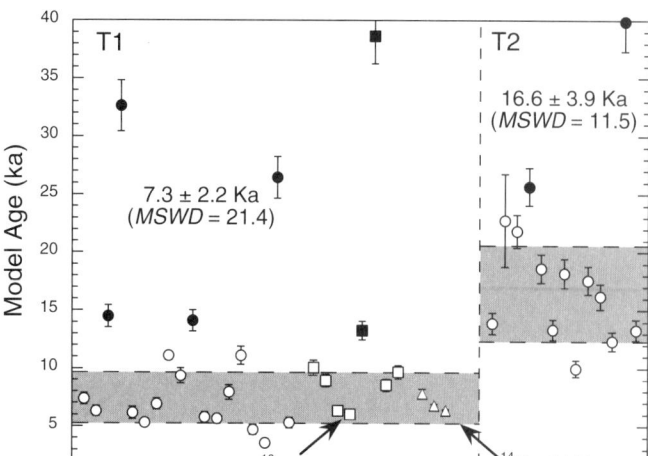

Figure 11. ¹⁰Be exposure model age determinations at Cherchen He (Mériaux et al., 2004). The average values are simple unweighted means with errors equal to one standard deviation for surface samples. The 1σ brackets on the average ages are shown as the shaded regions on the figure. ¹⁰Be ages of samples included in the average are shown as open symbols; filled symbols are for ages that were considered as outliers. Surface samples are shown as open and filled circles, and depth-corrected ages of subsurface samples are shown as open and filled squares. Radiocarbon ages are shown as open triangles. Surface samples yield an average age of 7.3 ± 2.2 ka for T1 and an average of 16.6 ± 3.9 ka for T2. Sample TU3-71, 71.8 ± 4.6 ka, has not been plotted in order to improve the resolution in the age range of interest.

with the rate at Cherchen He suggests that the longer-term measurement is similarly unaffected by erosion and/or inheritance. This issue will be discussed further below.

The eastern stretch (from 90°E to 97°E) of the Altyn Tagh fault strikes N70°E overall but has a complex geometry (Fig. 12) that has been described in detail (Mériaux et al., 2005). Between the western edge of the Qaidam basin and the Qilian range, the fault system comprises three principal parallel strands (Ge et al., 1992; Meyer et al., 1998; Peltzer et al., 1989; Van der Woerd et al., 2001). The southern strand or South Altyn Tagh fault follows the Qaidam side of the Altyn push-up, and then veers southeastward into the Tanghenan Shan. The northern strand follows the edge of the Tarim block from the Altyn to the Qilian Shan thrust fronts. East of the Qilian Shan front, the northern strand (North Altyn Tagh fault) continues with a N96°E strike north of the Hexi corridor. Finally, the central strand or Yema fault jogs southward from the North Altyn Tagh fault between Aksay and Subei and continues northeastward to the Taxue Shan thrust front.

The multiple strands and spatially related thrust faults that characterize the eastern Altyn Tagh fault provide mechanisms for transfer of slip from the main segment of the fault, and eastward-decreasing geomorphic slip rates do appear to confirm this effect. Millennial slip rates have been determined for the Altyn Tagh fault at three sites near Aksay (~94°E) along the piedmont of the Dangjin Shankou range in northeastern Tibet through a combination of offset measurements of fluvial channels and terrace risers

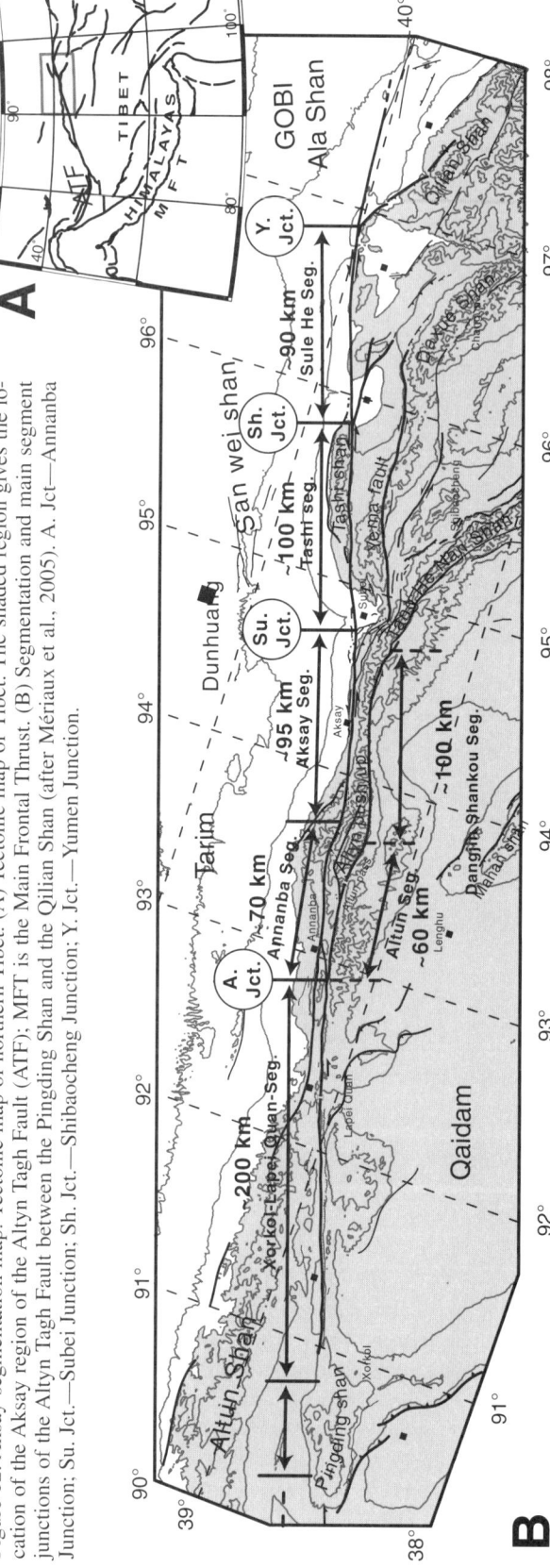

Figure 12. Aksay segmentation map. Tectonic map of northern Tibet. (A) Tectonic map of Tibet. The shaded region gives the location of the Aksay region of the Altyn Tagh Fault (ATF); MFT is the Main Frontal Thrust. (B) Segmentation and main segment junctions of the Altyn Tagh Fault between the Pingding Shan and the Qilian Shan (after Mériaux et al., 2005). A. Jct.—Annanba Junction; Su. Jct.—Subei Junction; Sh. Jct.—Shibaocheng Junction; Y. Jct.—Yumen Junction.

coupled with radiocarbon and ^{10}Be and ^{26}Al surface exposure dating (Mériaux et al., 2005). Cumulative offsets range from 20 m to 260 m and fall in distinct groups, indicative of climatically modulated regional landscape formation, and at least nine different surfaces have been defined based upon morphology, elevation, and dating. The abandonment ages of several of these surfaces are constrained by radiocarbon dating of subsurface charcoals. The majority of the samples are younger than ca. 14 ka, postdating the LGM. The end of the Early Holocene Optimum marks the boundary between the ages of the two main terrace levels at 5–6 ka. The radiocarbon ages typically coincide with the youngest cosmogenic ages for a particular surface. Surface exposure ages older than the radiocarbon dates are taken to represent diachronous terrace emplacement with a finite time interval between the onset of emplacement and abandonment. Explicit in this assumption is the possibility that some fraction of the offset of a terrace riser could have accumulated prior to abandonment of the underlying terrace. Slip rates are obtained by matching the riser offsets with both (1) the abandonment/emplacement ages of the terrace underlying the riser and (2) the abandonment ages of the terrace levels above (the "strath abandonment" and "strath emplacement" models described earlier), and provide bounding estimates on the slip rate (Fig. 13). Rates derived from channel offsets are determined using the age of the surface incised by the channel. Overall, slip-rate estimates using the abandonment age of the overlying level for fill terraces or channels and the emplacement of the underlying level for strath terraces give consistent results, with 32 determinations yielding an average Holocene rate of 17.8 ± 3.6 mm/yr (Fig. 13). As the Altyn Tagh fault is divided into a northern and southern branch at this longitude, the rate of the northern Altyn Tagh fault should thus be considered a minimum for the overall Altyn Tagh fault system. The Altyn Tagh fault slip rate at Aksay is 9 mm/yr less than the geomorphic rate obtained near Tura at ~87°E (26.9 ± 6.9 mm/yr), in keeping with the inference of an eastward-decreasing rate on the fault, due to increased thrusting to the south.

Regardless of any spatial variability in slip rate along the Altyn Tagh fault, the difference in millennial and geodetic slip rates is difficult to reconcile. For instance, GPS surveys across the central portion of the fault (~90°E) conducted in 1994, 1998, and 2002 yield a rate of 9 ± 4 mm/yr (Bendick et al., 2000; Wallace et al., 2004). Similarly, InSAR observation of the western Karakax Valley segment of the fault (82°E; Fig. 1) yields a slip rate of 5 ± 5 mm/yr (Wright et al., 2004). Both geodetic rates are substantially smaller than the millennial estimates. Wallace et al. (2004) conclude that the disparity is due to "systematic error that biases" geologic slip velocities to higher values. Given the agreement between radiocarbon, surface, and subsurface cosmogenic dating, it is not clear what these "inescapable" biases might be. Instead, we emphasize that the slip rates on the San Andreas fault at Wrightwood have varied by a factor of ~3.5 within the past 2000 yr (Weldon et al., 2004). Secular variations of this magnitude would reconcile millennial and geodetic rates on the Altyn Tagh fault, but the true nature

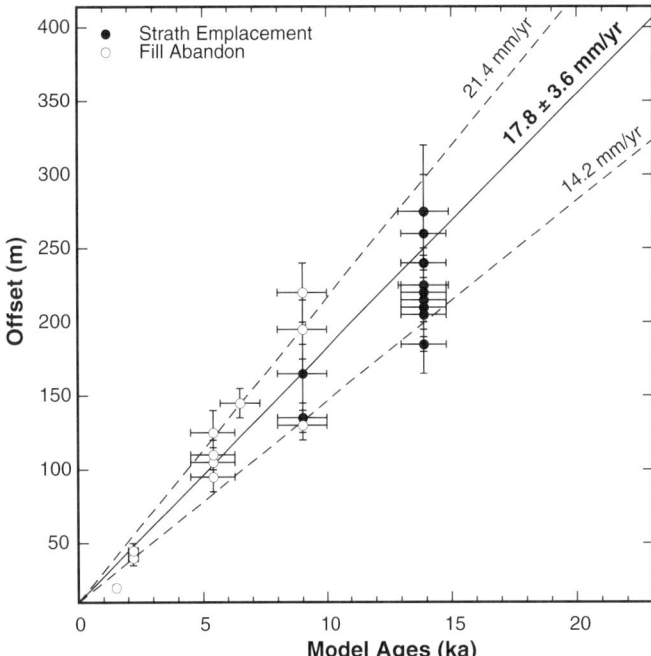

Figure 13. Aksay offset versus model age. Slip rate estimates are derived from using the age of the overlying surface for fill terraces and channels and the emplacement age of the underlying surface for strath terraces, yielding an average Holocene rate of 17.8 ± 3.6 mm/yr (after Mériaux et al., 2005).

and existence of this disparity must be confirmed by further millennial and decadal rate determinations.

The Karakorum Fault

The Karakorum fault is the main Quaternary right-lateral fault north of the Himalayas and trends roughly parallel to the western Himalayan range, extending from at least Mount Kailas to the Pamir, a length of >1200 km (Fig. 1). As the Altyn Tagh fault defines the northern edge of the Tibetan Plateau, the Karakorum fault defines at least the western portion of its southern border, and its slip rate has similar implications for the rheological character of the continental lithosphere and is a subject of active debate. Attempts to determine the rate over timescales ranging from the Oligo-Miocene to the Quaternary have produced disparate values ranging from 1 to 30 mm/yr (Banerjee and Burgmann, 2002; Brown et al., 2002a; Chevalier et al., 2005a; Lacassin et al., 2004; Liu, 1993; Murphy et al., 2002; Phillips et al., 2004; Wang et al., 2001; Wright et al., 2004).

Unlike the Altyn Tagh fault, the disparity in rate estimates on the Karakorum fault is not strictly a function of observational technique, however, and helps to point out an important distinction between estimates derived from geodesy and morphochronology. The InSAR observations of Wright et al. (2004) provide the most recent geodetic slip-rate estimate for the Karakorum fault. Using observations taken between 1992 and 1999, they obtain a rate of 1 ± 3 mm/yr. Based on cosmogenic dating, the geomorphic rate determined on a single strand of the Karakorum fault north of Bangong Lake is 4 ± 1 mm/yr (Brown et al., 2002a), in apparent agreement with the geodetic results. However, this geomorphic rate must be considered a minimum value as it only samples the northern of the two strands of the fault at this longitude, ~78°E (Fig. 1). While geodetic methods sample far-field deformation, capturing all of the motion distributed among the various strands of a fault system, the geomorphic rates are only representative of the motion on the strand on which the offset is observed. Motion accommodated on other strands within the system must be measured independently. A similar debate concerns the integrated Oligo-Miocene slip rate on the Karakorum fault. Phillips et al. (2004) estimate a total offset of 40–150 km to have accumulated on the fault west of Mount Kailas during the last ~15 m.y., yielding an integrated rate of 2–10 mm/yr. Lacassin et al. (2004) estimate a larger cumulative offset (~250 km) and earlier initiation of motion on this section of the fault (ca. 23 Ma), yielding a rate at the upper end of that proposed by Phillips et al. (2004). More importantly, Lacassin et al. (2004) propose that dextral slip along the Indus-Tsangpo suture east of Mount Kailas may have accommodated a similar cumulative offset, doubling the total offset and associated slip rate. Hence, the existing investigations of both the geologic and active slip rates on the Karakorum fault highlight the need for regional mapping of both active and geologic faults prior to smaller-scale analysis and sampling and that, other factors aside, the rates on a single strand of a complex fault system are necessarily minimum values.

Chevalier et al. (2005a) determined a millennial slip rate on one branch of the Karakorum fault at the Manikala glacial valley terminus, west of the Gar basin (32°2.529′N, 80°1.212′E; Fig. 14) using ^{10}Be surface exposure dating of dextrally offset moraines. The Manikala moraine complex lies at the base of the faulted Ayilari range front, which bounds the west side of the Gar Valley, a large pull-apart basin floored by marshland that hides other strands of the Karakorum fault system (Armijo et al., 1989) (Fig. 1). The dated moraines lie northeast of the U-shaped Manikala Valley, a glacial trough deeply entrenched into the range's igneous basement (Fig. 14). Within the till complex, two main groups of moraines are recognized (Fig. 14). All were emplaced by the Manikala Daer Glacier, whose terminus is today ~7 km upstream from the active fault trace. The morphology of the moraines indicates that they were formed during major advances of the glacier, and later abandoned when the glacier retreated upstream.

The relative ages of the moraine groups can be qualitatively assessed from their surface characteristics (Fig. 15). The M1 surface is rough and composed of chaotically distributed, imbricate blocks (as large as 3 m) surrounded by coarse debris. The smoother surface of M2 appears older with blocks (tens of centimeters to a meter) protruding above a mantle of smaller debris (Fig. 15). The morainic ridges thus appear to become younger from east to west, consistent with right-lateral motion on the fault.

The M2 moraine complex is divided into eastern and western sections (M2E and M2W, respectively) by a deep, beheaded, flat-floored, valley (labeled "PV" in Fig. 14) that is flanked by well-defined, lateral moraines. The crest of the lateral moraine east of PV is well preserved, and its eastern edge extends to the base of the faceted range front. There is no catchment on the mountain slope facing this valley, indicating that PV must correspond to a former channel of the Manikala Daer Glacier (Fig. 14). The youngest moraine group (M1 in Fig. 14) is the only one present on both sides of the Manikala outwash valley, displaying well-preserved terminal lobes and sharply defined ridge crests. The lateral limits of glacial incision in the basement upstream from the fault reach the base of the triangular facets on either side of the Manikala Valley. Once restored using satellite images, the M1 and M2E offsets are 220 ± 10 m and 1520 ± 50 m, respectively.

The peaks in the overall M1-M2 age distribution correspond to the coldest periods as derived from proxy paleotemperature records (e.g., SPECMAP δ^{18}O curve; Imbrie et al., 1984; ca. 19, 36, 151, and 182 ka) (Fig. 16), and hence, to maximal glacial advance. In particular, the younger M2 samples, 140 ± 5.5 ka, correspond roughly to the glacial maximum at the end of MIS 6 (150–140 ka), the older M1 subgroup, 40 ± 3 ka, to the cold period at the end of MIS 3 (ca. 40 ka), and the youngest M1 samples, 21 ± 1.0 ka, to the LGM (19 ka). The oldest ^{10}Be ages on M2E suggest that it was emplaced during the major glacial advance at the beginning of MIS 6, while the youngest ages on M2E are consistent with abandonment at ca. 140 ka at the beginning of the Eemian interglacial (Fig. 16). The bulk of the ages on the younger moraine, M1, are consistent with emplacement at ca. 40 ka. However, the younger ages on this surface suggest that it was not abandoned until the onset of post-LGM warming after ca. 20 ka.

Matching the 1520 ± 50 m offset of the M2E lateral moraine with the sample ages that approximate the end of the MIS 6 glacial maximum (140 ± 5.5 ka) yields an average slip rate of 10.9 ± 0.6 mm/yr (Fig. 17). Likewise, matching the 220 ± 10 m offset of M1 with the age of the M1 LGM samples (21 ± 1.0 ka) yields a rate of 10.5 ± 0.5 mm/yr, corresponding to a constant right-lateral slip rate of 10.7 ± 0.7 mm/yr on this segment of the Karakorum fault for the last 140,000 yr, a rate at least ten times greater than that obtained by InSAR (Wright et al., 2004). The total rate of displacement between southwestern Tibet and the western Himalayas is greater because the normal component of throw on the main fault must be taken into account, along with slip accommodated on other active fault strands within and on the opposite side of the Gar pull-apart basin. The disagreement between the various geomorphic and geodetic slip-rate observations on the Karakorum fault is a subject of continuing debate (cf. Brown et al., 2005; Chevalier et al., 2005b) and is unlikely to be resolved without continuing investigations in these remote areas.

Geomorphic Rates of Thrusting and Crustal Shortening

The region located northeast of the Qaidam basin, between the Altyn Tagh and Haiyuan faults (Fig. 1), is characterized by several large NW-SE-trending ranges (Tapponnier and Molnar, 1977) that have been interpreted to grow as ramp anticlines on

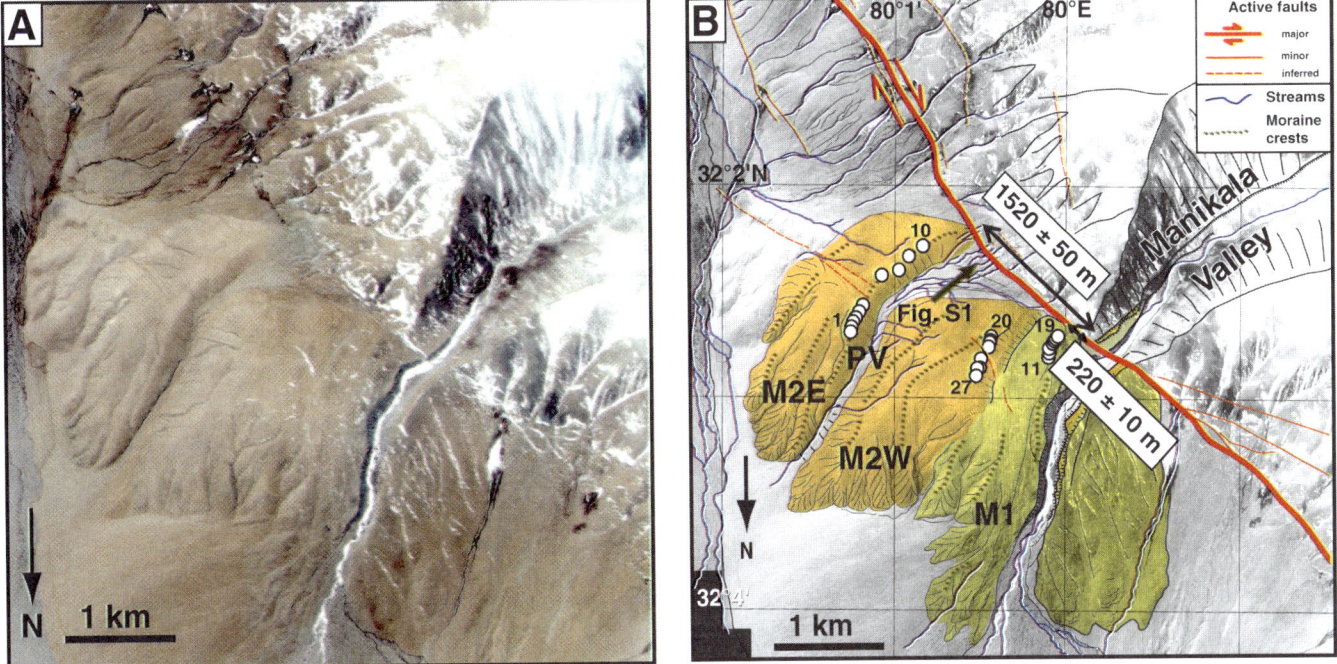

Figure 14. (A) IKONOS satellite image of abandoned Manikala Glacier moraines offset by Karakorum fault. Present-day Manikala Glacier outwash (lower right corner of image) is frozen and appears white in image. (B) Map of offset moraines (orange—M2; yellow—M1) and sample locations (circles with numbers). Note abandoned glacial channel (PV—paleovalley of Manikala Glacier) east of the Manikala outwash. Moraines M1 and M2E are offset ~220 and 1520 m, respectively.

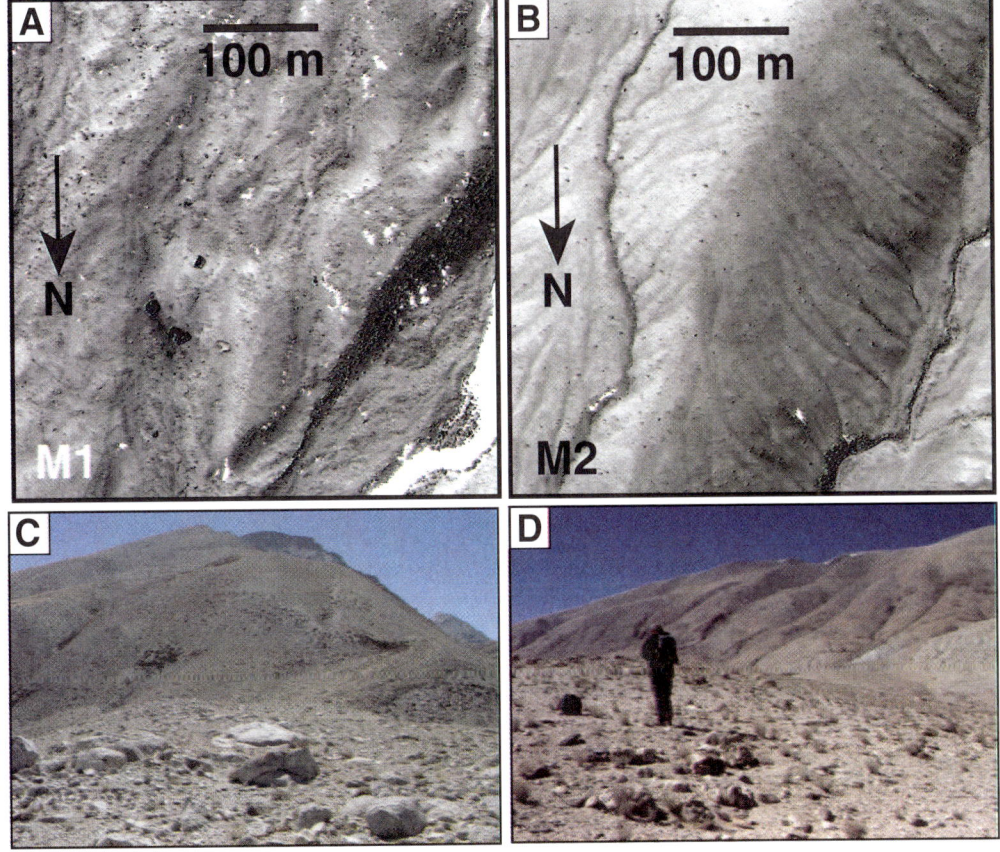

Figure 15. Manikala surface IKONOS images of lateral moraine crests M1 (A) and M2 (B) showing different surface morphology correlated with different age. Large blocks (dark) are visible on both moraines. C: Field photograph of the M1 surface. Large block in the center of the image is ~2 m across. D: Field photograph of the M2 surface (from Chevalier et al., 2005a).

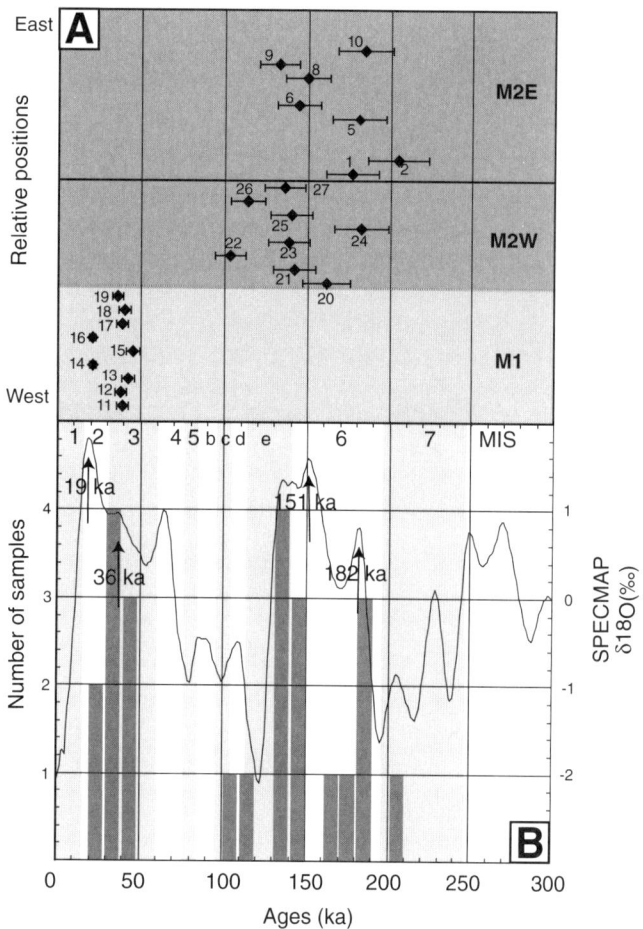

Figure 16. Manikala ages plus SPECMAP. A: ^{10}Be exposure ages of blocks sampled along M1 and M2 moraine crests. B: Comparison of age distribution (10 k.y. bins) with SPECMAP δ^{18}O proxy climate curve (δ^{18}O increases during glacial advances as ^{16}O is preferentially sequestered in the polar ice caps) shows a simple correlation of main moraine emplacement periods with coldest epochs of marine isotope stage MIS 6, MIS 3, and MIS 2 (LGM).

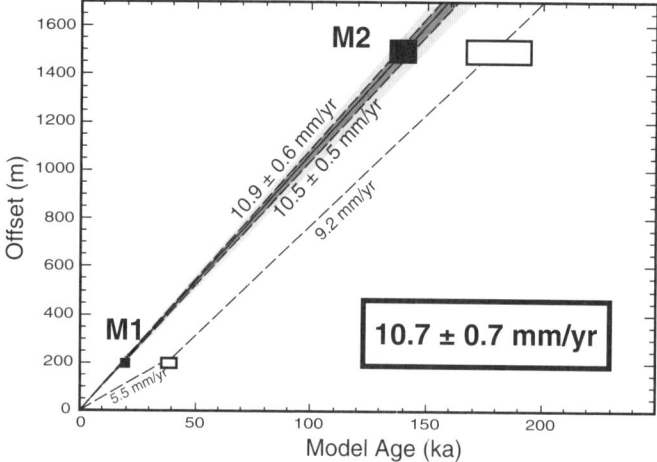

Figure 17. Rates obtained from the offset age relationships for the Manikala moraine complex (Chevalier et al., 2005a). An average rate of 10.7 ± 0.7 mm/yr is obtained from linking the 1520 ± 50 m and 220 ± 10 m offsets with the end of MIS 6 (ca. 140 ka) and MIS 2 (ca. 20 ka), respectively. This association yields a constant slip rate over the entire observation interval. A slip rate of 5.5 mm/yr is implied if the 220 ± 10 m is linked with the 40 ± 3 ka age of the older subgroup on M1. This low slip rate over the last ~40 k.y. requires a rate of 9.2 mm/yr between ca. 40 ka and ca. 180 ka to reconcile the rate obtained if the 1520 ± 50 m offset is linked with the 181 ± 14 ka age cluster on M2E.

active thrusts of crustal scale (Meyer et al., 1998; Tapponnier et al., 1990). The northeastward propagation of the left-lateral Altyn Tagh fault, and the regional northeastward elevation decrease (~3000 to 1200 masl), suggest a close relationship between sinistral faulting and thrusting in which sinistral motion along the Altyn Tagh fault is transferred to the active thrust faults (Métivier et al., 1998; Meyer et al., 1998; Peltzer et al., 1988). The rates of both sinistral slip on the Altyn Tagh fault and shortening across the thrusts should decrease due to the eastward transformation of lateral slip to shortening. In this region, Quaternary rates of uplift and crustal shortening have been quantitatively determined for the Tanghenan Shan thrust (Van der Woerd et al., 2001), the Yumen Shan thrust (Hetzel et al., 2002), and the Zhangye Shan thrust (Hetzel et al., 2004). All use the general strategy outlined in Figure 3; elevation profiles across the scarps and terraces are performed to obtain vertical offsets, coupled with either radiocarbon dating (Van der Woerd et al., 2001), cosmogenic dating (Hetzel et al., 2002), or a combination of cosmogenic and optically stimulated luminescence (OSL) dating (Hetzel et al., 2004).

The Tanghenan Shan, Yumen Shan, and Zhangye Shan thrusts occupy very different positions with respect to the eastern terminus of the Altyn Tagh fault and the growing thrust-bounded mountain ranges of the northeastern corner of Tibet. These differences appear to be reflected in the rates of uplift and shortening (Fig. 18). The Tanghenan Shan is one of the highest ranges (5500 m a.s.l.) north of the Qaidam basin, with a length of 300 km between the Altyn Tagh fault to the west and Hala Hu Lake to the east. Its structure and morphology are similar to the other actively growing ranges of this region, such as the Taxueh Shan and Qilian Shan. The Tanghenan Shan thrusts are northeast-vergent, the predominant trend in northeastern Tibet, and instead of being cut and displaced by strike-slip movement on the fault, as inferred by Wang (1997), the Tanghenan Shan (Fig. 18) appears to be a consequence of such movement (Meyer et al., 1998). North of Subei, the principal, northern branch of the Altyn Tagh fault keeps striking ~N70°E across the Subei basin (Ge et al., 1992; Meyer et al., 1996; Tapponnier et al., 1990), while the main Tanghenan Shan frontal thrust splays from it toward the southeast (Figs. 12 and 18). The southern branch of the Altyn Tagh fault, west and south of the Subei branch, follows the curved axis of the Tanghenan Shan. Such relationships between the sinistral faults, the range, and the thrust indicate that a fraction of the strike-slip motion is locally transformed into thrusting, folding, and mountain building. Southeast of Subei, the Tanghenan Shan frontal

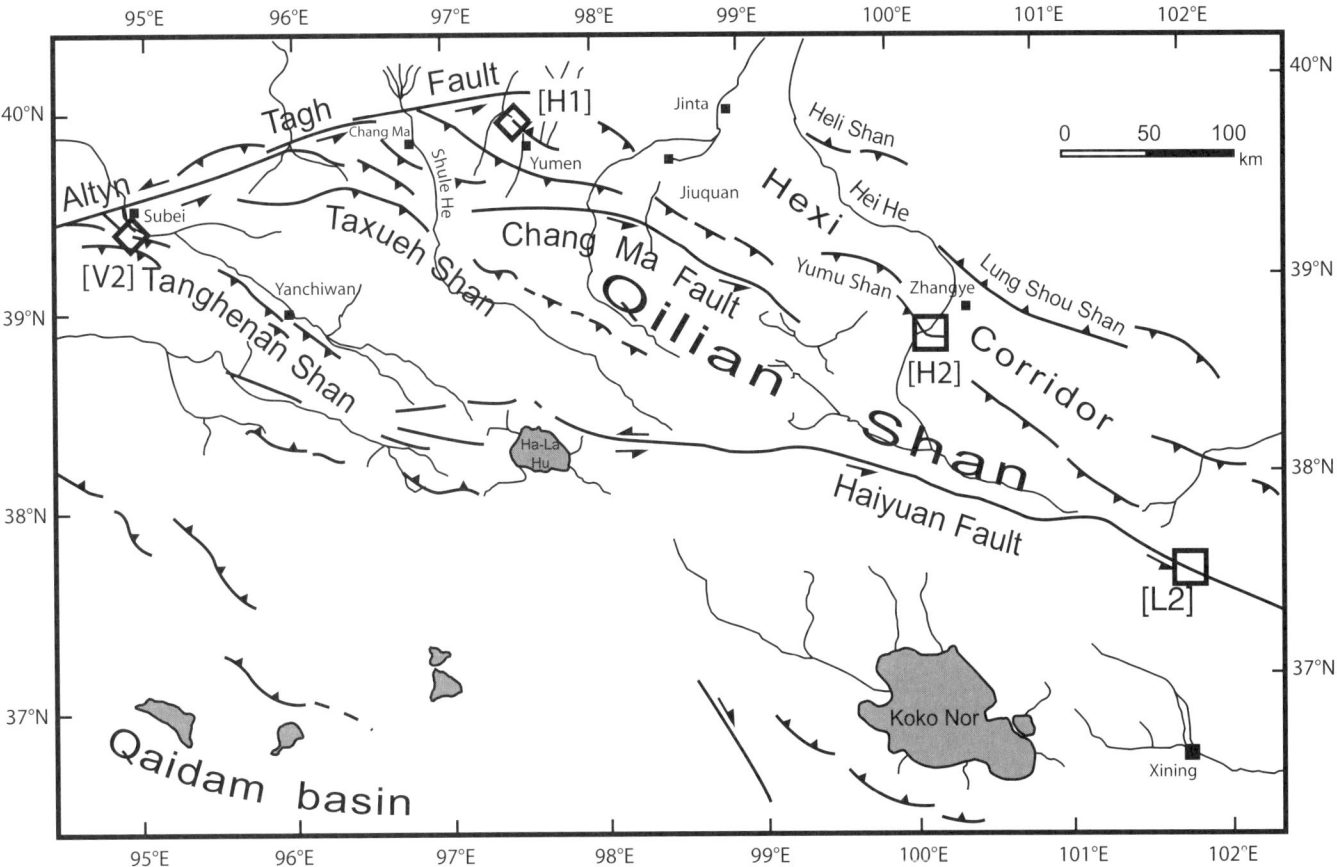

Figure 18. Seismotectonic map of northeast corner of Tibet modified from Tapponnier et al. (1990) and Meyer et al. (1998). Note linear N70°E-striking left-lateral Altyn Tagh fault at western terminations of Qilian, Taxueh, and Tanghenan Shan. N120°E-striking Tanghenan Shan veers counterclockwise to WSW before meeting with Altyn Tagh fault west of Subei. About five earthquakes with magnitude between 5 and 6.1 occurred along Tanghenan Shan since 1960. Fault plane solution for one event is from Molnar and Lyon Caen (1989). The nodal planes strike parallel to the range, compatible with north-vergent thrust. The boxes show the study areas of Van der Woerd et al. (2001) [V2], and Hetzel et al. (2002) [H1], Hetzel et al. (2004), [H2], and Lasserre et al. (2002) [L1].

thrust trace is displaced ~25 km northeastward by the sinistral Yema fault. Thrusting continues at least 150 km farther southeastward along the south side of the Yanchiwan basin (Meyer et al., 1998) (Fig. 18). All along the northern piedmont of the range, Neogene red beds, as well as the recent Quaternary terraces that unconformably cap them, are folded and uplifted by active north-vergent thrusts.

The rates of uplift in the Tanghenan Shan have been constrained by ^{14}C dating of organic remains collected on different strath terraces (Van der Woerd et al., 2001). Most of the fans and terraces in the southern part of the Subei basin were emplaced after the Last Glacial Maximum, many of them during the Early Holocene Optimum (9–5 ka). At one site, the minimum vertical throw of 34 ± 2 m of a surface dated at 8411 ± 530 yr B.P. provides a minimum vertical uplift rate of 4.1 ± 0.5 mm/yr. The maximum plausible uplift, deduced from stream incision and fold geometry, of the oldest terrace surface (115 ± 15 m), whose probable age is 15–18 ka, places an upper bound on the uplift rate of 7 ± 2 mm/yr. The thrust geometry at depth, and the cumulative shortening (9–12 km) across the imbricate thrust-wedge, were deduced from balancing two sections logged in the field and are consistent with a shortening rate of ~5 mm/yr and an onset of thrusting along the south side of the Subei basin at ca. 3 ± 1 Ma.

While the junction between the Altyn Tagh fault and the Tanghenan Shan thrust is ~300 km west of the eastern terminus of the fault, the Yumen Shan thrust lies ~10 km east of the Qilian Shan at roughly the same longitude as the fault's eastern terminus. Unlike the Tanghenan Shan thrust, which bounds a major mountain range, the Yumen Shan thrust, with opposite vergence, does not dip under a high mountain range (Meyer et al., 1998). While the Yumen Shan thrust may not be directly connected to the Altyn Tagh fault, it is arguably the easternmost thrust that is spatially associated with the Altyn Tagh fault. Hetzel et al. (2002) dated five uplifted terraces (up to ~60 m above the active river) using ^{10}Be, ^{26}Al, and ^{21}Ne cosmogenic dating, and demonstrate that the terraces at Yumen Shan record a protracted history of deposition and uplift extending to MIS 6 with ages ranging from ca. 40 to 170 ka. The uplift rate obtained is bounded between 0.3

and 0.4 mm/yr and is consistent with constant uplift and shortening over the observation interval (Fig. 19).

Projected along the regional strike of the Qilian Shan, the Zhangye Shan thrust's position relative to the terminus of the Altyn Tagh fault is roughly equivalent to that of the Yumen Shan thrust (Fig. 18). However, unlike the Yumen Shan thrust, the Zhangye Shan thrust is north-vergent and lies ~250 km southeast of the terminus of the Altyn Tagh fault, roughly midway between the Altyn Tagh and Haiyuan faults (Fig. 18). Hetzel et al. (2004) used a combination of ^{10}Be surface exposure and OSL dating of alluvial gravels and overlying loess deposits, respectively, to constrain the ages of terraces uplifted by 55–60 m and 25–30 m in the hanging wall of the thrust. The highest terrace yielded an age of 90 ± 11 ka and the lower terrace 31 ± 5 ka, constraining the long-term uplift rate to 0.6–0.9 mm/yr. Assuming a dip of 40°–60°, they obtain a shortening rate of ~1 mm/yr.

While there are numerous additional thrusts between the Qaidam basin and the Gobi-Ala Shan platform that have not been quantitatively dated, the results from those described here are in reasonable agreement with existing GPS data for the region (Chen et al., 2000, 2004; Zhang et al., 2004). For instance, as observed by Hetzel et al. (2004), two GPS estimates of shortening between the Qaidam basin and Hexi corridor yield shortening rates of 5.4 ± 1.8 mm/yr (Chen et al., 2004) and 8.1 ± 2.9 mm/yr (Chen et al., 2000), consistent with the combined rates from the Tanghenan Shan, Zhangye Shan, and Yumen Shan (assuming a dip of 40°–60° for the Yumen Shan). However, the rates of shortening for the individual faults decrease to the east, as demonstrated by the rates for the north-vergent Tanghenan Shan and Zhangye Shan thrusts. Interestingly, Meyer et al. (1998) estimated a shortening rate of 2.1 ± 0.6 mm/yr for a thrust in the piedmont of the Taxueh Shan. Taken together with the shortening rates for the Tanghenan Shan and Zhangye Shan thrusts, the Taxueh Shan rate, though not confirmed by quantitative dating, would yield an eastwardly decreasing shortening rate consistent with models for the propagation of the Altyn Tagh fault and asymmetric growth of the plateau via associated thrusting (Métivier et al., 1998; Meyer et al., 1998).

"SYSTEMATIC" UNDERESTIMATION OF SLIP RATES: EFFECTS OF EROSION

Perhaps the most striking feature of the assembled slip-rate data for central Asia is the disparity between geomorphic and geodetic slip-rate estimates on the Altyn Tagh and Karakorum faults; geomorphic rates are at least a factor of three or more higher than geodetic measurements. "Systematic" overestimation of the rate by geomorphic methods has been called upon to explain this disparity, but the underlying reasons for this overestimation remain unexplained (cf. Wallace et al., 2004). The effects most likely to skew cosmic-ray exposure ages of depositional surfaces are (1) predepositional exposure to cosmic rays, (2) postdepositional contamination of surfaces by younger material, and (3) erosion.

If unaccounted for, then predepositional exposure ("inheritance") will result in an apparent surface exposure age that is older than the true age of abandonment. Inheritance can be constrained by subsurface sampling (cf. Anderson et al., 1996; Hancock et al., 1999; Mériaux et al., 2004; Perg et al., 2001; Repka et al., 1997), and the convergence of surface and subsurface cosmogenic ages and radiocarbon dates from young surfaces (younger than 15 ka) at Cherchen He indicates that this approach is valid for fluvial/alluvial surfaces in Tibet (Mériaux et al., 2004). Overestimating the surface age would result in underestimates of the true slip rate, and this is not what is observed in Asia.

Postdepositional contamination can produce ages that are younger than the true age of abandonment. The most likely mode of postdepositional surface contamination in the settings discussed above is reoccupation of an abandoned fluvial terrace due to flooding, which is most likely for young terraces that are not separated either vertically or laterally from the active stream. Such an example has been described along the Kunlun fault by Van der Woerd et al. (1998). A distinct cluster of young ages (ca. 250 yr B.P.) was observed on a terrace where the majority of the samples had ages of ca.1500 yr B.P. The surface was only 1.5 m above that active river. As this surface ages, the relative difference between the abandonment and "flood" components will decrease, having a smaller overall effect on the average age of the surface. Also, as the terrace is farther removed from the site of the active river by tectonic activity and incision, it is less likely to be contaminated. More generally, subsurface sampling eliminates this effect, and in most cases, the subsurface cosmogenic

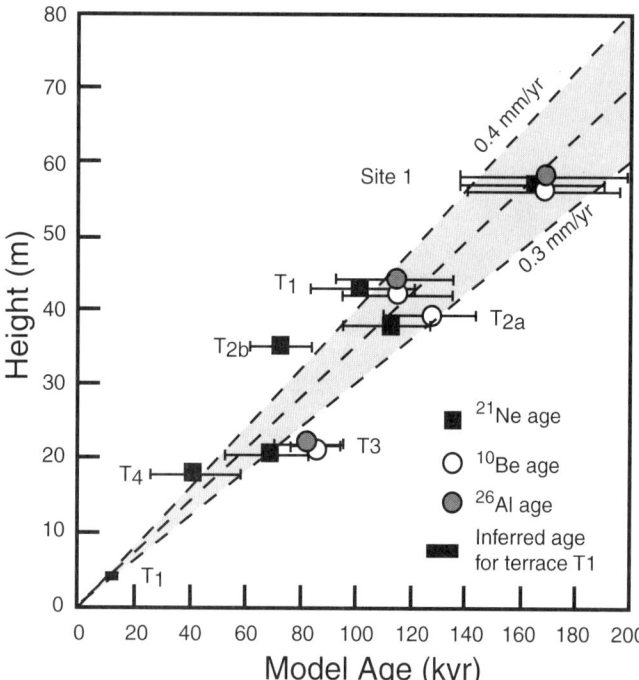

Figure 19. Plot of height above baseline versus model ages of the different terraces. Error bars indicate 2σ values (Hetzel et al., 2002). Three dashed lines indicate constant rates of river incision and tectonic uplift of 0.3, 0.35, and 0.4 mm/yr.

ages and radiocarbon data agree with the youngest surface ages (Mériaux et al., 2004, 2005).

Erosion does have the potential to decrease the apparent age of the surface, and the ages reported above are generally "model ages" that assume zero erosion rate. The effect of erosion on model ages becomes important when the amount of material removed approaches the depth at which the nuclide production rate drops to ~50% of the surface value (~50 cm). As a result, the effect of erosion on model ages is most pronounced for older surfaces (ca. 100 ka; Fig. 20). For instance, an erosion rate of 8 m/m.y. would yield a zero erosion rate model age of ca. 66 ka for a sample with a true exposure age of ca. 100 ka. Similarly, a rate of 6 m/m.y. would reduce the model age of a ca. 250 ka sample to ca. 110 ka. These erosion rates are similar to the maximum mean, alpine *bedrock* summit erosion rate, 7.6 ± 3.9 m/m.y., determined for four ranges in the western United States (Small et al., 1997). Model ages for younger samples are less perturbed simply because they haven't existed long enough to remove the requisite amount of material under reasonable conditions of erosion. Nevertheless, factor of two variations in age can be produced if the effect of erosion is ignored for older samples.

In theory, the ^{26}Al/^{10}Be ratio can be used to constrain erosion rates. However, this is only the case for very old samples with negligible erosion rates (Bierman and Caffee, 2001; Nishiizumi et al., 1991). Erosion has an effect similar to increasing the decay constant, and samples experiencing rapid erosion reach steady state before the ^{26}Al/^{10}Be ratio can deviate significantly from the production ratio. As a result, erosion rates are difficult to constrain using only ^{10}Be-^{26}Al systematics. In the absence of a direct method of assessing the effects of erosion on model ages, we appeal to other observations: (1) agreement with radiocarbon ages, (2) correlations between climate history and landscape formation, and (3) constancy of observed slip rates with time.

The agreement between cosmogenic dates and radiocarbon data has been established in a number of investigations in central Asia. For instance, Van der Woerd et al. (2002b) have shown that the same slip rate is obtained for the Kunlun fault using either ^{14}C or ^{10}Be-^{26}Al ages. At Cherchen He and Aksay, radiocarbon and ^{10}Be ages for the same surface are in good agreement (Mériaux et al., 2004, 2005). These surfaces are typically younger than ca. 15 ka, and erosion is not expected to be a major contributor, controlling the cosmogenic nuclide concentration. The comparison is ultimately limited in applicability by the temporal range of the radiocarbon method.

It is generally appreciated that landscape evolution is modulated by climate change (e.g., Bull, 1991), and that the ages of landscape features should be generally correlative with local and global climate history. The majority of terraces dated in central Asia postdate the Last Glacial Maximum (Mériaux et al., 2005). While fewer older surfaces have been dated, those that have appear to correlate with climate records. For instance, at Manikala Valley on the Karakorum fault, the bulk of ages older than 100 ka correlate with the cool periods of MIS 6 and MIS 5e (Fig. 16), and have been interpreted to indicate the onset of moraine emplacement during MIS 6 and retreat at the end of MIS 5e. Similarly, the major glacial features at the Sulamu Tagh site on the central Altyn Tagh fault also appear to have been abandoned during glacial retreat at the end of MIS 5e (Mériaux et al., 2004). Fluvial terrace emplacement and abandonment at Yumen Shan may also correlate with climate, with the two oldest terraces yielding model ages that are arguably consistent with warming trends during MIS 6 (ca. 170 ka) and the strong warming trend at the end of MIS 5e (ca. 120 ka). The presence of MIS 5e and MIS 6 landscape formation events suggests synchronous landscape evolution spanning the plateau from the northern boundary of the Himalayas to the Qilian Shan. If these model ages were reduced by erosion and the true age of these samples was significantly older, representing older glacial cycles, then we would have to explain the apparent, continent-wide absence of MIS 5e and MIS 6 landscape features. Hence, the correlation suggests that the ages of the landscape features used to constrain geomorphic slip rates have not been significantly affected by erosion.

The linearity of the offset versus model age data for both strike-slip and thrust faults (Figs. 7, 9, 13, 17, 19, and 21) in itself constitutes a strong argument in favor of low erosion rates with negligible effect on the model ages. Over timescales of 10^5 years, it is expected that, in general, large active strike-slip faults should

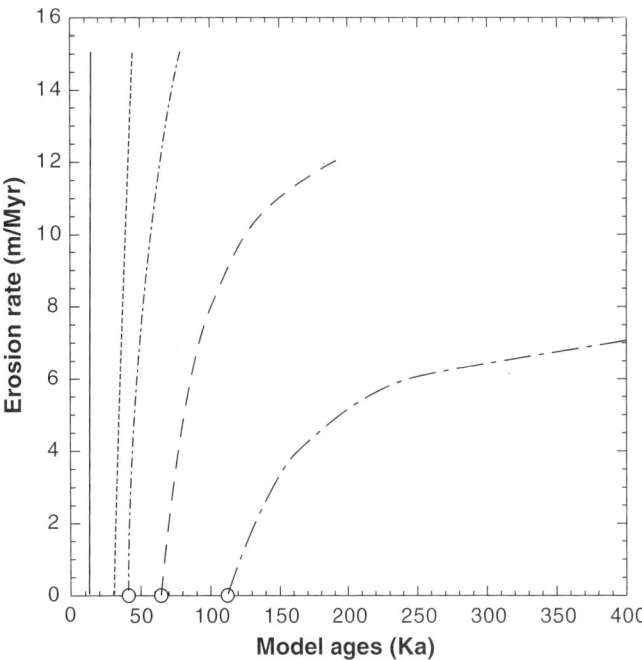

Figure 20. Effects of erosion on model ages. The zero erosion rate model ages for samples with ^{10}Be concentrations similar to those from the Sulamu Tagh site along the central Altyn Tagh fault (Mériaux et al., 2004) are shown along the abscissa. The curves indicate the true exposure age of a sample with these model ages for different erosion rates. Samples with model ages younger than ca. 60 ka are not significantly affected by erosion, but model ages for older samples can be significantly perturbed.

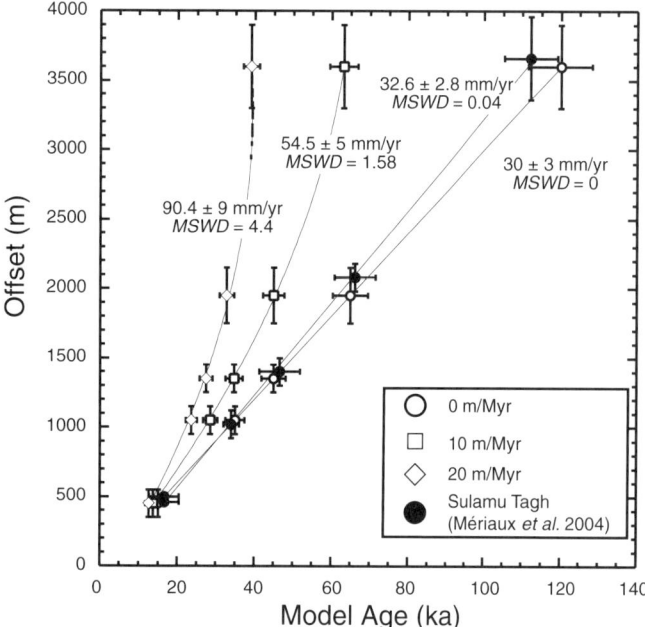

Figure 21. Effects of erosion rate on the apparent slip rate for a fault with a constant slip rate over time. For a fixed set of offset measurements, we assume a slip rate of 30 mm/yr with offsets, ages, and errors similar to those observed at Sulamu Tagh (Mériaux et al., 2004). For the 10 and 20 m/m.y. erosion rate models, we assume that the age is the same as in the zero erosion rate model, but recalculate the nuclide abundance for the different erosion rates. A "zero erosion rate model age" is then derived from these "erosionally decreased" nuclide concentrations. The zero erosion rate model ages generated for different erosion rates are then plotted against the observed offsets. The effect of erosion in decreasing the model ages is most apparent for the oldest samples, and as erosion rate increases, the slip-rate plots depart from linearity and the apparent slip rate increases dramatically. MSWD—mean square of weighted deviation.

slip at a constant rate. For instance, the slip rate on the Kunlun fault has been shown to remain constant over the last 40,000 yr (Van der Woerd et al., 2000, 2002b). Importantly, the slip rate obtained from the largest offset on the Kunlun fault (~400 m) was constrained by radiocarbon dating and, hence, not influenced by erosion. If erosion had played a significant role, the offset versus model age trends at high erosion rates should depart from linearity (Fig. 21). Rates obtained from larger, older offsets would be higher than the true value, as the effect of erosion on model ages is greatest for older samples (Fig. 20); the magnitude of the departure would decrease with offset. Attempts to apply linear fits to offset versus "erosion-influenced" model age data will result in higher slip rates and decreasing "goodness of fit" (as evidenced by increasing MSWD = mean square of weighted deviation) as the erosion rate increases. For instance, assuming a set of offset and age uncertainties similar to those on the central Altyn Tagh fault (Mériaux et al., 2004), a constant slip rate of 30 mm/yr yields a MSWD of ~0. Erosion rates of 10 m/m.y. and 20 m/m.y. yield offset versus model age fits with MSWD = 1.58 and 4.4, respectively, and linear fits yield nonzero intercepts (Fig. 21). Linear regression of the six offset-model age pairs from the Sulamu Tagh site yields a slip rate of 32.6 ± 2.8 mm/yr with MSWD = 0.04 (Mériaux et al., 2004). Regression of the combined Sulamu Tagh and Cherchen He offset versus model age data (nine pairs) yields a slip rate of 31.7 ± 1.9 mm/yr with MSWD = 0.14 (Mériaux et al., 2004). The clear linearity of these trends is a strong argument against significant erosional effects on the model ages and supports these model age–derived slip-rate estimates.

Based on the analysis above, we contend that the ca. 0–250 ka ages of depositional features used in constraining geomorphic slip-rate determinations in central Asia have not been significantly affected by erosion. This is somewhat surprising. Had these samples experienced high alpine bedrock erosion rates (Small et al., 1997), a model age–true age disparity should have been observed. The obvious conclusion is that these samples simply have not experienced such high rates, and/or do not erode like bedrock surfaces. In support of the former, the sites studied are clearly sites of deposition, not degradation, so the expectation is that erosion is not important. To the extent that they do occur, the processes of surface degradation for alluvium may be different from those that control bedrock erosion over the relevant timescale. Two surface degradation processes that clearly do occur are (1) the spallation of the surface pebbles, cobbles, and boulders that form fresh the alluvial surfaces, and (2) headward stream erosion. For instance, at Cherchen He, a section through the ~6000 yr old terrace shows that it is composed of pebbles with diameters of ~50 cm, while the surface is composed of shattered pebbles, sand, and minor loess (Fig. 22). Pebbles with sizes similar to those observed in the subsurface are scarce. As radiocarbon ages and cosmogenic ages agree, this process of initial degradation does not seem to have affected the surface chronology within current resolution. As this process happens in less than 6000 yr, it would be even more difficult to resolve for older samples, and perhaps represents the effective starting surface morphology for cosmic-ray exposure. The other process that does occur is headward erosion of ephemeral streams and rills. This process can be seen along the edges of the Manikala moraines (Fig. 14). If samples are taken on portions of the surface that are not currently experiencing this process, then it should have no effect on the ages obtained. Neither spallation of surface pebbles (at least after the initial period of surface smoothing) nor headward stream erosion need produce the same effect as erosion, which is to take samples from depth and bring them closer to the surface, thereby underestimating the time-integrated nuclide production rate.

CONCLUSIONS

The high elevation of the Tibetan Plateau along with its generally arid environment results in high cosmogenic nuclide production rates and preservation of millennial offsets. The close association of high topography and tectonic activity has also resulted in climatically generated landscape features along active faults, providing a variety of offset markers. Together these features make central Asia an ideal place for the application of tectonic morphochronology using cosmic-ray exposure dating. Slip rates on both strike-slip and thrust faults have been quantified in the time interval 2–250,000 yr, realizing the extended temporal range of this technique in extending the observation interval relative to radiocarbon dating. Where data permit, comparison of radiocarbon and cosmogenic dating generally agree and support the assertion that erosion, inheritance, and postdepositional contamination have only negligible effects on the surface age determination. Subsurface dating and multiple age determinations on surface samples are required to support this contention.

A striking feature of the relation between offset *and* age is its linearity, an observation that supports constant slip rates over the observed interval. Equally striking are the relatively low geodetic rates relative to the geomorphic slip-rate determinations for the Altyn Tagh and Karakorum faults. Data from other sources suggest that faults may exhibit temporal variations in slip velocity at timescales shorter than sampled by geomorphic measurements (cf. Ambraseys, 1971; Bennett et al., 2004; Peltzer et al., 2000; Weldon et al., 2004). Like the North and East Anatolian faults, the Altyn Tagh and Karakorum faults represent a conjugate fault pair on a continental scale. If the disparate geodetic and geomorphic rates were the result of transient strain accumulation, as has been suggested for the conjugate Garlock and Blackwater faults in southern California (Peltzer et al., 2000), then one fault should have a geodetic rate greater than the millennial rate. Secular variations in slip rate produced by such a mechanism are not consistent with the present data (both the Karakorum fault and Altyn Tagh fault geomorphic rates are higher than geodetic), arguing against this mechanism as the source of this disagreement. The geodetic and geomorphic rates for the Kunlun fault are in reasonable agreement, however, suggesting that there is no fundamental reason for disagreement between these methods in spite of very different observation intervals. Similarly, the shortening rates obtained by geodetic and geomorphic measurements across the Qilian Shan are also in good agreement. Where large disparities exist, there appears to be no easy path to reconciliation other than continuing investigations, including the analysis of additional geomorphic sites along these faults, better control of tropospheric effects on InSAR- and GPS-derived rates, and deployment of permanent GPS stations sufficiently far from the faults to eliminate contamination due to interseismic strain.

The disparity between geodetic and geomorphic rates notwithstanding, the geomorphic rates in northern Tibet support kinematic models that invoke block rheology and strain localization within the continental lithosphere (Peltzer and Tapponnier, 1988) coupled with intracontinental subduction and asymmetric growth of the northeastern edge of the plateau (Meyer et al., 1998; Tapponnier, 2001). For instance, the ~25 mm/yr rate on the central Altyn Tagh fault (Mériaux et al., 2004) is consistent with that required by block kinematic models (Avouac and Tapponnier, 1993). The eastward decrease in the Altyn Tagh fault slip rate between Tura (Mériaux et al., 2004) and Aksay (Mériaux et al., 2005) along with the eastwardly decreasing rates of shortening on thrust faults near the terminus of the Altyn Tagh fault (Hetzel et al., 2002, 2004; Van der Woerd et al., 2001) is consistent with the eastward propagation of the Altyn Tagh fault and the partitioning of slip onto genetically related thrust faults (Meyer et al., 1998; Tapponnier, 2001).

In spite of recent efforts (Brown et al., 2002b; Chevalier et al., 2005a), the southern boundary of Tibet remains largely unexplored with respect to geomorphic rate determinations, and assessment of these data in the context of the various kinematic models of Asian deformation is premature. For instance, the geomorphic rate of 10 mm/yr at the Manikala Valley site (Chevalier et al., 2005a) is still lower than the ~30 mm/yr required by block

Figure 22. Surface and subsurface samples at Cherchen He on the Altyn Tagh fault near Tura. The exposed active terrace riser (A) shows that the subsurface is composed of pebbles on the order of 50 cm in diameter and below. The surface (B) is largely composed of shattered pebbles, sand, and minor loess. In spite of the surface processing, radiocarbon, depth-corrected subsurface, and surface cosmogenic ages all agree.

kinematic models (Avouac and Tapponnier, 1993; Peltzer and Saucier, 1996). Whether this is the result of slip on other, as-yet uninvestigated, strands of the fault remains to be established, as are potential along-strike variations in velocity. In a similar connection, the rate of motion along the Karakorum-Jiali fracture zone has not been quantitatively constrained by cosmogenic dating. The role of the Karakorum-Jiali fracture zone in transferring slip from the Karakorum fault to strike-slip faults in eastern Asia, e.g., the Jiali fault, is a critical element of rigid block models of Asian deformation. Given the distributed nature of these features and the more subdued topography, acquisition of these data presents a formidable challenge.

ACKNOWLEDGMENTS

This work was performed under the auspices of the U.S. Department of Energy by University of California Lawrence Livermore National Laboratory under contract W-7405-Eng-48 under the sponsorship of the Laboratory Directed Research and Development program (report no. UCRL-BOOK-209322). The work was also supported by Institut National des Sciences de l'Univers, Centre National de la Recherche Scientifique (Paris, France), through programs Imagerie et Dynamique de la Lithosphère and Intérieur de la Terre, and by the China Earthquake Administration and the Ministry of Lands and Resources (Beijing, China).

REFERENCES CITED

Ambraseys, N.N., 1971, Value of historical record of earthquakes: Nature, v. 232, p. 375–379, doi: 10.1038/232375a0.

Anderson, R.S., Repka, J.L., and Dick, G.S., 1996, Explicit treatment of inheritance in dating depositional surfaces using in situ ^{10}Be and ^{26}Al: Geology, v. 24, p. 47–51, doi: 10.1130/0091-7613(1996)024<0047:ETOIID>2.3.CO;2.

Armijo, R., Tapponnier, P., and Tonglin, H., 1989, Late Cenozoic right-lateral strike-slip faulting in southern Tibet: Journal of Geophysical Research, B, Solid Earth and Planets, v. 94, no. B3, p. 2787–2838.

Avouac, J.P., and Peltzer, G., 1993, Active tectonics in southern Xinjiang, China—Analysis of terrace riser and normal-fault scarp degradation along the Hotan-Qira fault system: Journal of Geophysical Research, B, Solid Earth and Planets, v. 98, no. B12, p. 21,773–21,807.

Avouac, J.P., and Tapponnier, P., 1992, Kinematic model of active deformation in central Asia: Comptes Rendus de l'Academie des Sciences Serie Ii, v. 315, no. 13, p. 1791–1798.

Avouac, J.P., and Tapponnier, P., 1993, Kinematic model of active deformation in central Asia: Geophysical Research Letters, v. 20, no. 10, p. 895–898.

Banerjee, P., and Burgmann, R., 2002, Convergence across the northwest Himalaya from GPS measurements: Geophysical Research Letters, v. 29, no. 13, Art. No. 1652.

Bendick, R., Bilham, R., Freymueller, J., Larson, K., and Yin, G.H., 2000, Geodetic evidence for a low slip rate in the Altyn Tagh fault system: Nature, v. 404, no. 6773, p. 69–72, doi: 10.1038/35003555.

Bennett, R.A., Friedrich, A.M., and Furlong, K.P., 2004, Codepent histories of the San Andreas and San Jacinto fault zones from inversion of fault displacement rates: Geology, v. 32, p. 961–964, doi: 10.1130/G20806.1.

Bierman, P.R., and Caffee, M.W., 2001, Slow rates of rock surface erosion and sediment production across the Namib Desert and escarpment, southern Africa: American Journal of Science, v. 301, p. 326–358.

Bilham, R., Larson, K., Freymueller, J., Jouanne, F., LeFort, P., Leturmy, P., Mugnier, J.L., Gamond, J.F., Glot, J.P., Martinod, J., Chaudury, N.L., Chitrakar, G.R., Gautam, U.P., Koirala, B.P., Pandey, M.R., Ranabhat, R., Sapkota, S.N., Shrestha, P.L., Thakuri, M.C., Timilsina, U.R., Tiwari, D.R., Vidal, G., Vigny, C., Galy, A., and deVoogd, B., 1997, GPS measurements of present-day convergence across the Nepal Himalaya: Nature, v. 386, no. 6620, p. 61–64, doi: 10.1038/386061a0.

Brown, E.T., Bendick, R., Bourles, D., Gaur, V., Molnar, P., Raisbeck, G.M., and Yiou, F., 2002a, Slip rates of the Karakorum fault, Ladakh, India, determined using cosmic ray exposure dating of debris flows and moraines: Journal of Geophysical Research, B, Solid Earth and Planets, v. 107, Art. No. 2192.

Brown, E.T., Bendick, R., Bourles, D.L., Gaur, V., Molnar, P., Raisbeck, G.M., and Yiou, F., 2002b, Slip rates of the Karakorum fault, Ladakh, India, determined using cosmic ray exposure dating of debris flows and moraines: Journal of Geophysical Research, B, Solid Earth and Planets, v. 107, no. B9.

Brown, E.T., Molnar, P., and Bourles, D.L., 2005, Comment on "Slip-rate measurements on the Karakorum fault may imply secular variations in fault motion": Science, v. 309, no. 5739, p. 1326, doi: 10.1126/science.1112508.

Bull, W.B., 1991, Geomorphic responses to climatic change: New York, Oxford University Press, 326 p.

Burchfiel, B.C., King, R.W., Royden, L.H., Wang, E., Chen, Z., Zhang, X., and Zhao, J., 1998, GPS results from the entire eastern part of the Tibetan Plateau and its adjacent foreland and their tectonic interpretation: Eos (Transactions, American Geophysical Union), v. 79, p. 45.

Chen, Q.Z., Freymueller, J.T., Wang, Q., Yang, Z.Q., Xu, C.J., and Liu, J.N., 2004, A deforming block model for the present-day tectonics of Tibet: Journal of Geophysical Research, B, Solid Earth and Planets, v. 109, no. B1.

Chen, Z., Burchfiel, B.C., Liu, Y., King, R.W., Royden, L.H., Tang, W., Wang, E., Zhao, J., and Zhang, X., 2000, Global positioning system measurements from eastern Tibet and their implications for India/Eurasia intercontinental deformation: Journal of Geophysical Research, B, Solid Earth and Planets, v. 105, no. B7, p. 16,215–16,227, doi: 10.1029/2000JB900092.

Chevalier, M.-L., Ryerson, F.J., Tapponnier, P., Finkel, R.C., Van der Woerd, J., Haibing, L., and Qing, L., 2005a, Slip-rates measurements on the Karakorum fault may imply secular variations in fault motion: Science, v. 307, p. 411–414, doi: 10.1126/science.1105466.

Chevalier, M.L., Ryerson, F.J., Tapponnier, P., Finkel, R.C., Van Der Woerd, J., Li, H.B., and Liu, Q., 2005b, Response to comment on "Slip-rate measurements on the Karakorum fault may imply secular variations in fault motion": Science, v. 309, no. 5739, p. 1326, doi: 10.1126/science.1112629.

Cowgill, E., Yin, A., Harrison, T.M., and Wang, X.F., 2003, Reconstruction of the Altyn Tagh fault based on U-Pb geochronology: Role of back thrusts, mantle sutures, and heterogeneous crustal strength in forming the Tibetan Plateau: Journal of Geophysical Research, B, Solid Earth and Planets, v. 108, no. B7.

Cui, Z., and Yang, B., 1979, On the Tuouohu-Maqu active fault zone: Northwestern Seismological Journal, v. 1, p. 57–61.

Deng, Q.D., 1986, Variations in the geometry and amount of slip on the Haiyuan (Nanxihaushan) fault zone, China, and the surface rupture of the 1920 Haiyuan earthquake: Earthquake source mechanics, in Das, S., Boatwright, J, and Scholz, C.H., eds.: Washington, D.C., American Geophysical Union, Geophysical Monograph 37, p. 169–182.

England, P., and Houseman, G., 1986, Finite strain calculations of continental deformation, 2: Comparison with the India-Asia collision zone: Journal of Geophysical Research, B, Solid Earth and Planets, v. 91, no. B3, p. 3664–3676.

England, P., and Molnar, P., 1997a, Active deformation of Asia: From kinematics to dynamics: Science, v. 278, no. 5338, p. 647–650, doi: 10.1126/science.278.5338.647.

England, P., and Molnar, P., 1997b, The field of crustal velocity in Asia calculated from Quaternary rates of slip on faults: Geophysical Journal International, v. 130, no. 3, p. 551–582.

Gaudemer, Y., Tapponnier, P., Meyer, B., Peltzer, G., Guo, S.M., Chen, Z.T., Dai, H.G., and Cifuentes, I., 1995, Partitioning of crustal slip between linked, active faults in the eastern Qilian Shan, and evidence for a major seismic gap, the Tianzhu Gap, on the western Haiyuan fault, Gansu (China): Geophysical Journal International, v. 120, no. 3, p. 599–645.

Ge, S., Shen, G., Wei, R., Ding, G., and Wang, Y., 1992, Active Altun fault zone monograph: Beijing, China, State Seismological Bureau of China, 319 p.

Gehrels, G.E., Yin, A., and Wang, X.F., 2003, Detrital-zircon geochronology of the northeastern Tibetan Plateau: Geological Society of America Bulletin, v. 115, p. 881–896, doi: 10.1130/0016-7606(2003)115<0881:DGOTNT>2.0.CO;2.

Gilley, L.D., Harrison, T.M., Leloup, P.H., Ryerson, F.J., Lovera, O.M., and Wang, J.H., 2003, Direct dating of left-lateral deformation along the Red

River shear zone, China and Vietnam: Journal of Geophysical Research, v. 108, p. 2127, doi: 10.1029/2001JB001726.
Gu, G., Lin, Tinghaung, and Shi, Zhenliang, 1989, Catalogue of Chinese earthquakes (1831 BC–1969 AD): Beijing, China, Science Press, 872 p.
Hancock, G.S., Anderson, R.S., Chadwick, O.A., and Finkel, R.C., 1999, Dating fluvial terraces with Be-10 and Al-26 profiles: Application to the Wind River, Wyoming: Geomorphology, v. 27, no. 1-2, p. 41–60, doi: 10.1016/S0169-555X(98)00089-0.
Herquel, G., Tapponnier, P., Wittlinger, G., Mei, J., and Danian, S., 1999, Teleseismic shear wave splitting and lithospheric beneath and across the Altyn Tagh fault: Geophysical Research Letters, v. 26, no. 21, p. 3225–3228, doi: 10.1029/1999GL005387.
Hetzel, R., Niedermann, S., Tao, M., Kubik, P.W., Ivy-Ochs, S., Gao, B., and Strecker, M.R., 2002, Low slip rates and long-term preservation of geomorphic features in central Asia: Nature, v. 417, no. 6887, p. 428–432, doi: 10.1038/417428a.
Hetzel, R., Tao, M.X., Stokes, S., Niedermann, S., Ivy-Ochs, S., Gao, B., Strecker, M.R., and Kubik, P.W., 2004, Late Pleistocene/Holocene slip rate of the Zhangye thrust (Qilian Shan, China) and implications for the active growth of the northeastern Tibetan Plateau: Tectonics, v. 23, no. 6, article TC6006, doi: 10.1029/2004TC001653.
Imbrie, J., Hays, J.D., Martinson, D.G., McIntyre, A., Mix, A.C., Morley, J.J., Pisias, N.G., Prell, W.L., and Shackleton, N.J., 1984, The orbital theory of Pleistocene climate: Support from a revised chronology of the marine delta ^{18}O record, in Berger, A., et al., eds., Milankovitch and climate, Part I: Boston, Reidel, p. 269–305.
Kidd, W.S.F., and Molnar, P., 1988, Quaternary and active faulting observed on the 1985 Academia-Sinica Royal Society geotraverse of Tibet: Royal Society of London Philosophical Transactions, ser. A, Mathematical Physical and Engineering Sciences, v. 327, no. 1594, p. 337–363.
Lacassin, R., Valli, F., Arnaud, N., Leloup, P.H., Paquette, J.L., Haibing, L., Tapponnier, P., Chevalier, M.L., Guillot, S., Maheo, G., and Xu, Z.Q., 2004, Large-scale geometry, offset and kinematic evolution of the Karakorum fault, Tibet: Earth and Planetary Science Letters, v. 219, no. 3-4, p. 255–269, doi: 10.1016/S0012-821X(04)00006-8.
Larson, K.M., Burgmann, R., Bilham, R., and Freymueller, J.T., 1999, Kinematics of the India-Eurasia collision zone from GPS measurements: Journal of Geophysical Research, B, Solid Earth and Planets, v. 104, no. B1, p. 1077–1093, doi: 10.1029/1998JB900043.
Lasserre, C., Morel, P.H., Gaudemer, Y., Tapponnier, P., Ryerson, F.J., King, G.C.P., Metivier, F., Kasser, M., Kashgarian, M., Baichi, L., Taiya, L., and Daoyang, Y., 1999, Postglacial left slip rate and past occurrence of M ≥ 8 earthquakes on the western Haiyuan fault, Gansu, China: Journal of Geophysical Research, B, Solid Earth and Planets, v. 104, no. B8, p. 17,633–17,651, doi: 10.1029/1998JB900082.
Lasserre, C., Gaudemer, Y., Tapponnier, P., Meriaux, A.S., Van der Woerd, J., Yuan, D.Y., Ryerson, F.J., Finkel, R.C., and Caffee, M.W., 2002, Fast late Pleistocene slip rate on the Leng Long Ling segment of the Haiyuan fault, Qinghai, China: Journal of Geophysical Research, B, Solid Earth and Planets, v. 107, no. B11, Art. No. 2276.
Li, L., and Jia, Y., 1981, Characteristics of the deformation band of the 1937 Tuosuohu earthquake (M 7.5) in Qinhai: Northwestern Seismological Journal, v. 3, p. 61–65.
Liu, Q., 1993, Paléoclimat et Contraintes Chronologiques sur les Mouvemonts Récents dans l'Ouest du Tibet: Failles du Karakourm et de Longmu Co-Gozha Co, Lacs en Pull-Apart de Loungmu Co et de Sumxi Co [Ph.D. thesis]: Paris, L'Universite Paris VII, 360 p.
Massonnet, D., Rossi, M., Carmona, C., Adragna, F., Peltzer, G., Feigl, K., and Rabaute, T., 1993, The displacement field of the Landers earthquake mapped by radar interferometry: Nature, v. 364, no. 6433, p. 138–142, doi: 10.1038/364138a0.
Mériaux, A., Ryerson, F.J., Tapponnier, P., Van der Woerd, J., Finkel, R., Xu, X., Xu, Z., and Caffee, M.W., 2004, Rapid slip along the central Altyn Tagh fault: Morphochronologic evidence from Cherchen He and Sulamu Tagh: Journal of Geophysical Research, v. 109, B06401, doi: 10.1029/2003JB002558.
Mériaux, A.S., Tapponnier, P., Ryerson, F.J., Xu, X.W., King, G., Van der Woerd, J., Finkel, R.C., Li, H.B., Caffee, M.W., Xu, Z.Q., and Chen, W.B., 2005, The Aksay segment of the northern Altyn Tagh fault: Tectonic geomorphology, landscape evolution, and Holocene slip rate: Journal of Geophysical Research, B, Solid Earth and Planets, v. 110, no. B4.
Métivier, F., Gaudemer, Y., Tapponnier, P., and Meyer, B., 1998, Northeastward growth of the Tibet Plateau deduced from balanced reconstruction of two depositional areas: The Qaidam and Hexi Corridor basins, China: Tectonics, v. 17, no. 6, p. 823–842, doi: 10.1029/98TC02764.
Meyer, B., 1991, Mécanismes des grands tremblements de terre et du raccourcissement crustal oblique au bord nord-est du Tibet [Ph.D. thesis]: Paris, Université de Paris VII.
Meyer, B., Tapponnier, P., Gaudemer, Y., Peltzer, G., Guo, S.M., and Chen, Z.T., 1996, Rate of left-lateral movement along the easternmost segment of the Altyn Tagh fault, east of 96 degrees E (China): Geophysical Journal International, v. 124, no. 1, p. 29–44.
Meyer, B., Tapponnier, P., Bourjot, L., Metivier, F., Gaudemer, Y., Peltzer, G., Shunmin, G., and Zhitai, C., 1998, Crustal thickening in Gansu-Qinghai, lithospheric mantle subduction, and oblique, strike-slip controlled growth of the Tibet Plateau: Geophysical Journal International, v. 135, no. 1, p. 1–47, doi: 10.1046/j.1365-246X.1998.00567.x.
Molnar, P., and Lyon Caen, H., 1989, Fault plane solutions of earthquakes and active tectonics of the Tibetan Plateau and its margins: Geophysical Journal International, v. 99, no. 1, p. 123–153.
Murphy, M.A., Yin, A., Kapp, P., Harrison, T.M., Manning, C.E., Ryerson, F.J., Ding, L., and Guo, J.H., 2002, Structural evolution of the Gurla Mandhata detachment system, southwest Tibet: Implications for the eastward extent of the Karakoram fault system: Geological Society of America Bulletin, v. 114, p. 428, doi: 10.1130/0016-7606(2002)114<0428:SEOTGM>2.0.CO;2.
Nishiizumi, K., Kohl, C.P., Arnold, J.R., Klein, J., Fink, D., and Middleton, R., 1991, Cosmic ray produced ^{10}Be and ^{26}Al in Antarctic rocks: Exposure and erosion history: Earth and Planetary Science Letters, v. 104, p. 440–454, doi: 10.1016/0012-821X(91)90221-3.
Peltzer, G., and Rosen, P., 1995, Surface displacements of the 17 May 1993 Eureka Valley, California, earthquake observed by SAR interferometry: Science, v. 268, no. 5215, p. 1333–1336.
Peltzer, G., and Saucier, F., 1996, Present-day kinematics of Asia derived from geologic fault rates: Journal of Geophysical Research, B, Solid Earth and Planets, v. 101, no. B12, p. 27,943–27,956, doi: 10.1029/96JB02698.
Peltzer, G., and Tapponnier, P., 1988, Formation and evolution of strike-slip faults, rifts, and basins during the India-Asia collision—An experimental approach: Journal of Geophysical Research, B, Solid Earth and Planets, v. 93, no. B12, p. 15,085–15,117.
Peltzer, G., Tapponnier, P., Gaudemer, Y., Meyer, B., Guo, S.M., Yin, K.L., Chen, Z.T., and Dai, H.G., 1988, Offsets of late Quaternary morphology, rate of slip, and recurrence of large earthquakes on the Chang Ma fault (Gansu, China): Journal of Geophysical Research, B, Solid Earth and Planets, v. 93, no. B7, p. 7793–7812.
Peltzer, G., Tapponnier, P., and Armijo, R., 1989, Magnitude of late Quaternary left-lateral displacements along the north edge of Tibet: Science, v. 246, p. 1285–1289.
Peltzer, G., Hudnut, K.W., and Feigl, K.L., 1994, Analysis of coseismic surface displacement gradients using radar interferometry—New insights into the Landers earthquake: Journal of Geophysical Research, B, Solid Earth and Planets, v. 99, no. B11, p. 21,971–21,981, doi: 10.1029/94JB01888.
Peltzer, G., Crampe, F., and King, G., 1999, Evidence of nonlinear elasticity of the crust from the Mw 7.6 Manyi (Tibet) earthquake: Science, v. 286, no. 5438, p. 272–276, doi: 10.1126/science.286.5438.272.
Peltzer, G., Crampe, F., Hensley, S., and Rosen, P.A., 2000, Transient strain accumulation in the eastern California shear zone: Eos (Transactions, American Geophysical Union), v. 81, p. F1244.
Peltzer, G., Crampe, F., Hensley, S., and Rosen, P., 2001a, Transient strain accumulation and fault interaction in the eastern California shear zone: Geology, v. 29, p. 975–978, doi: 10.1130/0091-7613(2001)029<0975:TSAAFI>2.0.CO;2.
Peltzer, G., Crampe, F., and Rosen, P., 2001b, The Mw 7.1, Hector Mine, California earthquake: Surface rupture, surface displacement field, and fault slip solution from ERS SAR data: Comptes Rendus de l'Academie Des Sciences: Paris, Sciences De La Terre et des Planetes, v. 333, no. 9, p. 545–555.
Perg, L.A., Anderson, R.S., and Finkel, R.C., 2002, Use of a new Be-10 and Al-26 inventory method to date marine terraces, Santa Cruz, California, USA: Reply: Geology, v. 29, p. 879–882, doi: 10.1130/0091-7613(2001)029<0879:UOANBA>2.0.CO;2.
Perg, L.A., Anderson, R.S., and Finkel, R.C., 2003, Use of cosmogenic radionuclides as a sediment tracer in the Santa Cruz littoral cell, California, United States: Geology, v. 31, p. 299–302, doi: 10.1130/0091-7613(2003)031<0299:UOCRAA>2.0.CO;2.
Phillips, R.J., Parrish, R.R., and Searle, M.P., 2004, Age constraints on ductile deformation and long-term slip rates along the Karakoram fault zone,

Ladakh: Earth and Planetary Science Letters, v. 226, no. 3-4, p. 305–319, doi: 10.1016/j.epsl.2004.07.037.

Repka, J.L., Anderson, R.S., and Finkel, R.C., 1997, Cosmogenic dating of fluvial terraces, Fremont River, Utah: Earth and Planetary Science Letters, v. 152, no. 1-4, p. 59–73, doi: 10.1016/S0012-821X(97)00149-0.

Ritts, B.D., and Biffi, U., 2000, Magnitude of post–Middle Jurassic (Bajocian) displacement on the Altyn Tagh fault, northwest China: Geological Society of America Bulletin, v. 112, p. 61–74, doi: 10.1130/0016-7606(2000)112<0061:MOPMJB>2.3.CO;2.

Shen, F., Royden, L.H., and Burchfiel, B.C., 2001a, Large-scale crustal deformation of the Tibetan Plateau: Journal of Geophysical Research, B, Solid Earth and Planets, v. 106, no. B4, p. 6793–6816, doi: 10.1029/2000JB900389.

Shen, Z.K., Wang, M., Li, Y.X., Jackson, D.D., Yin, A., Dong, D.N., and Fang, P., 2001b, Crustal deformation along the Altyn Tagh fault system, western China, from GPS: Journal of Geophysical Research, B, Solid Earth and Planets, v. 106, no. B12, p. 30,607–30,621, doi: 10.1029/2001JB000349.

Sieh, K.E., and Jahns, R.H., 1984, Holocene activity of the San Andreas fault at Wallace Creek, California: Geological Society of America Bulletin, v. 95, p. 883–896, doi: 10.1130/0016-7606(1984)95<883:HAOTSA>2.0.CO;2.

Small, E.E., Anderson, R.S., Repka, J.L., and Finkel, R., 1997, Erosion rates of alpine bedrock summit surfaces deduced from in situ ^{10}Be and ^{26}Al: Earth and Planetary Science Letters v. 150, p. 413–425.

Sobel, E.R., Arnaud, N., Jolivet, M., Ritts, B.D., and Brunel, M., 2001, Jurassic to Cenozoic exhumation of the Altyn Tagh range, northwest China, constrained by $^{40}Ar/^{39}Ar$ and apatite fission track thermochronology, in Hendrix, M.S., and Davis, G.A., eds., Paleozoic and Mesozoic tectonic evolution of central Asia: From continental assembly to intracontinental deformation: Geological Society of America Memoir 194, p. 247–267.

Tapponnier, P., and Molnar, P., 1977, Active faulting and tectonics in China: Journal of Geophysical Research, v. 82, no. 20, p. 2905.

Tapponnier, P., and Molnar, P., 1979, Active faulting and Cenozoic tectonics of the Tien Shan, Mongolia, and Baykal regions: Journal of Geophysical Research, v. 84, no. B7, p. 3425.

Tapponnier, P., Meyer, B., Avouac, J.P., Peltzer, G., Gaudemer, Y., Guo, S.M., Xiang, H.F., Yin, K.L., Chen, Z.T., Cai, S.H., and Dai, H.G., 1990, Active thrusting and folding in the Qilian-Shan, and decoupling between upper crust and mantle in northeastern Tibet: Earth and Planetary Science Letters, v. 97, no. 3-4, p. 382–403, doi: 10.1016/0012-821X(90)90053-Z.

Tapponnier, P., Zhiqin, X., Roger, F., Meyer, B., Arnaud, N., Wittlinger, G., and Jingsui, Y., 2001, Oblique stepwise rise and growth of the Tibet Plateau: Science, v. 294, p. 1671–1677, doi: 10.1126/science.105978.

Van der Woerd, J.W., Ryerson, F.J., Tapponnier, P., Gaudemer, Y., Finkel, R., Meriaux, A.S., Caffee, M., Zhao, G.G., and He, Q.L., 1998, Holocene left-slip rate determined by cosmogenic surface dating on the Xidatan segment of the Kunlun fault (Qinghai, China): Geology, v. 26, p. 695–698, doi: 10.1130/0091-7613(1998)026<0695:HLSRDB>2.3.CO;2.

Van der Woerd, J., Ryerson, F.J., Tapponnier, P., Meriaux, A.S., Gaudemer, Y., Meyer, B., Finkel, R.C., Caffee, M.W., Zhao, G.G., and Xu, Z.Q., 2000, Uniform slip-rate along the Kunlun fault: Implications for seismic behaviour and large-scale tectonics: Geophysical Research Letters, v. 27, no. 16, p. 2353–2356, doi: 10.1029/1999GL011292.

Van der Woerd, J., Xu, X., Li, H.B., Tapponnier, P., Meyer, B., Ryerson, F.J., Meriaux, A.S., and Xu, Z.Q., 2001, Rapid active thrusting along the northwestern range front of the Tanghe Nan Shan (western Gansu, China): Journal of Geophysical Research, B, Solid Earth and Planets, v. 106, no. B12, p. 30,475–30,504, doi: 10.1029/2001JB000583.

Van der Woerd, J., Meriaux, A.-S., Klinger, Y., Ryerson, F.J., Gaudemer, Y., and Tapponnier, P., 2002a, The November 14th, 2001, Mw = 7.8 Kokoshili earthquake in northern Tibet (Qinghai Province, China): Seismological Research Letters, v. 73, p. 125–135.

Van der Woerd, J., Tapponnier, P., Ryerson, F.J., Meriaux, A.S., Meyer, B., Gaudemer, Y., Finkel, R.C., Caffee, M.W., Zhao, G.G., and Xu, Z.Q., 2002b, Uniform postglacial slip-rate along the central 600 km of the Kunlun fault (Tibet), from Al-26, Be-10, and C-14 dating of riser offsets, and climatic origin of the regional morphology: Geophysical Journal International, v. 148, no. 3, p. 356–388, doi: 10.1046/j.1365-246x.2002.01556.x.

Wallace, K., Yin, G.H., and Bilham, R., 2004, Inescapable slow slip on the Altyn Tagh fault: Geophysical Research Letters, v. 31, no. 9, doi: 10.1029/2004GL019724.

Wang, E., 1997, Displacement and timing along the northern strand of the Altyn Tagh fault zone, northern Tibet: Earth and Planetary Science Letters, v. 150, no. 1-2, p. 55–64.

Wang, Q., Zhang, P.Z., Freymueller, J.T., Bilham, R., Larson, K.M., Lai, X., You, X.Z., Niu, Z.J., Wu, J.C., Li, Y.X., Liu, J.N., Yang, Z.Q., and Chen, Q.Z., 2001, Present-day crustal deformation in China constrained by global positioning system measurements: Science, v. 294, no. 5542, p. 574–577, doi: 10.1126/science.1063647.

Wang, Q., Zhang, P.Z., Niu, Z.J., Freymueller, J.T., Lai, X., Li, Y.X., Zhu, W.Y., Liu, J.N., Bilham, R., and Larson, K.M., 2002, Present-day crustal movement and tectonic deformation in China continent: Science in China, ser. D, Earth Sciences, v. 45, no. 10, p. 865–874.

Washburn, Z., Arrowsmith, J.R., Forman, S.L., Cowgill, E., Wang, X.F., Zhang, Y.Q., and Chen, Z.L., 2001, Late Holocene earthquake history of the central Altyn Tagh fault, China: Geology, v. 29, p. 1051–1054, doi: 10.1130/0091-7613(2001)029<1051:LHEHOT>2.0.CO;2.

Weldon, R., Scharer, K., Fumal, T., and Biasi, G., 2004, Wrightwood and the earthquake cycle: What a long recurrence record tells us about how faults work: GSA Today, v. 14, no. 9, p. 4–10, doi: 10.1130/1052-5173(2004)014<4:WATECW>2.0.CO;2.

Weldon, R.J., and Sieh, K.E., 1985, Holocene rate of slip and tentative recurrence interval for large earthquakes on the San Andreas fault, Cajon Pass, southern California: Geological Society of America Bulletin, v. 96, p. 793–812, doi: 10.1130/0016-7606(1985)96<793:HROSAT>2.0.CO;2.

Wittlinger, G., Tapponnier, P., Poupinet, G., Mei, J., Danian, S., Herquel, G., and Masson, F., 1998, Tomographic evidence for localized lithospheric shear along the Altyn Tagh fault: Science, v. 282, no. 5386, p. 74–76, doi: 10.1126/science.282.5386.74.

Wittlinger, G., Vergne, J., Tapponnier, P., Farra, V., Poupinet, G., Jiang, M., Su, H., Herquel, G., and Paul, A., 2004, Teleseismic imaging of subducting lithosphere and Moho offsets beneath western Tibet: Earth and Planetary Science Letters, v. 221, no. 1-4, p. 117–130.

Wright, T.J., Parsons, B., England, P.C., and Fielding, E.J., 2004, InSAR observations of low slip rates on the major faults of western Tibet: Science, v. 305, no. 5681, p. 236–239, doi: 10.1126/science.1096388.

Xu, X.W., Tapponnier, P., Van Der Woerd, J., Ryerson, F.J., Wang, F., Zheng, R.Z., Chen, W.B., Ma, W.T., Yu, G.H., Chen, G.H., and Meriaux, A.S., 2005, Late Quaternary sinistral slip rate along the Altyn Tagh fault and its structural transformation model: Science in China, ser. D, Earth Sciences, v. 48, no. 3, p. 384–397.

Yang, J., Xu, Z., Zhang, J., Chu, C.-Y., Zhang, R., and Liou, J.G., 2001, Tectonic significance of early Paleozoic high-pressure rock in the Altun-Qaidam-Qilian Mountains, northwest China, in Hendrix, M.S., and Davis, G.A., eds., Paleozoic and Mesozoic tectonic evolution of central Asia: From continental assembly to intracontinental deformation: Geological Society of America Memoir 194, p. 151–170.

Yin, A., Harrison, T.M., Ryerson, F.J., Chen, W.J., Kidd, W.S.F., and Copeland, P., 1994, Tertiary structural evolution of the Gangdese thrust system, southeastern Tibet: Journal of Geophysical Research, B, Solid Earth and Planets, v. 99, no. B9, p. 18,175–18,201, doi: 10.1029/94JB00504.

Yin, A., Rumelhart, P.E., Butler, R., Cowgill, E., Harrison, T.M., Foster, D.A., Ingersoll, R.V., Zhang, Q., Zhou, X.Q., Wang, X.F., Hanson, A., and Raza, A., 2002, Tectonic history of the Altyn Tagh fault system in northern Tibet inferred from Cenozoic sedimentation: Geological Society of America Bulletin, v. 114, p. 1257–1295, doi: 10.1130/0016-7606(2002)114<1257:THOTAT>2.0.CO;2.

Yue, Y.J., Ritts, B.D., and Graham, S.A., 2001, Initiation and long-term slip history of the Altyn Tagh fault: International Geology Review, v. 43, no. 12, p. 1087–1093.

Zhang, P., Molnar, P., Burchfiel, B.C., Royden, L., Wang, Y., Deng, Q., and Song, F., 1988, Bounds on the Holocene slip-rate on the Haiyuan fault, north-central China: Quaternary Research, v. 30, p. 151–164, doi: 10.1016/0033-5894(88)90020-8.

Zhang, P.Z., Shen, Z., Wang, M., Gan, W.J., Bürgmann, R., and Molnar, P., 2004, Continuous deformation of the Tibetan Plateau from global positioning system data: Geology, v. 32, p. 809–812.

Zhang, W.Q., Jiao, D.C., Zhang, P.Z., Molnar, P., Burchfiel, B.C., Deng, Q.D., Wang, Y.P., and Song, F.M., 1987, Displacement along the Haiyuan fault associated with the Great 1920 Haiyuan, China, earthquake: Bulletin of the Seismological Society of America, v. 77, no. 1, p. 117–131.

MANUSCRIPT ACCEPTED BY THE SOCIETY 11 APRIL 2006

Geological Society of America
Special Paper 415
2006

Using in situ–produced ^{10}Be to quantify active tectonics in the Gurvan Bogd mountain range (Gobi-Altay, Mongolia)

J.-F. Ritz[†]
R. Vassallo
Laboratoire Dynamique de la Lithosphère, CNRS-UMII UMR 5573, Université Montpellier II, Montpellier, France

R. Braucher
CEREGE, Europole Méditerranéen de l'Arbois, Aix-en-Provence, France

E.T. Brown
Large Lakes Observatory, University of Minnesota, Duluth, Minnesota 55812, USA

S. Carretier
Laboratoire des Mécanismes de Transfert en Géologie, IRD-CNRS-UNIV UMR 5563, Toulouse, France

D.L. Bourlès
CEREGE, Europole Méditerranéen de l'Arbois, Aix-en-Provence, France

ABSTRACT

This paper presents an updated synthesis of morphotectonic studies that quantify active tectonics along the Gurvan Bogd mountain range in the Mongolian Gobi-Altay, the site of one of the strongest historic intracontinental earthquakes (Mw 8.1) in 1957. Our goal was to determine the slip rate along the constituent fault segments and to estimate the return period of such large events. Along each segment, cumulative offsets were estimated from topographic surveys, and the ages of the offset markers were determined using cosmic-ray exposure dating. In this review, we reevaluate ^{10}Be data reported in previous publications using a chi-square inversion analysis of depth profiles and an updated scaling model for spatial production rate variations. We also discuss sampling strategies for dating alluvial fans in arid settings.

This study confirms the low horizontal and vertical slip rates within the massifs of the Gurvan Bogd mountain range for the Late Pleistocene–Holocene period, suggests that episodes of aggradation occurred near the times of major glacial-interglacial terminations (at ca. 15–20 ka and ca. 100–130 ka), and provides evidence for another much earlier aggradational episode, occurring before 400 ka. The Bogd fault has a maximum horizontal left-lateral slip rate of ~1.5 mm/yr, while reverse fault segments along the Gurvan Bogd fault system have vertical slip rates between 0.1 and 0.2 mm/yr. Characteristic dislocations observed along the Bogd fault suggest return periods of earthquakes similar to 1957 between 3000 and 4000 yr.

Keywords: Mongolia, Gurvan Bogd, active faults, slip rates, ^{10}Be.

[†]E-mail: ritz@dstu.univ-montp2.fr.

Ritz, J.-F., Vassallo, R., Braucher, R., Brown, E.T., Carretier, S., and Bourlès, D.L., 2006, Using in situ–produced ^{10}Be to quantify active tectonics in the Gurvan Bogd mountain range (Gobi-Altay, Mongolia), *in* Siame, L.L., Bourlès, D.L., and Brown, E.T., eds., In Situ–Produced Cosmogenic Nuclides and Quantification of Geological Processes: Geological Society of America Special Paper 415, p. 87–110, doi: 10.1130/2006.2415(06). For permission to copy, contact editing@geosociety.org. © 2006 Geological Society of America. All rights reserved.

INTRODUCTION

In intraplate domains, strain rates can be very low, and earthquake recurrence intervals may be thousands of years. It is therefore important to study active faulting over several seismic cycles. In such regions, long-term slip rates determined through dating surficial features that accumulated deformation over significant timescales provide a means of characterizing tectonic activity. However, until the mid-1990s, dating morphological features displaced along active faults was problematic. In arid domains, for example, the absence of organic material and fine-grained deposits often precluded radiocarbon or thermoluminescence dating, so ages of morphological markers were typically estimated by correlation with global and regional climatic events. The development of cosmic-ray exposure dating in the mid-1980s (Nishiizumi et al., 1986; Klein et al., 1986) provided the possibility to determine surface exposure age of quartz-rich detrital material.

In this paper, we present an updated synthesis of a series of morphotectonic studies (Ritz et al., 1995, 2003; Carretier, 2000; Carretier et al., 2002; Vassallo et al., 2005) that aimed to determine the long-term slip rates along the Gurvan Bogd fault system in the Gobi-Altay (Mongolia) using in situ–produced ^{10}Be (Fig. 1). These studies were based on five months of fieldwork during seven expeditions. The Gurvan Bogd fault system is within the easternmost extent of the Mongolian Gobi-Altay, and in 1957 was the site of one of the strongest intraplate earthquakes of the past century (Mw 8.1; Florensov and Solonenko, 1965; Kurushin et al., 1997). This earthquake generated more than 350 km of surface ruptures, principally along the east-west-trending left-lateral Bogd strike-slip fault. The Bogd fault was the site of one of the first studies (Ritz et al., 1995)—along with a site along the Owens Valley fault in California (Bierman et al., 1995)—that applied cosmic-ray exposure dates to estimate long-term slip rates. Because of the arid climate, the Gobi-Altay offers extraordinary preservation of the morphological markers and thus is well suited for cosmogenic dating.

The ^{10}Be studies in Gobi-Altay, as well as others in similar settings (e.g., Anderson et al., 1996; Repka et al., 1997; Van der Woerd et al., 1998, 2002; Hancock et al., 1999; Brown et al., 2002; Meriaux et al., 2004, 2005), showed that superficial samples often contain inherited ^{10}Be due to preexposure that can lead to overestimation of their exposure ages. Measurement of the distribution of ^{10}Be with depth and comparison with theoretically predicted exponential decreases in vertical profiles (e.g., Brown et al., 1992) provide a means of evaluating complex exposure histories. Therefore, after a first protocol consisting of sampling only the surface, a second protocol consisted of analyzing the distribution of ^{10}Be at depth along soil pits dug into the upper two meters of studied markers. Below this depth and for deposits younger than a few 100 ka, negligible ^{10}Be is produced. Assuming that there was little temporal variability in the inherited component, the concentration profiles tend toward an asymptotic value at depth that indicates average inheritance, and may be used to correct the surface age (Burbank and Anderson, 2001, p. 50–51).

The distribution of ^{10}Be at depth is also a function of the erosion rate of the surface (Brown et al., 1995). Therefore, the knowledge of this parameter improves the precision of age calculations. When the erosion rate cannot be estimated, an assumption of no erosion is generally made to obtain minimum exposure ages (e.g., Brown et al., 1992; Ritz et al., 1995). The divergence between minimum and real exposure ages increases with the age of the surface, especially for surfaces older than ca. 100 ka. Indeed, the evolution of the ^{10}Be concentration of a sample with time, at a given depth and for a given production rate, firstly increases linearly and then tends to a steady-state equilibrium, which is reached more or less rapidly depending on the erosion rate (Brown et al., 1991). Using a novel chi-square inversion analysis of depth profiles (Siame et al., 2004), we reevaluate the cosmic-ray exposure data obtained for the Gurvan Bogd fault system in eastern Gobi-Altay reported in previous publications (Ritz et al., 1995, 2003; Carretier, 2000; Vassallo et al., 2005). We applied to all of them the same scaling model for spatial production rate variation calculations (Stone, 2000). This allows a discussion of contributions of morphological and tectonic processes in active fault systems.

TECTONIC SETTING

The Gobi-Altay mountain range in Mongolia and its continuation to the northwest, the Altay mountain range, represent the northernmost active compressional belt in central Asia (Molnar and Tapponnier, 1975; Tapponnier and Molnar, 1979) (Fig. 1). Western Mongolia and its immediately surrounding areas were the site of four M 8 earthquakes during the twentieth century, and thus may be considered among the most active intracontinental regions (e.g., Baljinnyam et al., 1993; Schlupp, 1996; Bayasgalan, 1999).

In 1957 the most recent of these earthquakes, the Gobi-Altay earthquake, ruptured the eastern part of the Valley-of-Lakes fault (called the Bogd fault), a Paleozoic structure that was reactivated during the Cenozoic (Florensov and Solonenko, 1965). The following year, a Mongolian-Russian expedition provided an outstanding description of ground surface effects of the earthquake at the epicentral zone (Florensov and Solonenko, 1965). Baljinnyam et al. (1993) revisited some of the piercing points of the surface breaks, and Kurushin et al. (1997) furnished an updated and thorough description of the entire rupture area. The main rupture of more than 260 km of left-lateral strike slip occurred along the Bogd fault, to the north of the Ih Bogd (3957 m) and Baga Bogd (3590 m) massifs (Fig. 1). The average horizontal displacement ranged between 3 and 4 m with a maximum section of offsets up to 5–7 m (Kurushin et al., 1997). An additional 100 km of reverse faulting, distributed on five secondary segments, ruptured simultaneously with the Bogd fault during the 1957 earthquake (Ölziyt, Gurvan Bulag, Toromhon, Dalan Türüü, and Hetsüü). These fault segments correspond mainly to thrust faults found at the base of ridges and low hills or "forebergs" (Florensov and Solonenko, 1965; Kurushin et al., 1997; Owen et al., 1999; Bayasgalan et al., 1999b) that are shortening structures associated with the main Bogd strike-slip fault.

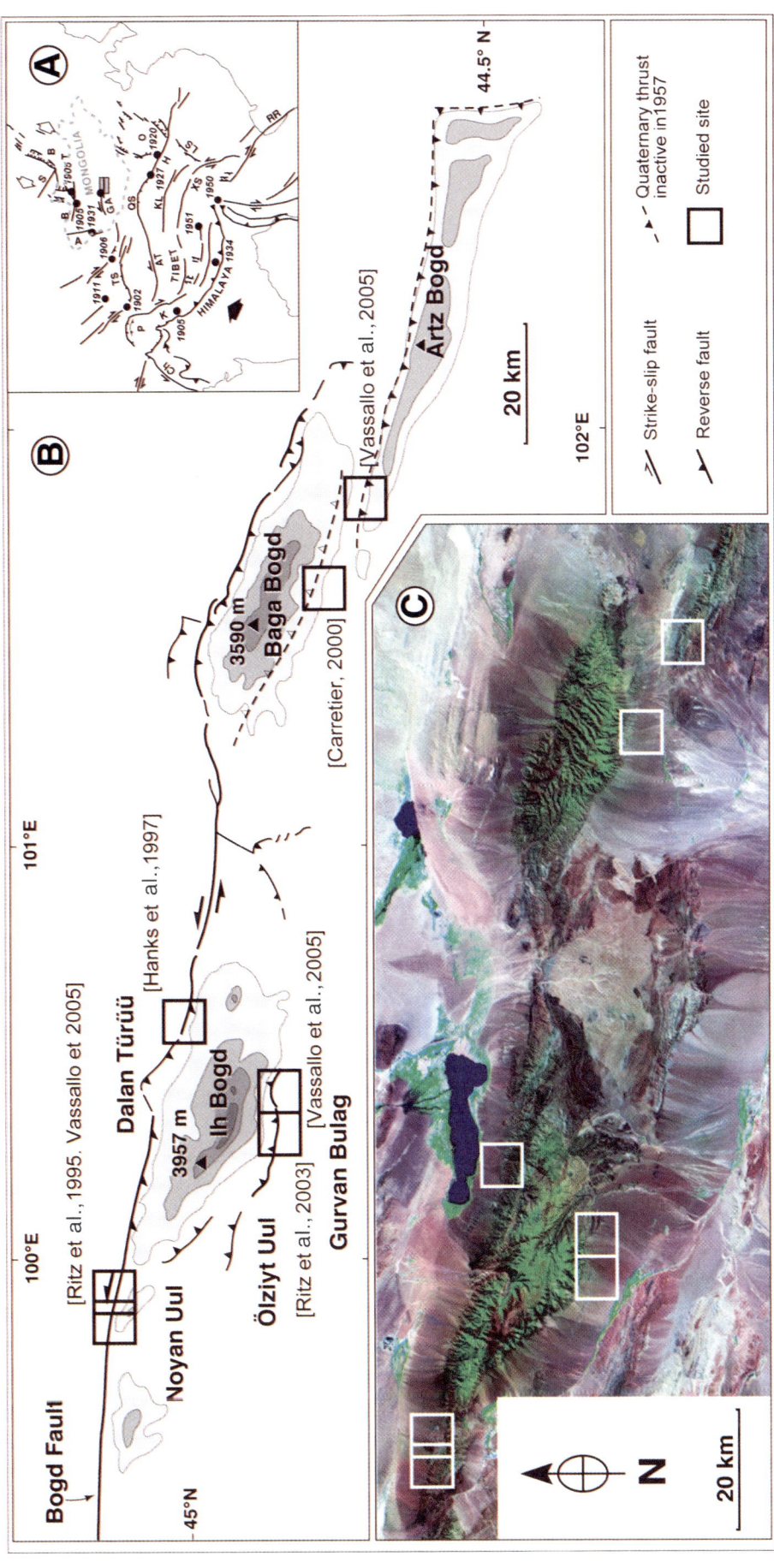

Figure 1. A: Simplified map of Quaternary faults in central Asia and M ≥ 8 earthquakes (solid dots) recorded during the past century (modified after Molnar and Qidong, 1984). B—Baikal; S—Sayan; T—Tsetserleg; B—Bolnay; A—Altay; GA—Gobi-Altay; TS—Tien Shan; P—Pamir; AT—Altyn Tagh; QS—Qilian Shan; O—Ordos; H—Haiyuan; KL—Kunlun; LS—Lungmen Shan; XS—Xian Shui; RR—Red River; K—Karakorum; Ch—Chaman. B: Sketch map of the 1957 Gobi-Altay rupture. Dashed lines are fault ruptures that did not break in 1957. Open squares are sites where slip-rate estimations were made using in situ–produced ^{10}Be dates. A and B modified after Ritz et al., 2003. C: Landsat image showing the Gurvan Bogd mountain range and the studied sites.

The topography of the region appears to be tightly associated with the geometry, the kinematics, and the distribution of ruptures (Fig. 1). Along the Ih Bogd and Baga Bogd massifs, left-lateral slip is associated with a vertical reverse component. These two massifs are bounded by oblique reverse faults along their northern flanks and pure reverse faults along their southern flanks. They can thus be considered as rigid pop-up structures resulting from transpressional deformations within restraining bends along the Bogd strike-slip fault (Kurushin et al., 1997; Cunningham et al., 1996, 1997; Cunningham, 1998). The broad flatness of the summit plateau of the Ih Bogd massif (Fig. 2), an elevated remnant of an ancient erosional surface, also suggests that the bounding faults have had similar long-term slip rates. The difference in height between the surface and the bounding faults is ~2000 m.

Morphological analysis of offset streams, ridges, or alluvial fans along the Gurvan Bogd fault system permits estimation of cumulative displacements. This allows evaluation of the late Quaternary slip rates along the various fault segments involved in building the mountain range. Vertical slip rates along the reverse faults allow estimation of the time since initiation of relief uplift, while the horizontal slip rate combined with analysis of coseismic displacements along the Bogd strike-slip fault allows calculation of the recurrence interval of large earthquakes in the area.

FIELDWORK AND METHODS

Field Site Selection

Soviet-Mongolian aerial photographs (1:35,000 scale) taken in 1958 were used to define surfaces and to choose sites for detailed field studies, selecting those with the best-preserved morphotectonic markers (typically fan surfaces displaced by fault movements [Fig. 3] and their associated surface features, mainly debris flows). Based on their general appearance, the studied morphotectonic markers can be categorized into groups with qualitatively decreasing age. The oldest markers are rounded ridges corresponding to remnants of alluvial surfaces that have been reincised by the drainage network. Boulders on such features are entirely embedded in the surface. Inset in these ridges, two to three flat surfaces are recognizable: The older, flatter surfaces contain deeply weathered granite boulders generally embedded in finer material (debris flows); the younger surfaces are broader and can be covered by broad debris-flow deposits characterized by dense boulder fields (according to the lithology of the bedrock within the upstream drainage basin). The well-preserved granite boulders (1 m on average, up to 3 m) on the different surfaces often show well-developed desert varnish coatings, some of which having petroglyphs reported to be 3000 yr old (Florensov and Solonenko, 1965) (Fig. 4).

Measurement of Offsets

Topography of the surfaces was surveyed by measuring cross sections or digital elevation models (DEMs) with kinematic GPS using two receivers. One was used as a base, its antenna fixed on a tripod; the other one was mobile, its antenna attached to a hand-carried pole. Both receivers recorded positioning data (from at least four satellites) at intervals of 1, 3, 5, or 10 seconds depending on the length of profiles or the surface of the DEM and the need for finer topographic details. Positioning data were processed after or during (by means of a radio connection between receivers) the survey depending on the type of kinematic GPS stations. Remeasurements of starting points for profiles or DEM indicated horizontal and vertical uncertainties on the order of 1 cm. Displacements and associated uncertainties were calculated using mathematical parameterizations developed by Hanks et al. (1984) from the work of Bucknam and Anderson (1979).

Dating Morphotectonic Markers

Cosmic-ray exposure dating of the morphological markers was performed using in situ–produced ^{10}Be (e.g., Brown et al., 1991; Bierman, 1994; Siame et al., 2000). The alluvial fans cut by the Gurvan Bogd fault system show variations in morphology and degree of preservation. Our sampling strategy was developed to minimize the effects of exposure prior to deposition and of postdeposition erosional processes. Some of the surfaces show dense boulder fields preserved at the surface, whereas some others—because they are older and more eroded or because they are composed of material with other lithologies—do not contain large boulders. Our sampling strategies evolved over time. At the beginning of the study in 1993, the working assumption was that the occurrence of large granite boulder flows represented the effect of strong erosional events that reworked massive quantities of slope material from the upstream drainage basins (Ritz et al., 1995, 2003). In addition, because these debris flows represented intense erosional events, it was thought that deposition would be rapid, and little cosmogenic nuclide accumulation would occur during transport. Under these conditions, the concentration of in situ cosmogenic ^{10}Be would be directly related to the time when alluviation ended, or when subsequent incision led to the abandonment of the fan surface.

However, during evaluation of first ages obtained from surface samples (Ritz et al., 1995), we realized that some of these surface boulders were likely to contain ^{10}Be due to prior exposure. This inherited ^{10}Be leads to the overestimation of exposure ages. In subsequent fieldwork, we collected samples for ^{10}Be analysis in depth profiles in fan surfaces in order to evaluate the potential role of prior exposure (Carretier, 2000; Ritz et al., 2003; Vassallo et al., 2005). Observations of the evolution of ^{10}Be with depth may be compared with theoretically predicted exponentially decreasing profiles (e.g., Brown et al., 1992, 1995) to develop strategies for corrections for prior exposure.

To estimate exposure ages from boulders embedded in alluvial fans as well as both denudation rates and exposure ages from depth profiles, the following equation was used:

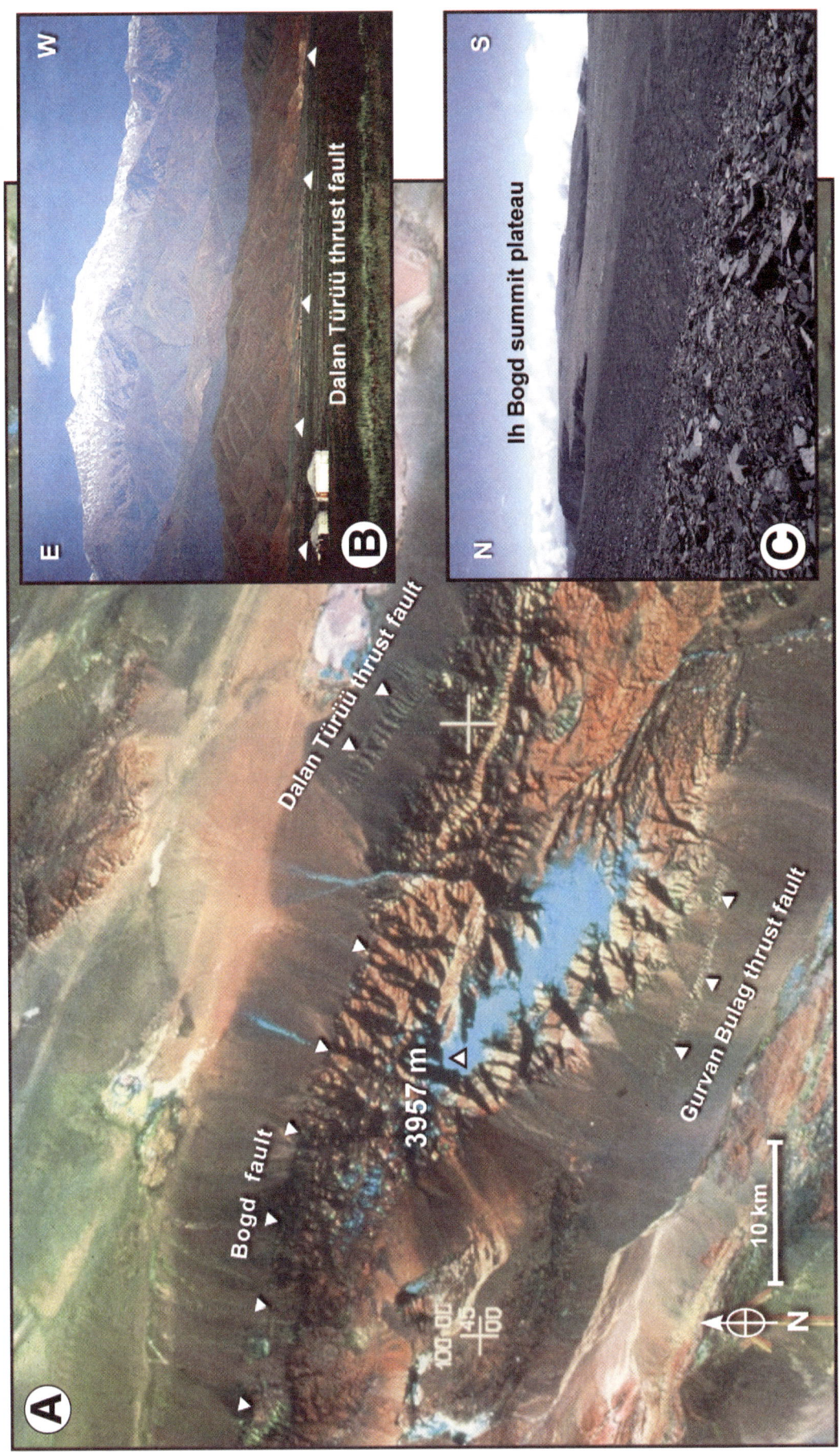

Figure 2. A: SPOT image of Ih Bogd massif (note the flat summit surface covered by a thin snow cap). B: Ih Bogd massif seen from the Dalan Türüü thrust fault. C: View toward the east of Ih Bogd summit surface corresponding to an old erosional surface.

Figure 3. Examples of offset alluvial fan surfaces. A: Along the Bogd left-lateral strike-slip fault. B: Along the Gurvan Bulag thrust fault at Noyan Uul.

Figure 4. Large boulder (~1 m diameter) exposed at the surface of a fan, well embedded in finer material. Note the petroglyph representing a ram drawn on the heavy desert varnish coating. Note also the absence of desert varnish at the bottom of the boulder, attesting to the erosion of the finer material by wind deflation.

$$C(x,\varepsilon,t) = C_{inh} \cdot e^{-\lambda t} + \frac{P_0 \cdot p_n}{\frac{\varepsilon}{\Lambda_n} + \lambda} \cdot e^{-\frac{x}{\Lambda_n}} \left[1 - e^{-t\left(\frac{\varepsilon}{\Lambda_n} + \lambda\right)} \right] + \frac{P_0 \cdot p_{\mu s}}{\frac{\varepsilon}{\Lambda_{\mu s}} + \lambda}$$

$$\cdot e^{-\frac{x}{\Lambda_{\mu s}}} \left[1 - e^{-t\left(\frac{\varepsilon}{\Lambda_{\mu s}} + \lambda\right)} \right] + \frac{P_0 \cdot p_{\mu f}}{\frac{\varepsilon}{\Lambda_{\mu f}} + \lambda} \cdot e^{-\frac{x}{\Lambda_{\mu f}}} \left[1 - e^{-t\left(\frac{\varepsilon}{\Lambda_{\mu f}} + \lambda\right)} \right] \quad (1)$$

where $C(x, \varepsilon, t)$ is the ^{10}Be concentration function of depth x (g/cm^2), erosion rate ε (g/cm^2/yr) and exposure time t (yr); Λ_n, $\Lambda_{\mu s}$, and $\Lambda_{\mu f}$ are the effective apparent attenuation lengths (g/cm^2), for neutrons, slow muons, and fast muons, respectively; p_n, $p_{\mu s}$, and $p_{\mu f}$ are the relative contributions to the ^{10}Be production rate of the three reactions ($p_n + p_{\mu s} + p_{\mu f} = 100\%$); P_0 is the production rate at the surface taken from Stone (2000); and C_{inh} represents the ^{10}Be concentration potentially acquired by the sample during exposure to cosmic rays prior to emplacement in their sampling position. All calculations were performed using attenuation lengths of 150, 1500, and 5300 g/cm^2 with associated relative contributions to the total surface production rate of 97.85%, 1.50%, and 0.65% for neutrons, slow muons and fast muons, respectively. These values are based on field-calibrated measurements (Braucher et al., 2003).

Recently, Siame et al. (2004) showed that measurement of ^{10}Be concentrations along a depth profile allows estimation of both exposure time and erosion rate using a chi-square inversion that minimizes the function

$$Chi - square = \sum_{i=1}^{n} \left[\frac{C_i - C_{(x_i,\varepsilon,t)}}{\sigma_i} \right]^2, \quad (2)$$

where C_i is the measured ^{10}Be concentration at depth x_i, $C(x_i, \varepsilon, t)$ is the theoretical ^{10}Be concentration determined using equation 1, σ_i is the analytical uncertainty at depth i, and n is the total number of samples in the profile. Chi-square inversion allows us to determine the Q value, an estimator of the "goodness of fit" (e.g., Press et al., 1996). The model is considered a good one if the Q value is greater than 0.001. A lower value can result from oversimplifying model assumptions (such as constant erosion rate through time or same inheritance for all samples), or from too large analytical uncertainties.

For each profile, we assume a null inheritance and an inheritance of 0.15 M atoms/g to evaluate the effects of inheritance on age determination. The second value corresponds to the typical concentration measured in samples at depths greater than 1.5 m. This concentration would correspond to a residence time of 5000 yr at the average elevation of the major drainage basins feeding the alluvial fans. For both values, we generated a chi-square contour diagram versus erosion rate and time. The minimum chi-square value defines the best-fit erosion-time couple. A 1σ confidence interval contour is determined, whose projection on erosion and time axes determines 1σ uncertainties associated with the best-fit couple. However, these chi-square contour diagrams usually display a vertical trend parallel to the time axis for large ages. If the best-fit solution lies in this portion of the diagram, the best-fit model corresponds to a steady state, for which the determination of the age is impossible. In this case, calculation only yields an erosion rate and a minimum age, which corresponds to the minimum chi-square value for no erosion.

For the surficial samples of each studied surface, a mean exposure age, weighted to account for the variable analytical uncertainties for each sample, was computed. To perform this calculation, we did not include outlying data that were significantly different (when considering 2σ uncertainty intervals) from the mean value of the main data cluster for a given feature. Such outliers are interpreted as being associated either with reworked material with significant predepositional exposure (highest values), or with material exposed to postdepositional processes (lowest values). The uncertainty for the age of each surface was then estimated by the difference between this weighted mean age and the highest and lowest considered ages (error bars included), since we do not know whether scatter is primarily due to variation in inheritance (predepositional exposure) or erosion/shielding (postdepositional processes). Because the erosion rate affecting surficial boulders may be significantly different (lower) than the erosion rate estimated for the corresponding surfaces, the exposure ages of boulders were calculated assuming no erosion and are therefore minimum ages. These minimum ages allow us to check the validity of the ages obtained from the depth profiles. We use preferably ages obtained from the depth profiles because they give more accurate ages. The ages obtained from surficial samples are used only if ages given by depth profiles are poorly constrained.

Samples were analyzed by accelerator mass spectrometer at the Tandétron AMS facility, Gif-sur-Yvette, France (Raisbeck et al., 1987, 1994) or the Lawrence Livermore National Laboratory AMS facility, Livermore, California, USA (Davis et al., 1990), after isolation of quartz and chemical preparation of Be targets (Ritz et al., 2003). One goal of this paper is to homogenize and synthesize all data gathered in the region, having the ^{10}Be concentrations all normalized with reference to National Institute of Standards and Technology Standard Reference materials (NIST SRM) 4325 using its certified ^{10}Be/^{9}Be ratio of $(26.8 \pm 1.4) \times 10^{-12}$ (Tables 1–4). All exposure age calculations use production rates from Stone (2000).

TABLE 1. RESULTS OF THE ^{10}Be ANALYSIS, NOYAN UUL

Sample	Surface	P0 (at/g/yr)	10Be (Mat/g)	Uncertainty (Mat/g)	Minimum age (yr)	Uncertainty (yr)
Noyan Uul						
D VI 1	S1	21.1	1.42	0.11	6.99E+04	6.85E+03
D VI 3	S1	21.1	1.30	0.17	6.39E+04	9.19E+03
D VI 4	S1	21.1	1.04	0.11	5.09E+04	6.19E+03
D VI 5	S1	21.1	1.37	0.14	6.74E+04	7.98E+03
D VII 1	S2	21.1	1.10	0.14	5.39E+04	7.59E+03
D VII 2	S2	21.1	1.56	0.17	7.69E+04	9.56E+03
D VII 3	S2	21.1	1.83	0.12	9.05E+04	8.04E+03
D VII 4	S2	21.1	1.65	0.14	8.14E+04	8.46E+03
D VII 5	S2	21.1	1.55	0.12	7.6E+04	7.48E+03
D IV 1	R1	21.1	1.30	0.09	6.39E+04	5.85E+03
D IV 2	R1	21.1	1.13	0.08	5.54E+04	5.14E+03
D IV 3	R1	21.1	1.22	0.10	5.99E+04	6.08E+03
D IV 4	R1	21.1	0.97	0.09	4.75E+04	5.25E+03

Soil pit	Surface	Depth (cm)	10Be (Mat/g)	Uncertainty (Mat/g)
NU42	S1	205	0.34	0.06
NU45	S1	160	0.63	0.14
NU47	S1	110	0.45	0.07
NU49	S1	80	0.32	0.04
NU51	S1	20	1.11	0.16
NU52A	S1	0	1.85	0.27

P0 (at/g/yr)	Density (g/cm^3)
18.2	2.0

TABLE 2A. CENTRAL GURVAN BULAG, WESTERN FAN

Sample	Surface	P0 (at/g/yr)	10Be (Mat/g)	Uncertainty (Mat/g)	Minimum age (ka)	Uncertainty (ka)
Gurvan Bulag (West)						
GB96-9	S2	25.0	3.05	0.10	1.29E+05	8.85E+03
GB96-10	S2	25.0	3.19	0.10	1.35E+05	9.12E+03
GB96-11	S2	25.0	3.17	0.08	1.34E+05	8.67E+03
GB96-12	S2	25.0	2.99	0.07	1.26E+05	8.15E+03
GB96-13 (65 cm)	S2	25.0	1.18	0.03		
GB96-14 (90 cm)	S2	25.0	0.84	0.03		
GB96-13[†]	S2	25.0	2.81	0.31	1.18E+05	1.47E+04
GB96-14[†]	S2	25.0	2.79	0.32	1.17E+05	1.50E+04
GB96-15[‡]	S2	25.0	2.10	0.05	8.77E+04	5.66E+03
Mo93-AII1	S3	25.0	0.28	0.03	1.17E+04	1.42E+03
Mo93-AII2	S3	25.0	0.29	0.04	1.21E+04	1.79E+03
Mo93-AII3	S3	25.0	0.43	0.04	1.75E+04	1.95E+03
Mo93-AII4	S3	25.0	0.59	0.05	2.44E+04	2.52E+03
Mo93-AII5[‡]	S3	25.0	1.02	0.08	4.22E+04	4.16E+03
Mo95-11	S3	25.0	0.37	0.03	1.53E+04	1.54E+03
Mo95-12	S3	25.0	0.29	0.02	1.20E+04	1.09E+03
Mo95-13	S3	25.0	0.32	0.02	1.32E+04	1.14E+03
Mo95-14	S3	25.0	0.42	0.05	1.73E+04	2.30E+03
Mo95-15	S3	25.0	0.44	0.03	1.82E+04	1.62E+03
Mo95-16	S4	24.1	0.11	0.01	4.50E+03	5.03E+02
Mo95-17	S4	24.1	0.21	0.03	8.75E+03	1.22E+03
Mo95-18	S4	24.1	0.08	0.01	3.37E+03	5.88E+02
Mo95-19	S4	24.1	0.13	0.02	5.43E+03	7.92E+02
Mo95-20[‡]	S4	24.1	0.36	0.03	1.54E+04	1.52E+03
Mo95-21[‡]	S5	24.1	1.39	0.08	5.98E+04	4.84E+03
Mo95-22	S5	24.1	0.43	0.03	1.83E+04	1.83E+03
Mo95-23	S5	24.1	0.31	0.03	1.30E+04	1.42E+03
Mo95-24	S5	24.1	0.41	0.04	1.74E+04	1.88E+03
Mo95-25	S5	24.1	0.35	0.04	1.51E+04	1.83E+03

Soil pit	Surface	Depth (cm)	10Be (Mat/g)	Uncertainty (Mat/g)
GB96-1	S2	40–45	1.21	0.09
GB96-2	S2	55–60	0.89	0.06
GB96-3	S2	90	0.78	0.05
GB96-4	S2	110–115	0.50	0.04
GB96-5	S2	130	0.51	0.04
GB96-6	S2	145–148	0.44	0.04
GB96-7A	S2	160	0.25	0.09
GB96-7B	S2	160	0.21	0.02
GB96-8A	S2	170	0.18	0.02
GB96-8B	S2	172	0.19	0.03

P0 (at/g/yr)	Density (g/cm^3)
25.0	2.0

[†]Data normalized to the surface.
[‡]Outlying data not included in calculations (see text).

TABLE 2B. CENTRAL GURVAN BULAG, EASTERN FAN

Soil pits	Surface	Depth (cm)	^{10}Be (Mat/g)	Uncertainty (Mat/g)
Gurvan Bulag (East)				
IBSA23	S2	140	0.54	0.07
IBSA25	S2	100	0.73	0.09
IBSA26	S2	70	1.37	0.19
IBSA28	S2	30	1.79	0.20
IBSA30	S2	0	2.54	0.27
IBSB31	S3	200	0.16	0.04
IBSB33	S3	160	0.38	0.05
IBSB35	S3	120	0.15	0.03
IBSB37	S3	80	0.37	0.05
IBSB39	S3	40	0.50	0.07
IBSB41	S3	0	0.70	0.10

P0 (at/g/yr)	Density (g/cm³)
23.3	2.0

TABLE 3. SOUTH BAGA BOGD

Sample	Surface	P0 (at/g/yr)	^{10}Be (Mat/g)	Uncertainty (Mat/g)	Minimum age (yr)	Uncertainty (yr)
South Baga Bogd						
BBS97-1†	S3	27.6	1.68	0.04	6.47E+04	4.20E+03
BBS97-2	S3	27.6	0.52	0.05	1.85E+04	2.02E+03
BBS97-3	S3	27.6	0.63	0.02	2.23E+04	1.57E+03
BBS97-4	S3	27.6	0.53	0.05	1.86E+04	2.03E+03
BBS97-5†	S3	27.6	1.40	0.03	5.26E+04	3.38E+03

Soil pit	Surface	Depth (cm)	^{10}Be (Mat/g)	Uncertainty (Mat/g)
BBS97-18	S1	0	2.85	0.11
BBS97-15	S1	60	1.61	0.01
BBS97-12	S1	110	0.99	0.11
BBS97-11	S1	130	0.94	0.12
BBS97-9	S1	160	0.39	0.09
BBS97-8	S1	185	0.56	0.04
BBS97-6	S1	220	0.22	0.03

P0 (at/g/yr)	Density (g/cm³)
27.6	2.0

†Outlying data not included in calculations (see text).

TABLE 4. NORTH ARTZ BOGD

Soil pit	Surface	Depth (cm)	^{10}Be (Mat/g)	Uncertainty (Mat/g)
North Artz Bogd				
ABW1	UNIT 1	200	0.42	0.10
ABW4	UNIT 1	160	0.74	0.09
ABW6	UNIT 1	120	1.23	0.14
ABW8	UNIT 1	80	2.41	0.32
ABW10	UNIT 2	40	3.29	0.41
ABW12B	UNIT 2	0	3.39	0.37

P0 (at/g/yr)	Density (g/cm³)
21.0	2.0

SUMMARY OF PREVIOUS RESULTS

The Bogd Fault: Noyan Uul

At Noyan Uul, immediately to the northwest of the Ih Bogd massif (location in Fig. 1), there are morphological features clearly displaying cumulative horizontal left-lateral displacements (Fig. 5). The east-southeast-trending fault scarp delimits two morphological domains: (1) to the south a mountainous area incised by gorges and deep ravines and (2) to the north an alluvial plain that dips gently (5°) northward. The drainage network clearly shows small-scale left-lateral strike-slip movements; along the fault an ~5 m offset is visible, and numerous small streams show left-lateral displacements. North of the fault scarp there are several generations of incised alluvial fans. The younger fans (S0) are cone-shaped in plan view. During deposition, they truncated parts of the older fans. The old fans (S1, S2) do not presently correspond to any upslope stream, indicating sinistral displacement of the alluvial plain relative to the mountainous domain. Ritz et al. (1995) studied the site where the misalignment of cones with respect to the drainage basin was clearest and where the left-lateral strike-slip offset was also manifested by the misalignment of large stream incisions in the old alluvial fans (D1, D2), which appear beheaded relative to streams upslope of the fault (Fig. 6). Simultaneous alignment of floodplain features (S2, D1, and D2) with upslope streams (U0, U1, and U2, respectively) requires compensation for a 220 ± 10 m horizontal offset, while alignment of surface S1 apex with the outlet of the upstream drainage basin requires compensation for 110 ± 10 m. The incision of valleys U1-D1 and U2-D2 in the hanging wall and the deposition of alluvial fans S0, S1, and S2 in the footwall can be interpreted as the result of enhanced stream power associated with a major regional climate change (Carretier et al., 1998). Analogously displaced fans are observed for several kilometers in both directions along the fault (Ritz, 2003).

Ritz et al. (1995) sampled S1 and S2 as well as old ridges (R1) that were interpreted as remnants of even older fan surfaces (Fig. 6). If we consider a constant slip rate through time, surface S2 should be twice as old as S1, and its boulders should have correspondingly higher ^{10}Be concentrations. Instead, Ritz et al. (1995) found that there were only small differences in ^{10}Be concentrations between the two surfaces, and that ^{10}Be concentrations of ridgetop boulders were lower than those associated with the stratigraphically younger surfaces. This suggested that the concentrations were approaching steady-state values on surfaces

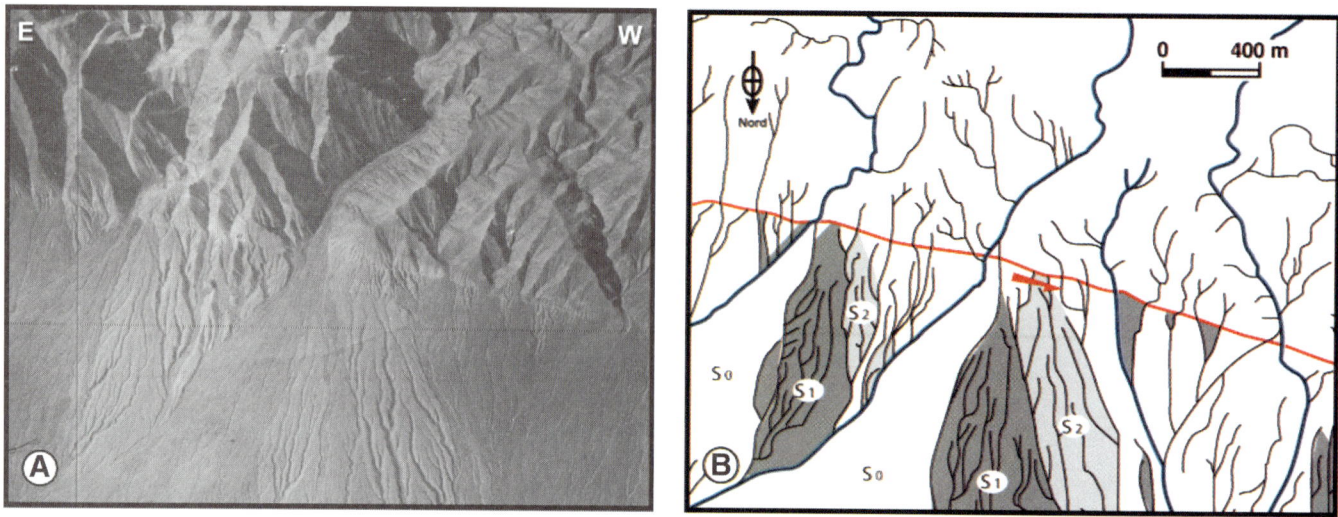

Figure 5. A and B: Aerial image and corresponding sketch map of two sequences of three alluvial fan surfaces (S0, S1, S2) shifted left-laterally along the Bogd strike-slip fault at Noyan Uul site. S0 (white) is the younger alluvial surface, S2 (light gray) is the older and S1 (dark gray) is the intermediate one. Main streams are underlined in blue.

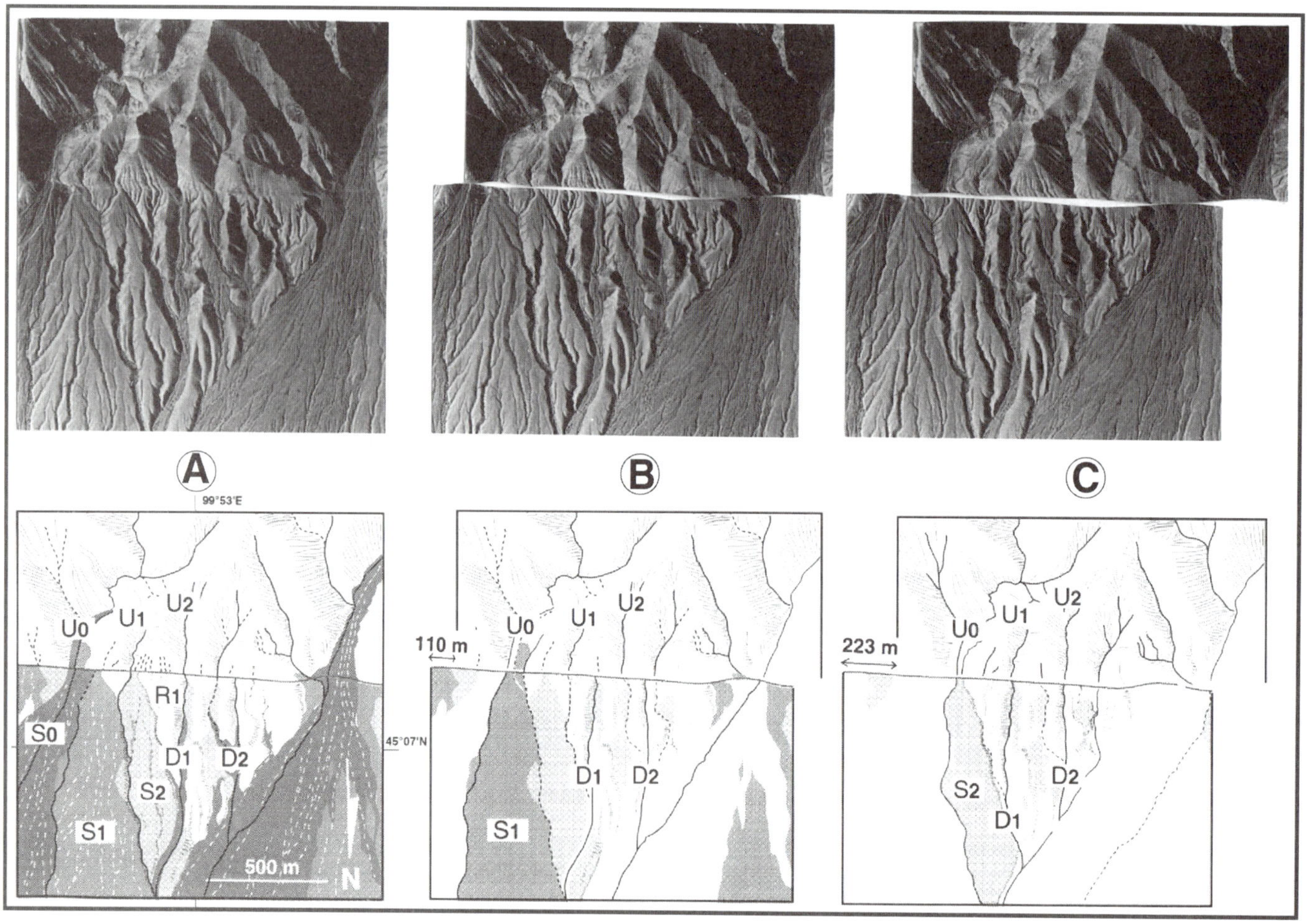

Figure 6. Reconstruction of history of alluvial fan deposition and erosion along the Bogd fault at Noyan Uul based on aerial photographs. A: Present day. B: Compensation for ~110 m horizontal offset. C: Offset compensation of ~220 m. Solid lines labeled U0, U1, and U2 designate major stream axes upstream of fault; D1 and D2 are major axes below the fault. Dotted lines indicate smaller river valleys. Shading represents relative age of depositional surfaces (S2, S1, and S0); lightest tones corresponding to oldest surfaces. Only present-day relicts of alluvial fans S1 and S2 are represented in B and C. After Ritz et al. (1995).

S1 and S2. Taking the apparent minimum age calculated for the youngest surface S1 (80 ka), Ritz et al. (1995) calculated a maximum horizontal slip rate of 1.2 mm/yr. Vassallo et al. (2005) reestimated the age of surface S1 at 125 ± 28 ka by analyzing the ^{10}Be distribution at depth—yielding an average inheritance of 0.20 ± 0.10 M atoms/g—and by estimating the erosion rate at 7 ± 1 m/m.y. This allowed the authors to calculate a left-lateral slip rate of 0.95 ± 0.29 mm/yr.

The Gurvan Bulag Thrust Fault

The Gurvan Bulag fault is a 23 km thrust that ruptured most recently in 1957, simultaneously with the Bogd strike-slip fault during the Mw 8.1 Gobi-Altay event (e.g., Kurushin et al., 1997) (Fig. 1). "Gurvan Bulag" means "three springs" in Mongolian; the 1957 earthquake changed local hydrology and dried up the springs. The morphology of the Gurvan Bulag thrust fault zone is described as a foreberg resulting from interaction between tectonics and fan dynamics (Bayasgalan et al., 1999b; Carretier et al., 2002). Flat, active surfaces are generally directly downstream of drainage basins where erosional and depositional rates are at a maximum, whereas hills are found in areas where erosion and deposition rates are lower, typically at the lateral margins of the fans. The foreberg is thus a system of inset surfaces that show clear cumulative vertical slips.

Ritz et al. (2003) and Vassallo et al. (2005) focused on two fans—termed western and eastern fans—within the central part of the Gurvan Bulag thrust fault, where the 1957 fault displacements and the cumulative deformation appear to be the largest (Fig. 7), and where the offset surfaces were the best preserved. Four markers could be distinguished from their relative elevations and surface characteristics (see Ritz et al., 2003, for more detailed description). S1 corresponds to upper, old eroded surfaces found within the hanging wall and represents elevated remnants of planar abrasive or depositional surfaces. S2 is an intermediate alluvial surface inset in S1 and is found in large patches extending on both sides of the fault. S3 is the youngest depositional surface, extending from the apex of the cones to the Gurvan Bulag foreberg, and appears to be inset in S2. Notice that the nomenclature of the surfaces, in terms of relative ages, does not correspond to that of Noyan Uul. In the western fan, as surface S3 approaches the fault zone, it overlies S2 and then dissipates before reaching the fault scarp. This termination of S3 deposits above the fault scarp is not observed within the eastern fan, where S3 is more deeply inset within the two older surfaces and crosses the fault scarp in broad channels (Fig. 7). Additional features are present only within the western fan: S4 corresponds to gullies cut in alluvial surfaces near the fault scarp, with local cones found in the footwall in front of them. Ritz et al. (2003) interpreted these features as local debris cones that accumulated at the toe of the fault scarp simultaneously with the incision of S4 gullies into older hanging-wall alluvial surfaces.

Within the western fan, Ritz et al. (2003) estimated a minimum vertical offset for S1 of 31 ± 1 m (profile 7, Fig. 8) and a mean vertical offset for surface S2 in the studied area at 19.8 ± 1.9 m (profiles 5–7, Fig. 8). Estimation of vertical slip for surface S3 was more uncertain because within the western fan S3 terminates before reaching the fault scarp and is not found at the footwall. Because of this, Ritz et al. (2003) proposed two extreme scenarios depending on whether the cumulative offset occurred before or after S3 deposition. If most of the cumulative offset occurred before S3 deposition, the minimum vertical displacement for S3 was 6.5 ± 1.2 m, corresponding to the mean vertical displacement calculated from the gullies S4 incising surface S2 (in this scenario, the S3 debris flow covered the preexisting topography of S2 by overbank flow). In contrast, if the cumulative offset occurred after deposition of S3, the measured mean vertical separation of 17.3 ± 0.4 m would represent a maximum value for the vertical offset for S3. Finally, Ritz et al. (2003) estimated from topographic analysis that the mean 1957 vertical offset along the central part of the Gurvan Bulag thrust fault was 4.2 ± 0.3 m (profiles 1–4, Fig. 8). From the offset measurements and surface ages (131 ± 20 ka and 16 ± 20 ka for S2 and S3 surfaces, respectively), Ritz et al. (2003) determined the following vertical slip rates: 0.14 ± 0.03 mm/yr over the Late Pleistocene–Holocene and between 0.44 ± 0.11 mm/yr and 1.05 ± 0.25 mm/yr since the end of the Late Pleistocene.

Within the eastern fan, Vassallo et al. (2005) studied surfaces S2 and S3. The advantage of this site—although alluvial surfaces are not covered by boulder field as on the western adjacent fan—is that these surfaces are found in both the hanging wall and the footwall, allowing more accurate estimates of vertical offsets (profiles 8 and 9, Fig. 8). The older surface is vertically displaced by 16.0 ± 0.5 m and the younger surface by 5.0 ± 0.5 m. No estimation of the 1957 offset could be made because it was clear that within the eastern fan, the frontal part of the preexisting scarp collapsed during the 1957 event (Carretier et al., 2002, defined a gravity-controlled face). This suggests that estimates of the 1957 offset made by Ritz et al. (2003) or Kurushin et al. (1997) from topography within the western fan, 2 km farther west, are too large. Ten kilometers farther east, paleoseismological evidence indicates that the 1957 vertical offset was between 1 and 2 m, and suggests that the scarp height reported along the central part of the Gurvan Bulag thrust may represent the cumulative result of repeated fault ruptures (Prentice et al., 2002). From the offset measurements and surface ages (128 ± 13 ka and 22 ± 3 ka for S2 and S3 surfaces, respectively), Vassallo et al. (2005) determined the following vertical slip rates: 0.12 ± 0.03 mm/yr over the Late Pleistocene–Holocene and between 0.23 ± 0.05 mm/yr and 0.19 ± 0.05 mm/yr since the end of the Late Pleistocene.

The Southern Baga Bogd Thrust Fault

Similarly to the Ih Bogd massif, the southern flank of the Baga Bogd massif is bounded by a 50-km-long reverse fault that cuts through the alluvial deposits (Fig. 1). Within the eastern part of the massif, Carretier et al. (2002) identified three main geomorphic surfaces (S1, S2, and S3) (Fig. 9). These surfaces

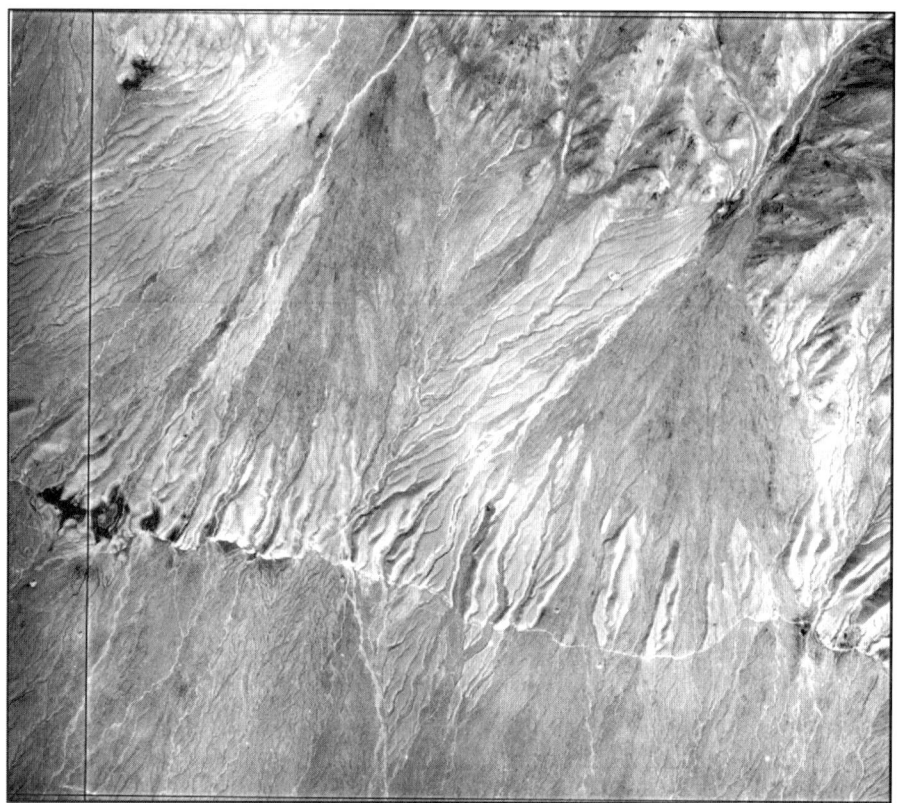
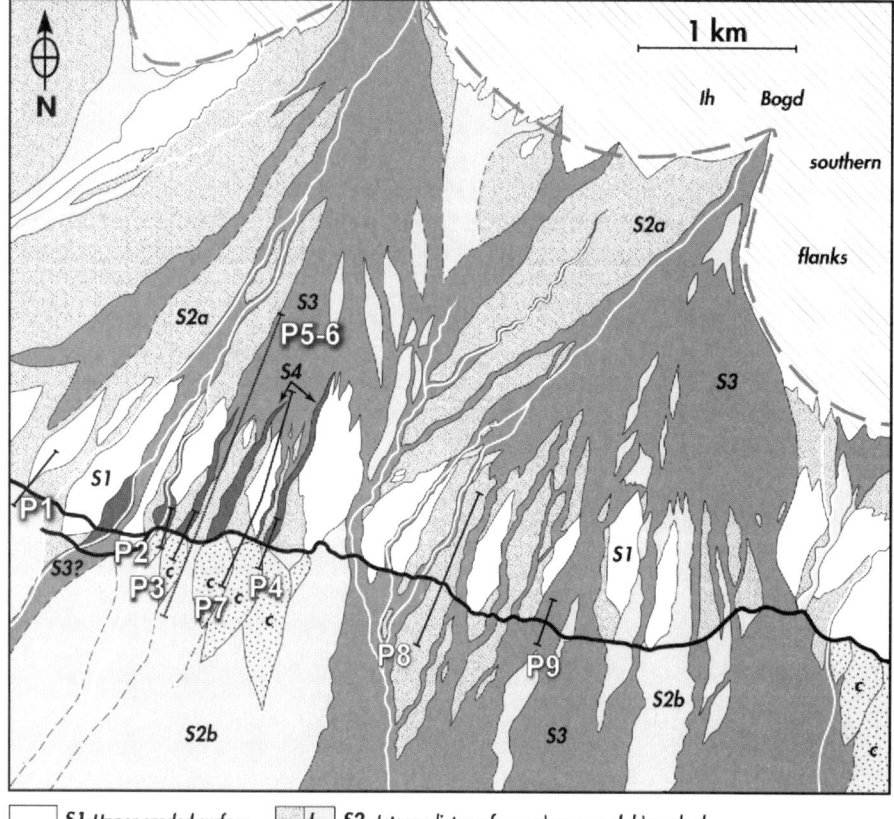

Figure 7. A: Aerial photograph of the central part of the Gurvan Bulag thrust fault cutting through two fans (see Fig. 1 for location). B: Corresponding morphological map of inset surfaces (modified after Ritz et al., 2003).

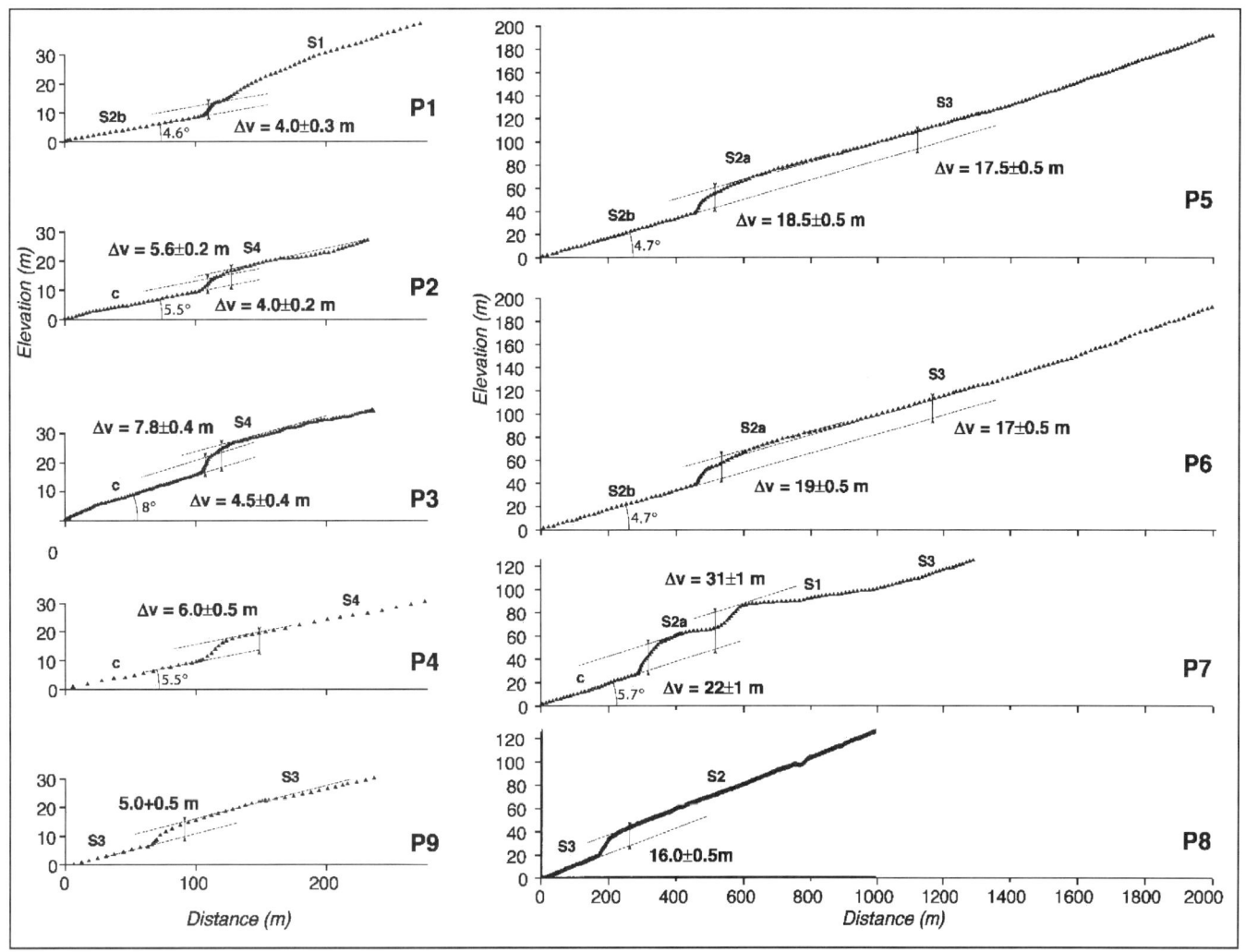

Figure 8. Topographic profiles across uplifted surfaces within the central part of the Gurvan Bulag thrust fault (see Fig. 7).

were not cut by thrusting in the 1957 earthquake (Florensov and Solonenko, 1965). The oldest (S1) is incised by dendritic drainage networks and is the highest recognizable alluvial fan surface uplifted by the reverse fault. The alluvial fan surface (S2) shows characteristics of both dendritic incision and bar-and-swales. The most recent alluvial fan surface (S3) has not been uplifted by the fault, indicating that thrust activity on this fault segment ceased between the depositions of S2 and S3. Carretier et al.'s (2002) survey of surface S1 indicated a vertical offset of 19 ± 0.5 m (Fig. 9C). They dug a soil pit in surface S1 and determined a surface age of 206 ± 50 ka, which yielded a long-term vertical slip rate of 0.10 ± 0.03 mm/yr.

The Artz Bogd Thrust Fault

The Artz Bogd thrust fault is a 75 km long fault bounding the Artz Bogd massif to the north (Fig. 1). The thrust fault is cutting through detrital slopes deposed at the piedmont of the massif (Bayasgalan et al., 1999b). Vassallo et al. (2005) studied the western termination of the fault (Fig. 10). As was the case for the western central fan studied by Ritz et al. (2003) within the Gurvan Bulag thrust fault, the surface that extends downslope of the fault scarp does not correspond to the surface that is vertically offset in the hanging wall. The hanging-wall surface was incised by the drainage network after its vertical displacement. Extension of the planar hanging-wall surface to the north indicates a vertical separation of 20.3 ± 0.5 m with respect to the footwall surface (Fig. 11A). Vassallo et al.'s (2005) analysis of the ^{10}Be concentration distribution along a soil pit dug in the hanging-wall surface shows the superposition of two depositional sequences consistently with the stratigraphy observed in the soil pit. The ages obtained for the lower and the upper layers were 360 ± 36 ka and 160 ± 16 ka, respectively. From the offset of the upper deposit (20.3 ± 0.5 m) and its exposure age, Vassallo et al. (2005) estimated a vertical slip rate of 0.13 ± 0.01 mm/yr.

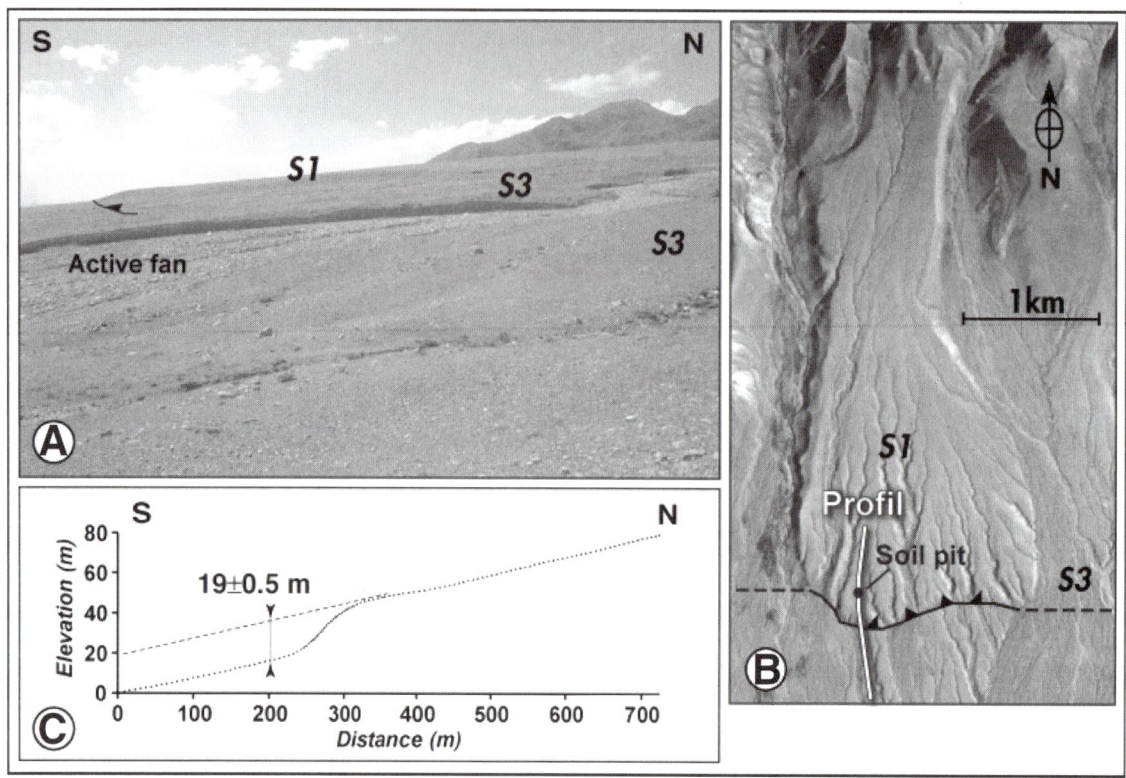

Figure 9. A: Fault scarp along the Southern Baga Bogd thrust fault. B: Corresponding site seen on aerial photograph (see Fig. 1 for location). C: Topographic profile across the fault scarp. Modified after Carretier (2000).

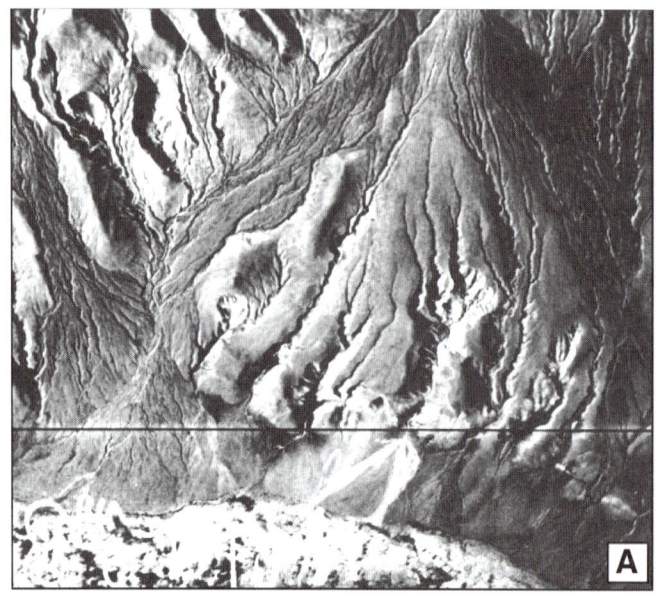

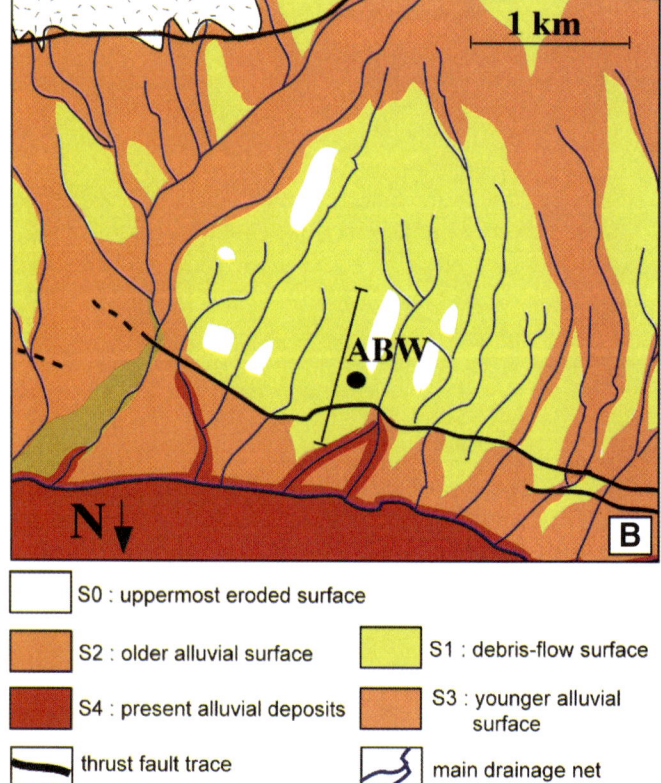

Figure 10. A: Aerial photograph of the western part of the Artz Bogd thrust fault (see Fig. 1 for location). B: Corresponding morphological map of inset surfaces. After Vassallo et al. (2005). ABW—Artz Bogd West soil pit.

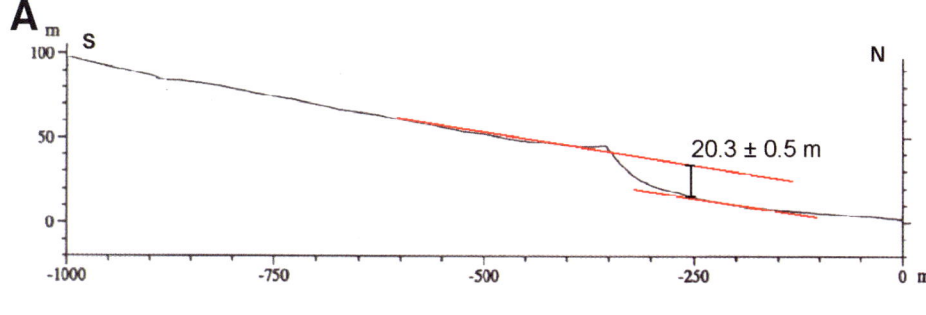

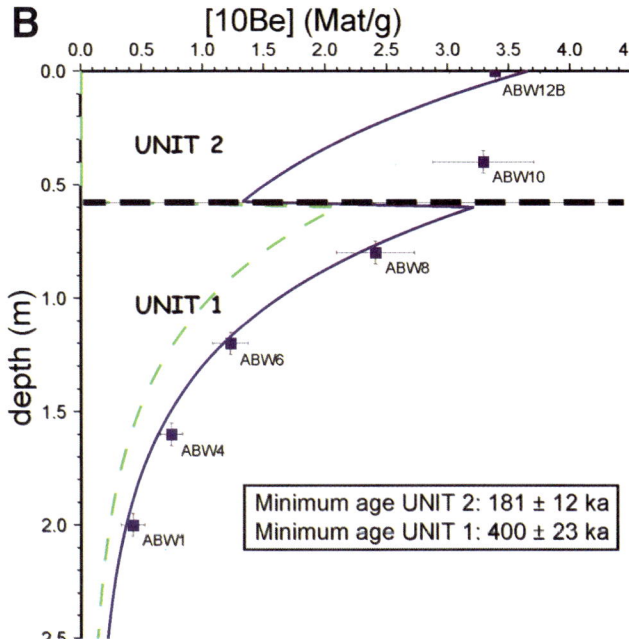

Figure 11. A: Topographic profile across the Artz Bogd fault scarp. B: Results of the ^{10}Be analysis. The ^{10}Be distribution in unit 1 before deposit of unit 2 (with removal of ~40 cm of unit 1) is in dashed line, the present distribution is in solid line. After Vassallo et al. (2005).

REEVALUATED ^{10}Be AGES

Profiles

We present results of the chi-square inversion for all the profiles in Figures 12–16 (see Tables 1–4). For the Artz Bogd site, because of the stratigraphic complexity of the deposits, we did not apply the chi-square model to the profile. However, the distribution of ^{10}Be at depth shows that there is little inherited ^{10}Be within the lower deposits (Fig. 11B). Considering no inheritance in the upper layer as well, and assuming no erosion for both units, we calculated minimum ages of 400 ± 23 ka and 181 ± 12 ka for the lower and upper depositional units, respectively.

For all profiles, except S2 in the eastern fan of Gurvan Bulag, the Q value is lower than 0.001. This is due to the scatter of samples with respect to the theoretical models, and also to the analytical uncertainties that vary significantly from one sample to another (see for instance the case of Noyan Uul, Fig. 12).

Models with the assumption of no inheritance indicate that all surfaces, except S2 at Gurvan Bulag, are at steady state. On the other hand, if we introduce an average inheritance of 0.15 M atoms/g, the models show patterns where the age of the surface is well constrained but the erosion rate for the best fit is zero, which is not realistic. Furthermore, as can be observed from the profiles or the chi-square values, the fits are not improved. Therefore, there are no mathematical reasons to prefer a model with inheritance to a model without inheritance. Nevertheless, for some surfaces (especially S3 in the eastern fan of Gurvan Bulag, the younger one) the value of the average inheritance chosen has a significant influence on the age estimation.

We found a constant optimum erosion rate value of 6 ± 1 m/m.y. for all the old surfaces, other than that at Noyan Uul. At this site, however, the scattering of samples relative to the best-fit model is large. Moreover, samples falling to the right of the best-fit curve (see Fig. 12) have greater uncertainties than other samples, diminishing their importance in the inversion process. If we do not take into account the analytical uncertainty associated with the sample at 1.6 m (lying on the right side of the model and having the largest uncertainty), the inversion leads to a best-fit erosion rate of 6 m/m.y. (Fig. 12).

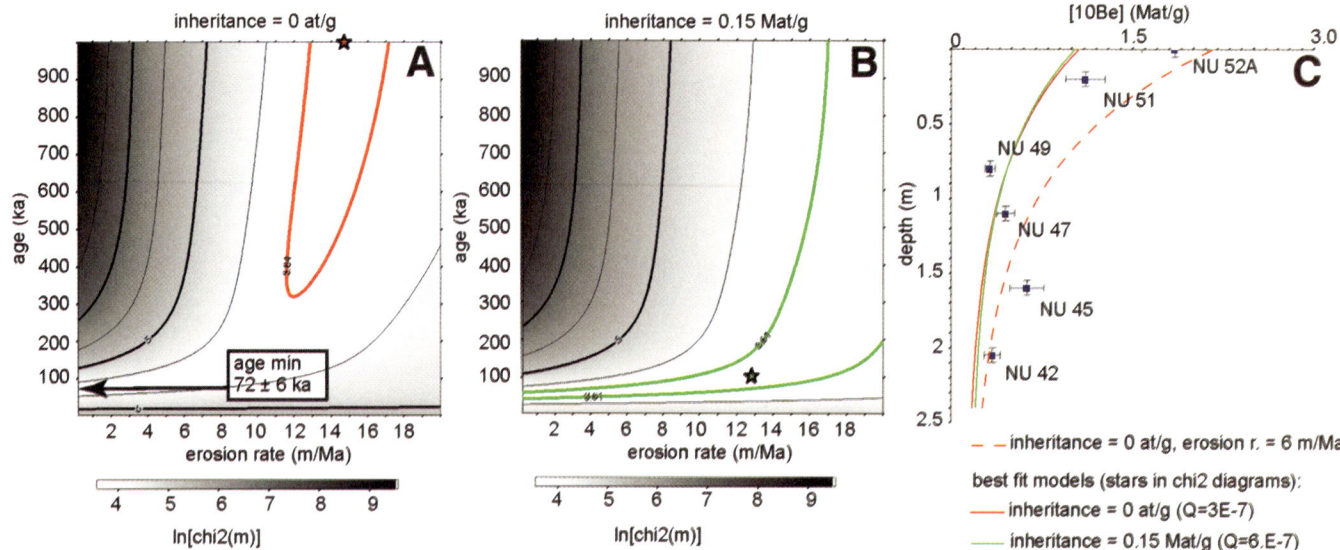

Figure 12. Results of the ¹⁰Be analysis on surface S1 in Noyan Uul. A and B: Plots of the results of the chi-square inversion. Stars represent the best-fit solutions, and colored curve defines the associated 1σ uncertainties. The inversion gives a minimum age of 72 ± 6 ka for the model with no inheritance and 50 ± 8 ka for the model with 0.15 M atoms/g of inheritance. C: Plot of the ¹⁰Be concentration of the samples along the depth profile and best-fit theoretical models issued from the chi-square inversion.

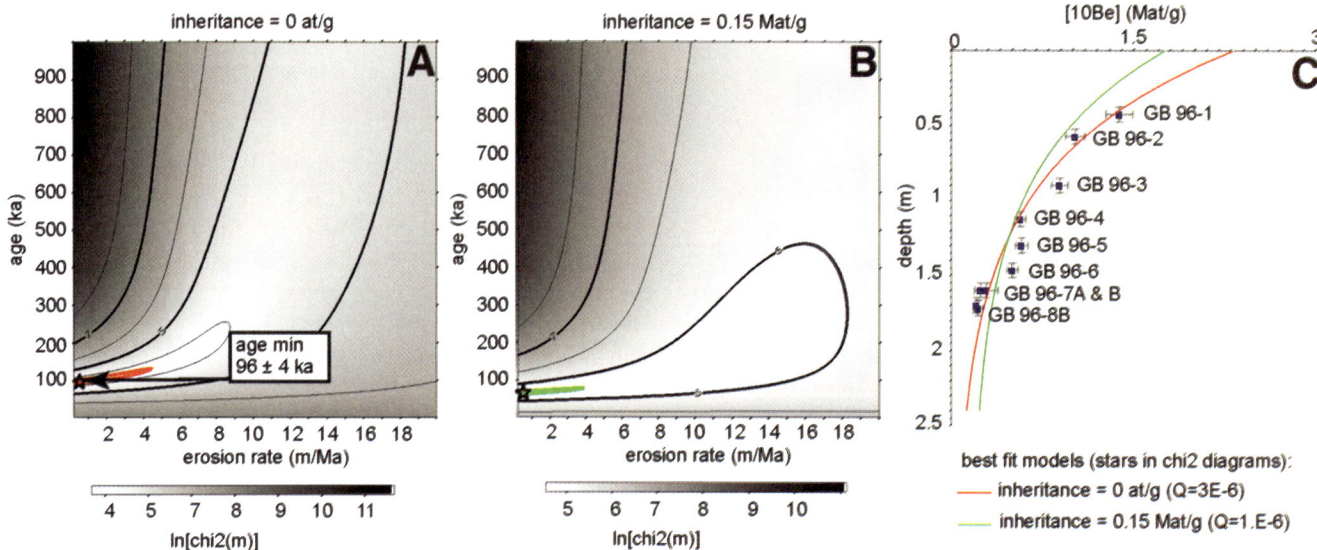

Figure 13. Results of the ¹⁰Be analysis on surface S2 in the western fan (central part of the Gurvan Bulag thrust fault). A and B: Plots of the results of the chi-square inversion, giving a minimum age of 96 ± 4 ka for the model with no inheritance and 66 ± 2 ka for the model with 0.15 M atoms/g of inheritance. C: Plot of the ¹⁰Be concentration of the samples along the depth profile and best-fit theoretical models issued from the chi-square inversion.

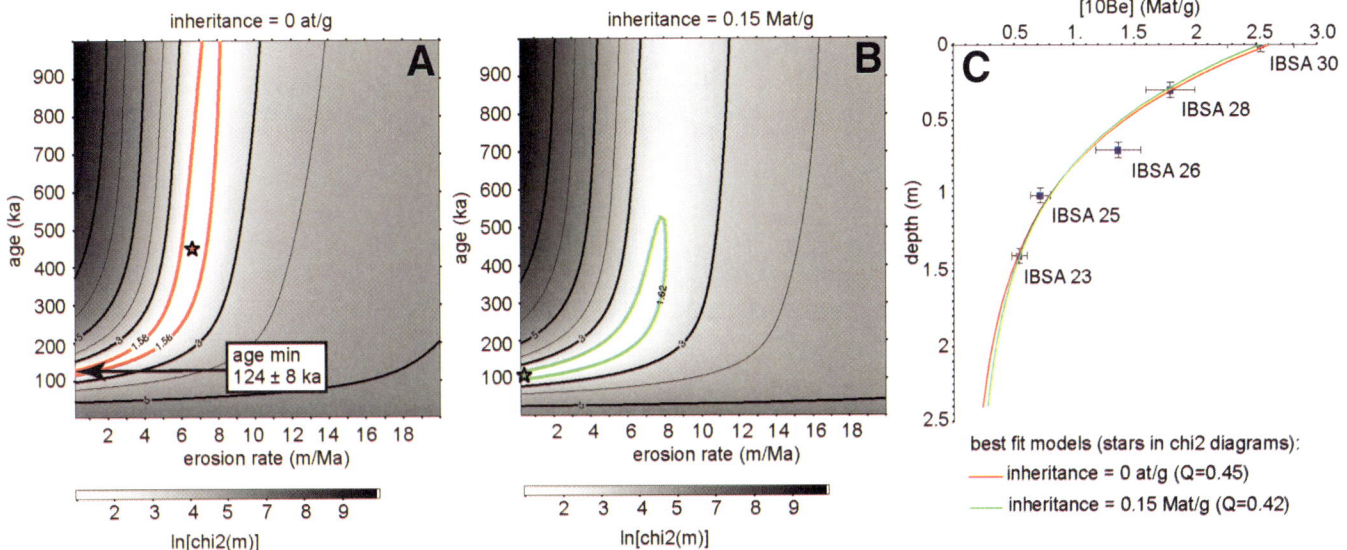

Figure 14. Results of the ^{10}Be analysis on surface S2 in the eastern fan (central part of the Gurvan Bulag thrust fault). A and B: Plots of the results of the chi-square inversion, giving a minimum age of 124 ± 8 ka for the model with no inheritance and 106 ± 6 ka for the model with 0.15 M atoms/g of inheritance. C: Plot of the ^{10}Be concentration of the samples along the depth profile and best-fit theoretical models issued from the chi-square inversion.

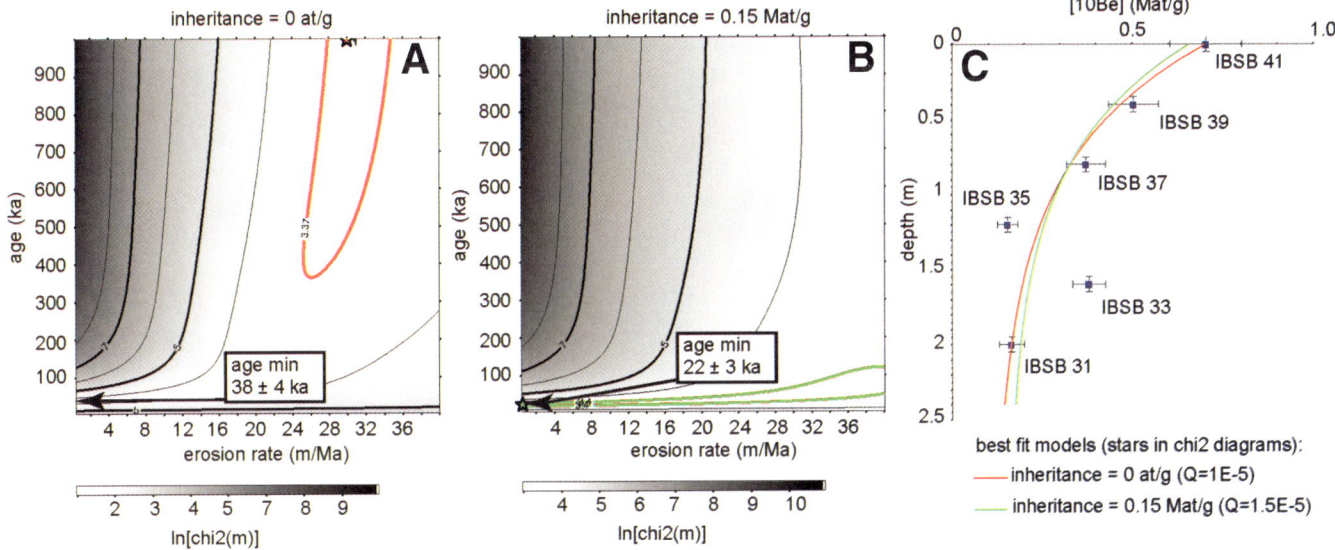

Figure 15. Results of the ^{10}Be analysis on surface S3 in the eastern fan (central part of the Gurvan Bulag thrust fault). A and B: Plots of the results of the chi-square inversion, giving a minimum age of 38 ± 4 ka for the model with no inheritance and 22 ± 3 ka for the model with 0.15 M atoms/g of inheritance. C: Plot of the ^{10}Be concentration of the samples along the depth profile and best-fit theoretical models issued from the chi-square inversion.

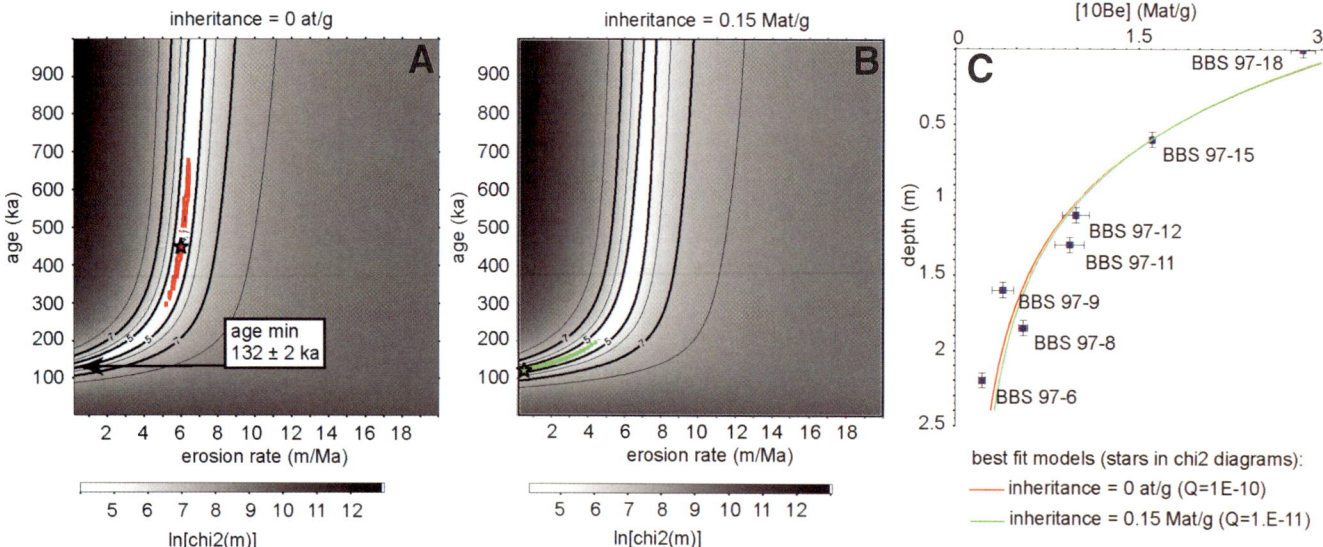

Figure 16. Results of the ^{10}Be analysis on surface S1 at the studied site (Baga Bogd thrust fault). A and B: Plots of the results of the chi-square inversion, giving a minimum age of 132 ± 2 ka for the model with no inheritance and 120 ± 2 ka for the model with 0.15 M atoms/g of inheritance. C: Plot of the ^{10}Be concentration of the samples along the depth profile and best-fit theoretical models issued from the chi-square inversion.

Samples with concentrations approaching steady-state values do not yield exposure ages. Thus, we chose to estimate the ages corresponding to models for no inheritance and no erosion. These ages a priori correspond to minimum ages, although increase of apparent age by inheritance is possible for the young surface S3 at Gurvan Bulag. On the time versus erosion rate diagrams, ages are given by the minimum of the chi-square value on the y-axis (erosion rate = 0). Uncertainties correspond to a confidence interval of 1σ. The minimum ages obtained are 72 ± 6 ka for S1 at Noyan Uul (Fig. 12); 96 ± 4 ka for S2 on the western fan at Gurvan Bulag (Fig. 13), 124 ± 8 ka for S2 (Fig. 14) and 38 ± 4 ka for S3 (Fig. 15) on the eastern fan at the same site; 132 ± 2 ka for S1 at the south of Baga Bogd (Fig. 16).

Surficial Boulders

The reevaluation of surface exposure ages using the ^{10}Be concentrations on surficial boulders (see the penultimate paragraph of the section "Dating Morphotectonic Markers" and Tables 1–3) gives the following results: At Noyan Uul, the reevaluation of the exposure age of the surface S1 gives 61.8 $^{+14.9}/_{-4.7}$ ka. At Gurvan Bulag, within the western fan, exposure ages for surfaces S2, S3, and S4 are 128.7 $^{+15.1}/_{-26.5}$ ka, 14.3 $^{+12.6}/_{-4.1}$ ka, and 4.6 $^{+5.4}/_{-1.8}$ ka, respectively. The age found on S3 was also found in samples collected at a depth ≥2 m in a recent debris flow S5 (15.6 $^{+3.7}/_{-4.0}$ ka) inset in the S4 surface reworking upstream and previously exposed material. At Baga Bogd, the reevaluation of the exposure ages for the surface S3 gives 20.2 $^{+3.7}/_{-3.7}$ ka.

Concluding Remarks on Reevaluated ^{10}Be Ages

For the old fans (surface S1 at Noyan Uul, surface S2 at Gurvan Bulag), minimum ages estimated using surficial boulders or depth profiles (with or without inheritance) are consistent for a given surface. On the other hand, the age of the young surfaces (S3 at Gurvan Bulag and at Baga Bogd) determined using surficial boulders is more consistent with the age given by the modeling of the depth profile of S3 at Gurvan Bulag that takes into account 0.15 M atoms/g of inheritance (Fig. 15B). Because exposure ages calculated from profiles on young surfaces are highly sensitive to inheritance, and considering the larger sample population of surficial boulders, we suggest an age of ca. 20 ka for S3.

RECALCULATION OF SLIP RATES

Taking into account the measured offsets and the reevaluated exposure ages (for all sites, we use ages obtained from the depth profiles except at Gurvan Bulag for surface S3, where we used the weighted mean age obtained from surficial samples collected within the western fan), we calculated fault slip rates (with uncertainties incorporating errors in ages as well as in offsets) during the Late Pleistocene–Holocene within the Gurvan Bogd fault system (Fig. 17).

At Noyan Uul, the left-lateral displacement of 110 ± 10 m and the minimum age of 72 ± 6 ka for surface S1 yield a maximum horizontal slip rate of the Bogd fault of 1.55 ± 0.26 mm/yr during the Late Pleistocene–Holocene.

At Gurvan Bulag, we estimated vertical slip rates over two periods of time, the past ~100–130 k.y. and the past ~20 k.y.,

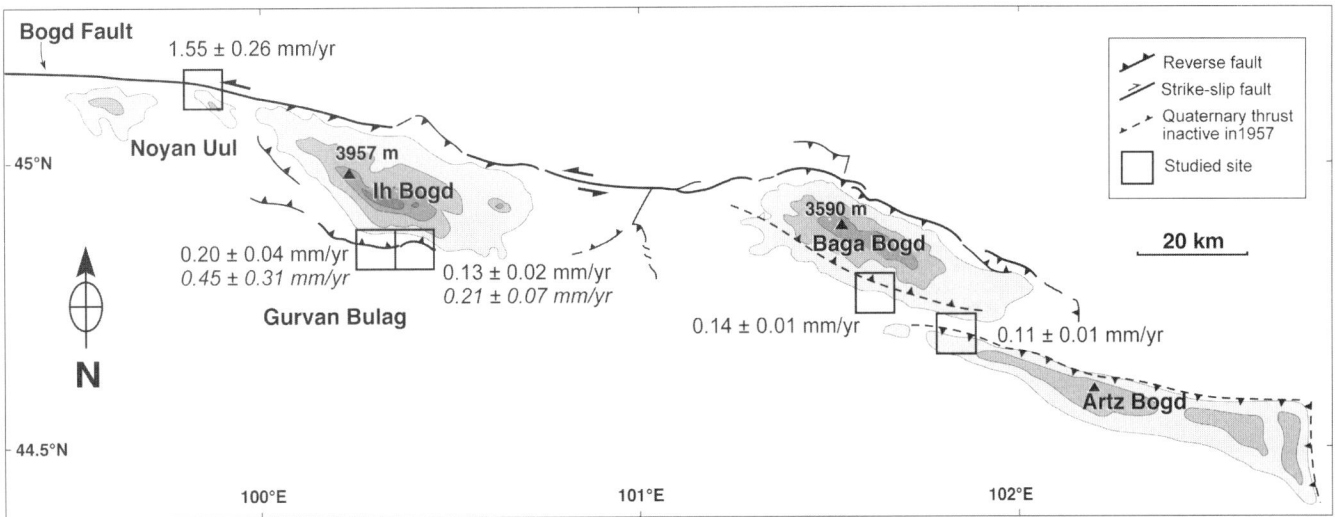

Figure 17. Reestimated maximum fault slip rates within the Gurvan Bogd fault system. All the slip rates are vertical ones, except at Noyan Uul, and calculated for the Late Pleistocene–Holocene. At Gurvan Bulag, the rates in italics are calculated for the past ~15–20 k.y.

from the two main offset surfaces S2 and S3 observed within the two fans. Because it cannot be determined when surfaces formed during the seismic cycle, the Late Pleistocene–Holocene slip rates were bracketed using the mean total offset of surfaces and the mean total offset of surfaces less the mean 1957 offset. On the basis of earlier work (Prentice et al., 2002), we estimated the mean 1957 offset at ~1.5 m within the western fan and ~1 m within the eastern fan. Dividing these bracketed offsets by the surface age yields upper and lower limits on the vertical slip rate. Within the western fan, over the past ~100 k.y., the slip-rate upper and lower limits are 0.21 ± 0.03 mm/yr and 0.19 ± 0.03 mm/yr, respectively. For the past ~20 k.y., the slip-rate upper and lower limits are 0.48 ± 0.28 mm/yr and 0.37 ± 0.23 mm/yr, respectively, considering an offset of 6.5 ± 1.2 m. They increase to 1.18 ± 0.55 mm/yr and 1.08 ± 0.51 mm/yr, respectively, assuming an offset of 17.3 ± 0.4 m. Within the eastern fan, vertical slip-rate upper and lower limits are 0.13 ± 0.01 mm/yr and 0.12 ± 0.01 mm/yr, respectively, for the past ~130 k.y., and 0.23 ± 0.05 mm/yr and 0.19 ± 0.05 mm/yr, respectively, for the past ~20 k.y.

Vertical slip rates estimated on both fans are consistent if we assume that within the western fan, the S3 debris flow was deposited on a preexisting offset morphology: the S2 surface that was already incised by gullies S4. Consequently the age of S4 should be the same as that of S3. This is not inconsistent with ages reported for S4; one of the five S4 samples (MO95-20) yields an age of 15.4 ± 1.5 ka, consistent with the age of the S3 debris flow (Ritz et al., 2003). It may have been deposited on S4 when the debris flow S3 was deposited atop surface S2. Under this scenario the younger exposure ages ($4.6^{+5.4}_{-1.8}$ ka) of the other four S4 samples would represent the results of complex exposure histories, including delivery to the surface by erosion of adjacent slopes.

At South Baga Bogd, the vertical offset of 19.0 ± 0.5 m measured across the scarp and the minimum age of 132 ± 2 ka given by the model for surface S1 yield a maximum vertical slip rate of 0.14 ± 0.01 mm/yr during the Late Pleistocene–Holocene.

At Artz Bogd, surface incision by the drainage network clearly postdates the vertical displacement (20.3 ± 0.5 m) of the deposits. For a minimum age of the upper deposit of 181 ± 12 ka, this yields a maximum vertical slip rate of 0.11 ± 0.01 mm/yr.

SLIP RATES AND RECURRENCE INTERVALS OF EARTHQUAKES WITHIN THE GURVAN BOGD FAULT SYSTEM

It is possible to compare the 1957 dislocation along the main left-lateral strike-slip Bogd fault with dislocations associated with the penultimate earthquake and earlier events. Our morphotectonic study of two sites demonstrates that features (offset streams or shutter ridges) offset during the 1957 event also show well-preserved pre-1957 offsets corresponding to one or more previous events. Digital elevation models of these sites show dislocations that are multiples of the 1957 dislocation. At Noyan Uul, we measured constant dislocation of 5.25 ± 0.25 m (Fig. 18), and at north of Ih Bogd, where the slip is oblique along the fault, we measured several horizontal components that were all multiples of 3 m (Ritz, 2003). This suggests that the successive dislocations along the Bogd fault have the same magnitude and can be defined as characteristic dislocations (e.g., Schwartz and Coppersmith, 1984; Sieh, 1996). Coupled with knowledge of average slip rate, this allows estimation of the return period. At Noyan Uul, for instance, dividing the characteristic dislocation (~5.25 m) by the maximum slip rate (1.55 ± 0.26 mm/yr) yields a minimum average recurrence interval of 3000–4000 yr.

At Gurvan Bulag, the amount of vertical offset for the 1957 is more difficult to establish (see above) and, a fortiori, we do not have estimates of earlier dislocations, so estimation of recurrence

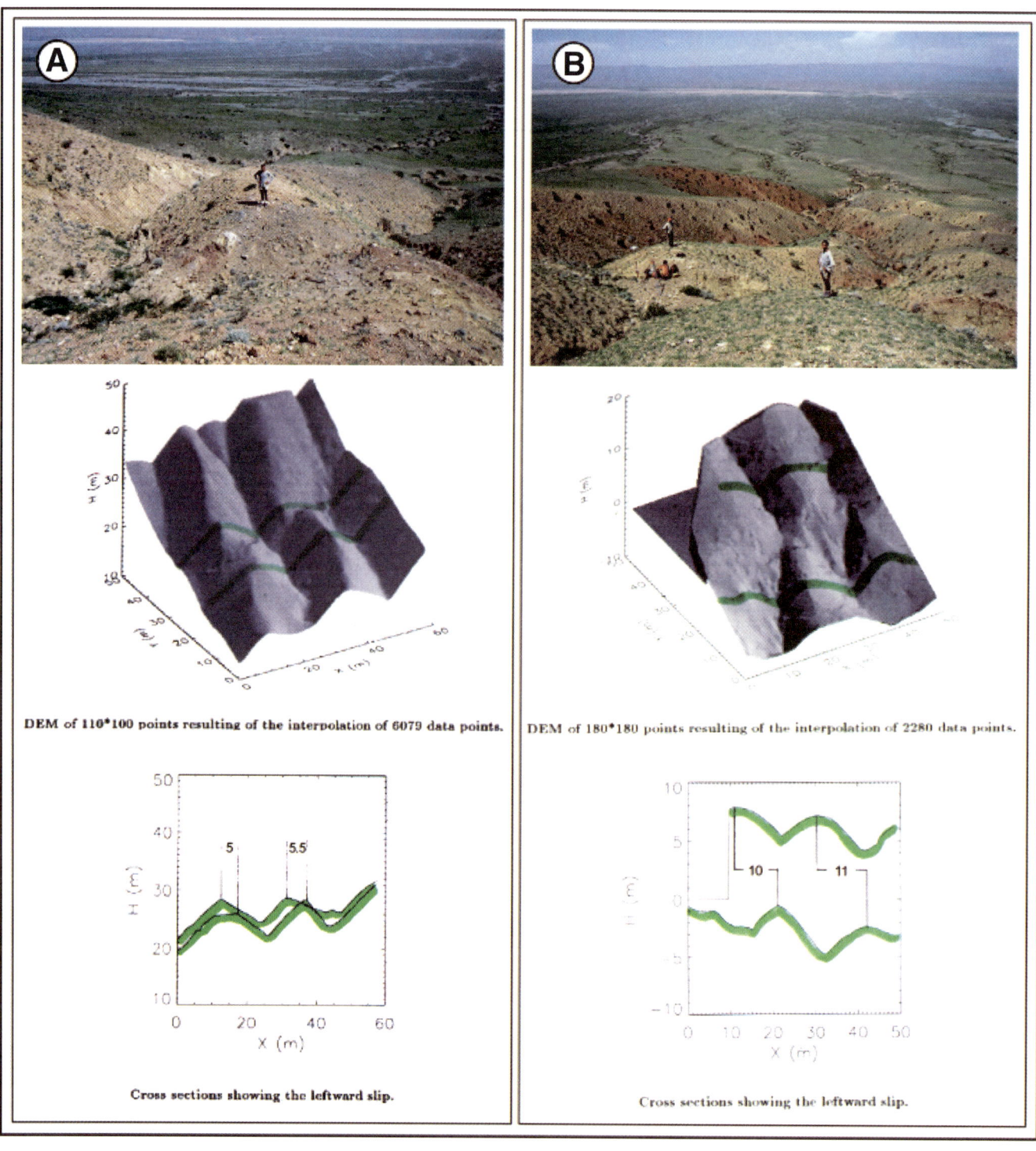

Figure 18. Examples of shutter ridges and streams along the Bogd fault at Noyan Uul with field picture, corresponding digital elevation model, and cross sections. A: 1957 dislocation. B: Dislocation twice the 1957 displacement.

intervals from dating offset markers is problematic. However, paleoseismic investigations (Prentice et al., 2002) indicate an average periodicity of ~3600 yr, similar to that reported for the Bogd fault. This suggests that the two faults may have ruptured simultaneously during earlier events, as they did in 1957, although the paleoseismic data clearly indicate that this was not the case for the penultimate event (Prentice et al., 2002). The hypothesis of a general pattern of simultaneous rupture is also supported by the overall correspondence between the 1957 fault patterns (geometry, kinematics, magnitudes of dislocations) and the topography of the Gurvan Bogd mountain range.

ALLUVIAL SURFACES DEPOSITION AND THEIR EVOLUTION THROUGH TIME IN GOBI-ALTAY: CONSEQUENCES IN TERMS OF SAMPLING STRATEGY

Compilation of morphotectonic studies of the Gurvan Bogd mountain range leads to several conclusions on the history and evolution of alluvial surfaces within the arid climate of Gobi-Altay. ^{10}Be dates suggest the occurrence of episodes of significant aggradation localized in time. Even though the old surface (S1) in Noyan Uul on the northern flank of the Ih Bogd massif appears younger than the two old surfaces in the southern flank (S2 at Gurvan Bulag and S1 at Baga Bogd), it is likely that the three surfaces are associated with the same climatic pulse. Indeed, the morphology on both sides clearly shows that episodes of alluviation are separated by long periods of drought. It is therefore difficult to imagine that the pulses that controlled the alluviation within the Gurvan Bogd massifs were different from one flank to another. Furthermore, Hanks et al. (1997) found another ca. 100 ka alluvial fan along the northern flank of the Ih Bogd massif. Therefore, we believe that the observed differences in the minimum ages are associated with greater postdepositional perturbation of surfaces on the northern flank. Despite the uncertainties inherent in cosmic-ray exposure ages, our results suggest that the two last pulses could have been contemporaneous with global climate changes at the terminations of marine isotope stages (MIS) 2 and 6, and can be interpreted as the effects of major alluvial events due to enhanced stream power reworking the material that accumulated in the slopes or in the drainage network of the upstream basins during drier and colder periods.

Taking into account the exposure ages of the surfaces and their morphologies, our study also enables us to outline the evolution of the geomorphic surfaces and their associated deposits (Fig. 19): Fan surfaces evolve from a bar-and-swale morphology characterized by a high-frequency/low-amplitude topographic signal totally covered by boulder fields (with different sizes of boulders) toward a low-frequency/high-amplitude topographic signal on which the number of standing boulders diminishes gradually. Eventually, the surfaces become flat with no more boulders remaining. This scenario suggests that the erosion rate of the boulders gradually catches up to the rate of removal of fine-grained material on the surface. In the Gurvan Bogd mountain range, this stage appears to be reached after ~100 k.y.

The foregoing observations allow development of sampling strategy for such context: In all cases, because of the potential of inheritance and of complex postdepositional history—especially when studying stepped markers—it is useful to study the distribution of ^{10}Be concentrations at depth, especially when the surface is young. This allows determination of a minimum surface age corrected for preexposure. To get closer to the true age, this protocol can be combined with a statistical analysis of surface concentrations on top of the remaining boulders—if any—that are well embedded in surfaces.

CONCLUSIONS

This article reviews age estimates of faulted morphological markers along the Gurvan Bogd fault system, and documents climatic and tectonic processes in eastern Gobi-Altay. These results suggest episodes of aggradation occurring at the times of major global climatic changes at ca. 15–20 ka and ca. 100–130 ka, and provide evidence for another much earlier aggradational episode occurring before 400 ka.

Dating alluvial surfaces and calculation of their offsets permitted quantification of slip rates along the fault segments bounding the Gurvan Bogd fault system. The main fault, the Bogd fault, has a maximum horizontal left-lateral slip rate of ~1.5 mm/yr during the Late Pleistocene–Holocene, consistent with the GPS measurements (Calais et al., 2003). Segments of reverse faulting along the Gurvan Bogd fault system have vertical slip rates between 0.1 and 0.2 mm/yr during the past ~100–130 k.y. At Gurvan Bulag, the activity of the fault appears to have increased slightly since ca. 15–20 ka. Characteristic dislocations observed along the Bogd fault suggest return periods of earthquakes similar to 1957 between 3000 and 4000 yr. If we extrapolate the Late Pleistocene–Holocene rates to a longer period of time, the uplift of the summit erosional surface of the Ih Bogd massif (the difference in height between the summit surface and the bounding faults being ~2000 m) would have begun between 10 and 20 Ma. When compared with the 0.2–0.3 mm/yr uplift rate of the Ih Bogd massif estimated from dating of strath terraces in the Bitut River (Vassallo et al., 2004), our results suggest that the thrust faults that we studied do not fully account for the uplift of the Gurvan Bogd massif. This is consistent with the suggestion of Bayasgalan et al. (1999a) that other thrust faults also contribute to the uplift.

ACKNOWLEDGMENTS

This paper is a review of several studies in Gobi-Altay since 1992. Seven expeditions involving a total of five months of fieldwork were undertaken in collaboration with the Center of Informatic and Remote Sensing; the Mongolian University of Science and Technology, Ulaanbataar; the Bullard Laboratories, Cambridge, UK; the U.S. Geological Survey, Menlo Park;

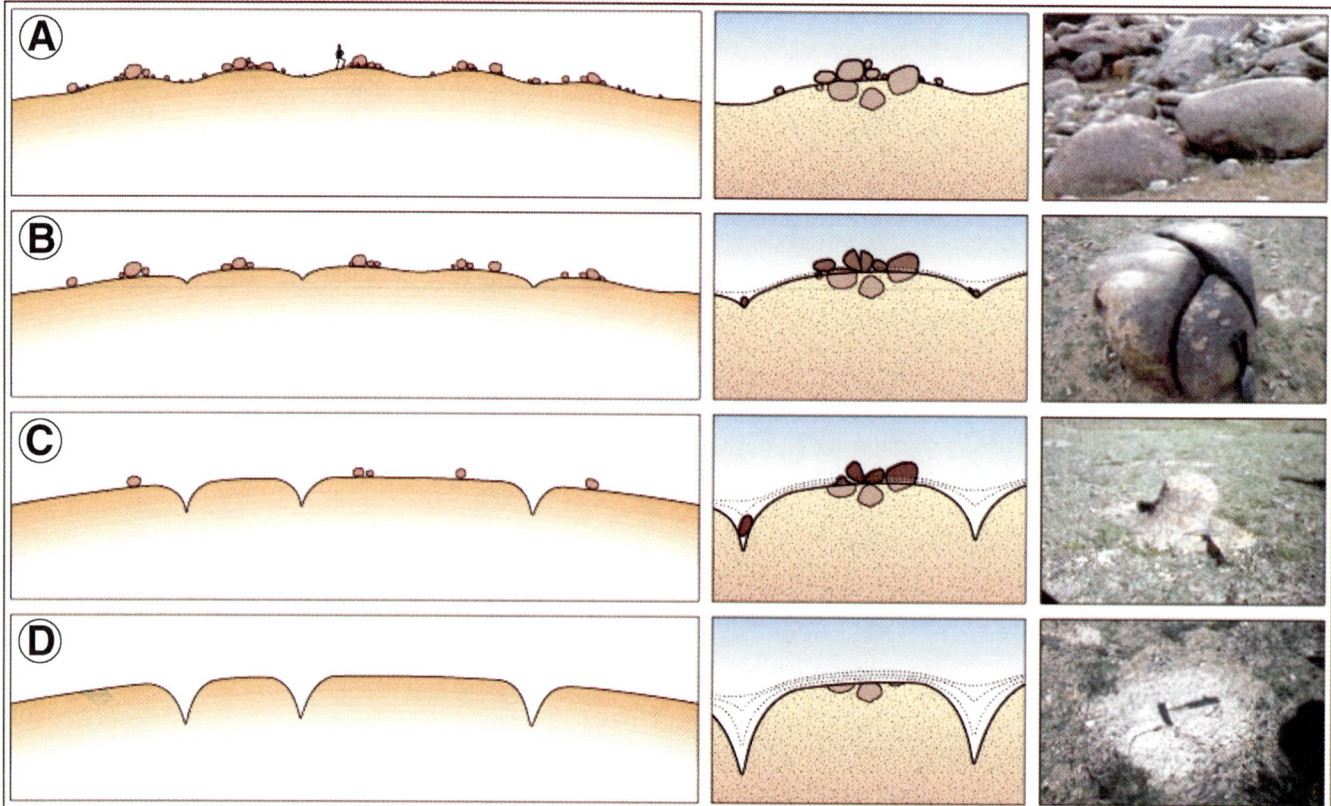

Figure 19. Scenario of the evolution of an alluvial fan surface abandoned after its displacement along a fault. A: "Bar-and-swale" morphology with largest boulders along bars. B: Smoothing of bars by collapse of boulders and diffusion of finer material (boulders are weathered and fissured). C: The surface gets gradually flatter and more incised (few hard-core boulders still stand on the surface). D: The surface is totally flattened with deep incisions (phantoms of boulders are eroded at the same erosion rate as the surface).

and the LLNL, Livermore. We thank again A. Bayasgalan, K. Berryman, E. Calais, J. Deverchères, B. Enhtuvshin, R. Finkel, P. Galsan, M. Ganzorig, T. Hanks, J. Jackson, K. Kendrick, H. Philip, C. Prentice, G. Raisbeck, A. Schlupp, D. Schwartz, M. Todbileg, and F. Yiou for fruitful discussions. Many thanks to Anne Delplanque for the drawings. We acknowledge L. Siame and an anonymous referee for their reviews that helped us to improve the original manuscript.

REFERENCES CITED

Anderson, R.S., Repka, J.L., and Dick, G.S., 1996, Explicit treatment of inheritance in dating depositional surfaces using in situ ^{10}Be and ^{26}Al: Geology, v. 24, p. 47–51, doi: 10.1130/0091-7613(1996)024<0047:ETOIID>2.3.CO;2.

Baljinnyam, I., Bayasgalan, A., Borisov, B.A., Cisternas, A., Dem'yanovich, M.G., Ganbaatar, L., Kochetkov, V.M., Kurushin, R.A., Molnar, P., Philip, H., and Vashchilov, Yu.Ya., 1993, Ruptures of major earthquakes and active deformation in Mongolia and its surroundings: Geological Society of America Memoir 181, 62 p.

Bayasgalan, A., 1999, Active tectonics of Mongolia [Ph.D. thesis]: Cambridge, UK, University of Cambridge, 182 p.

Bayasgalan, A., Jackson, J., Ritz, J.-F., and Carretier, S., 1999a, Field examples of strike-slip fault terminations in Mongolia and their tectonic significance: Tectonics, v. 18, p. 394–411, doi: 10.1029/1999TC900007.

Bayasgalan, A., Jackson, J., Ritz, J.-F., and Carretier, S., 1999b, "Forebergs", flowers structures, and the development of large intra-continental strike-slip fault: The Gurvan Bogd fault system in Mongolia: Journal of Structural Geology, v. 21, p. 1285–1302, doi: 10.1016/S0191-8141(99)00064-4.

Bierman, P.R., 1994, Using in situ produced cosmogenic isotopes to estimate rates of landscape evolution: A review from the geomorphic perspective: Journal of Geophysical Research, v. 99, p. 13,885–13,896, doi: 10.1029/94JB00459.

Bierman, P.R., Gillespie, A.R., and Caffee, M.W., 1995, Cosmogenic ages for earthquakes recurrence intervals and debris flow fan deposition, Owen valley, California: Science, v. 270, p. 447–450.

Braucher, R., Brown, E.T., Bourlès, D.L., and Colin, F., 2003, In situ produced ^{10}Be measurements at great depths: Implications for production rates by fast muons: Earth and Planetary Science Letters, v. 211, p. 251–258, doi: 10.1016/S0012-821X(03)00205-X.

Brown, E.T., Edmond, J.M., Raisbeck, G.M., Yiou, F., Kurz, M.D., and Brook, E.J., 1991, Examination of surface exposure ages of Antarctic moraines using in situ produced ^{10}Be et ^{26}Al: Geochimica et Cosmochimica Acta, v. 55, p. 2269–2283, doi: 10.1016/0016-7037(91)90103 C.

Brown, E.T., Brook, E.J., Raisbeck, G.M., Yiou, F., and Kurz, M.D., 1992, Effective attenuation lengths of cosmic rays producing ^{10}Be and ^{26}Al in quartz: Implication for exposure age dating: Geophysical Research Letters, v. 19, no. 4, p. 369–372.

Brown, E.T., Bourlès, D.L., Colin, F., Raisbeck, G.M., Yiou, F., and Desgarceaux, S., 1995, Evidence for muon-induced in situ production of ^{10}Be in near-surface rocks from the Congo: Geophysical Research Letters, v. 22, p. 703–706, doi: 10.1029/95GL00167.

Brown, E.T., Bendick, R., Bourlès, D.L., Gaur, V., Molnar, P., Raisbeck, G.M., and Yiou, F., 2002, Slip rates of the Karakorum fault, Ladakh, India, determined using cosmic ray exposure dating of debris flows and moraines: Journal of Geophysical Research, v. 107, no. B9, p. 2192, doi: 10.1029/2000JB000100.

Bucknam, R.C., and Anderson, R.E., 1979, Estimation of fault-scarp ages from a scarp-height-slope-angle relationship: Geology, v. 7, p. 11–14, doi: 10.1130/0091-7613(1979)7<11:EOFAFA>2.0.CO;2.

Burbank, D.W., and Anderson, R.S., 2001, Tectonic geomorphology: Malden, USA, Blackwell Science, 274 p.

Calais, E., Vergnolle, M., Sankov, V., Lukhnev, A., Miroshnitchenko, A., Amarjargal, S., and Dervèrchère, J., 2003, GPS measurements of crustal deformation in the Baikal-Mongolia area (1994–2002): Implications on current kinematics of Asia: Journal of Geophysical Research, v. 108, no. B10, p. 2501, doi: 10.1029/2002JB002373.

Carretier, S., 2000, Cycle sismique et surrection de la chaîne de Gurvan Bogd (Mongolie): Approche de la géomorphologie quantitative [Ph.D. thesis]: Université de Montpellier 2, 324 p.

Carretier, S., Lucazeau, F., and Ritz, J.-F., 1998, Approche numérique des intéractions entre climat, faille active et érosion: Compte Rendus Academie Sciences Paris, v. 326, p. 391–397.

Carretier, S., Ritz, J.F., Jackson, J., and Bayasgalan, A., 2002, Morphological dating of cumulative reverse fault scarp: Examples from the Gurvan Bogd fault system, Mongolia: Geophysical Journal International, v. 148, p. 256–277, doi: 10.1046/j.1365-246X.2002.01599.x.

Cunningham, D.W., 1998, Lithospheric controls on late Cenozoic construction of the Mongolian Altai: Tectonics, v. 17, p. 891–902, doi: 10.1029/1998TC900001.

Cunningham, D.W., Windley, B.F., Dorjnamjaa, D., Badamgarov, G., and Saandar, M., 1996, A structural transect across the Mongolian western Altai: Active transpressional mountain building in central Asia: Tectonics, v. 15, p. 142–156, doi: 10.1029/95TC02354.

Cunningham, D.W., Windley, B.F., Owen, L.A., Barry, T., Dorjnamjaa, D., and Badamgarav, J., 1997, Geometry and style of partitioned deformation within a late Cenozoic transpressional zone in the eastern Gobi Altai Mountains, Mongolia: Tectonophysics, v. 277, p. 285–306, doi: 10.1016/S0040-1951(97)00034-6.

Davis, J.C., Proctor, I.D., Southon, J.R., Caffee, M.W., Heikkinen, D.W., Roberts, M.L., Moore, T.L., Turtletaub, K.W., Nelson, D.E., Loyd, D.H., and Vogel, J.S., 1990, Lawrence Livermore National Laboratory–University of California Center for Accelerator Mass Spectrometry facility and research program: Nuclear Instruments and Methods in Physics Research, B52, p. 269–272.

Florensov, N.A., and Solonenko, V.P., editors, 1965, The Gobi-Altay earthquake: Washington, D.C., U.S. Department of Commerce, 424 p.

Hancock, G.S., Anderson, R.S., Chadwick, O.A., and Finkel, R.C., 1999, Dating fluvial terraces with ^{10}Be and ^{26}Al profiles: Application to the Wind River, Wyoming: Geomorphology, v. 27, p. 41–60, doi: 10.1016/S0169-555X(98)00089-0.

Hanks, T., Ritz, J.-F., Kendrick, K., Finkel, R.C., and Garvin, C.D., 1997, Uplift rates in a continental interior: Faulting offsets of a ~100 ka abandoned fan along the Bogd fault, southern Mongolia: Proceedings of the Penrose Conference on the Tectonics of Continental Interiors, 23–28 September 1997, Cedar City, Utah.

Hanks, T.C., Bucknam, R.C., Lajoie, K.R., and Wallace, R.E., 1984, Modification of wave-cut and faulting-controlled landforms: Journal of Geophysical Research, v. 89, p. 5771–5790.

Klein, J., Giegengack, R., Middleton, R., Sharma, P., Underwood, J.R., and Weeks, R.A., 1986, Revealing histories of exposure using in situ produced ^{26}Al and ^{10}Be in Libyan Desert Glass: Radiocarbon, v. 28, no. 2A, p. 547–555.

Kurushin, R.A., Bayasgalan, A., Ölziybat, M., Enkhtuvshin, B., Molnar, P., Bayarsayhan, C., Hudnut, K.W., and Lin, J., 1997, The surface rupture of the 1957 Gobi-Altay, Mongolia, earthquake: Geological Society of America Special Paper 320, 143 p.

Meriaux, A.S., Ryerson, F.J., Tapponnier, P., Van der Woerd, J., Finkel, R.C., Xiwei Xu, Zhiqin Xu, and Caffee, M.W., 2004, Rapid slip along the central Altyn Tagh fault: Morphochronologic evidence from Cherchen He and Sulamu Tagh: Journal of Geophysical Research, v. 109, B06401, doi: 10.1029/2003JB002556.

Meriaux, A.S., Tapponnier, P., Ryerson, F.J., Xu Xiwei, King, G., Van der Woerd, J., Finkel, R.C., Li Haibing, Caffee, M.W., Xu Zhiqin, and Chen Wenbin, 2005, The Aksay segment of the northern Altyn Tagh fault: Tectonic geomorphology, landscape evolution, and Holocene slip rate: Journal of Geophysical Research, v. 110, B04404, doi: 10.1029/2004JB003210.

Molnar, P., and Tapponnier, P., 1975, Cenozoic tectonics of Asia: Effects of a continental collision: Science, v. 189, p. 419–426.

Molnar, P., and D. Qidong, 1984, Faulting associated with large earthquakes and the average rate of deformation in Central and eastern Asia: Journal of Geophysical Research, v. 89, p. 6203–6227.

Nishiizumi, K., Lal, D., Klein, J., Middleton, R., and Arnold, J.R., 1986, Production of ^{10}Be and ^{26}Al by cosmic rays in terrestrial quartz in situ and implications for erosion rates: Nature, v. 319, p. 134–135, doi: 10.1038/319134a0.

Owen, L.A., Cunningham, D.W., Richards, B.W., Rhodes, E., Windley, B.F., Dorjnamjaa, D., and Badamgarav, J., 1999, Timing of formation of forebergs in the northeastern Gobi-Altai, Mongolia: Implications for mountain uplift rates and earthquake recurrence intervals: Geological Society [London] Journal, v. 156, p. 457–464.

Prentice, C., Kendrick, K., Berryman, K., Bayasgalan, A., Ritz, J.F., and Spencer, J.Q., 2002, Prehistoric ruptures of the Gurvan Bulag fault, Gobi Altay, Mongolia: Journal of Geophysical Research, v. 107, p. 2321, doi: 10.1029/2001JB000803.

Press, W.H., Teukolsky, S.A., Vetterling, W.T., and Flannery, B.P., 1996, Numerical recipes in Fortran 90—The art of parallel scientific computing: Cambridge, UK, Cambridge University Press, 1356 p.

Raisbeck, G.M., Yiou, F., Bourlès, D.L., Lestringuez, J., and Deboffe, D., 1987, Measurements of ^{10}Be and ^{26}Al with a Tandetron AMS facility: Nuclear Instruments and Methods in Physics Research, v. 29, p. 22–26, doi: 10.1016/0168-583X(87)90196-0.

Raisbeck, G.M., Yiou, F., Bourlès, D.L., Brown, E.T., Deboffle, D., Jouhanneau, P., Lestringuez, J., and Zhou, Z.Q., 1994, The AMS facility at Gif-sur-Yvette: Progress, perturbations and projects: Nuclear Instruments and Methods in Physics Research, v. 92, p. 43–46, doi: 10.1016/0168-583X(94)95972-2.

Repka, J.L., Anderson, R.S., and Finkel, R.C., 1997, Cosmogenic dating of fluvial terraces, Fremont River, Utah: Earth and Planetary Science Letters, v. 152, no. 1-4, p. 59–73.

Ritz, J.F., 2003, Analyse de la tectonique active en domaine continental: Approche morphotectonique et paléosismologique [HDR thesis]: Université de Montpellier 2, V1, 72 p.

Ritz, J.F., Brown, E.T., Bourlès, D.L., Philip, H., Schlupp, A., Raisbeck, G.M., Yiou, F., and Enkhtuvshin, B., 1995, Slip rates along active faults estimated with cosmic-ray-exposure dates: Application to the Bogd fault, Gobi-Altaï, Mongolia: Geology, v. 23, p. 1019–1022, doi: 10.1130/0091-7613(1995)023<1019:SRAAFE>2.3.CO;2.

Ritz, J.F., Bourlès, D., Brown, E.T., Carretier, S., Chery, J., Enhtuvushin, B., Galsan, P., Finkel, R.C., Hanks, T.C., Kendrick, K.J., Philip, H., Raisbeck, G., Schlupp, A., Schwartz, D.P., and Yiou, F., 2003, Late Pleistocene to Holocene slip rates for the Gurvan Bulag thrust fault (Gobi-Altay, Mongolia) estimated with ^{10}Be dates: Journal of Geophysical Research, v. 108, no. B3, p. 2162, doi: 10.1029/2001JB000553.

Schlupp, A., 1996, Néotectonique de la Mongolie Occidentale analysée à partir de données de terrain, sismologiques et satellitaires [Ph.D. thesis]: Strasbourg, University Louis Pasteur, 172 p.

Schwartz, D., and Coppersmith, K., 1984, Fault behavior and characteristic earthquakes: Examples from the Wasatch and San Andreas faults: Journal of Geophysical Research, v. 89, p. 5681–5698.

Siame, L., Braucher, R., and Bourlès, D., 2000, Les nucléides cosmogéniques produits in situ: de nouveaux outils en géomorphologie quantitative: Bulletin de la Société Géologique de France, v. 171, p. 383–396, doi: 10.2113/171.4.383.

Siame, L., Bellier, O., Braucher, R., Sébrier, M., Cushing, M., Bourlès, D., Hamelin, B., Baroux, E., de Voogd, B., Raisbeck, G., and Yiou, F., 2004, Local erosion rates versus active tectonics: Cosmic ray exposure modelling in Provence (south-east France): Earth and Planetary Science Letters, v. 7010, p. 1–21.

Sieh, K., 1996, The repetition of large-earthquake ruptures: Proceedings of the National Academy of Sciences of the United States of America, v. 93, p. 3764–3771, doi: 10.1073/pnas.93.9.3764.

Stone, J.O., 2000, Air pressure and cosmogenic isotope production: Journal of Geophysical Research, v. 105, no. B10, p. 23,753–23,759, doi: 10.1029/2000JB900181.

Tapponnier, P., and Molnar, P., 1979, Active faulting and Cenozoic tectonics of the Tien Shan, Mongolian and Baykal regions: Journal of Geophysical Research, v. 84, p. 3425–3459.

Van der Woerd, J., Ryerson, F.J., Tapponnier, P., Gaudemer, Y., Finkel, R., Meriaux, A.S., Caffee, M.W., Guoguang, Z., and Qunlu, H., 1998, Holocene left-slip rate determined by cosmogenic surface dating on the Xidatan segment of the Kunlun fault (Qinghai, China): Geology, v. 26, p. 695–698, doi: 10.1130/0091-7613(1998)026<0695:HLSRDB>2.3.CO;2.

Van der Woerd, J., Tapponnier, P., Ryerson, F.J., Meriaux, A.S., Meyer, B., Gaudemer, Y., Finkel, R., Caffee, M.W., Zhao, G.G., and Xu, Z.Q., 2002, Uniform postglacial slip rate along the central 600 km of the Kunlun fault (Tibet), from Al-26, Be-10, and C-14 dating of riser offsets, and climatic origin of the regional morphology: Geophysical Journal International, v. 148, p. 356–388, doi: 10.1046/j.1365-246x.2002.01556.x.

Vassallo, R., Ritz, J-F., Braucher, R., Jolivet, M., Larroque, C., Sue, C., Todbileg, M., Javhaa, D., and Bourlès, D.L., 2004, Timing and uplift rates in Cenozoic transpressional mountain ranges within the Mongolia-Siberia region, RST Strasbourg, September 2004, Abstract RSTGV-A-00210.

Vassallo, R., Ritz, J.-F., Braucher, R., and Carretier, S., 2005, Dating faulted alluvial fans with cosmogenic ^{10}Be in the Gurvan Bogd mountain (Gobi-Altay, Mongolia): Climatic and tectonic implications: Terra Nova, v. 17, p. 278–285, doi: 10.1111/j.1365-3121.2005.00612.x.

MANUSCRIPT ACCEPTED BY THE SOCIETY 11 APRIL 2006

Eroding the land: Steady-state and stochastic rates and processes through a cosmogenic lens

Arjun M. Heimsath[†]
Department of Earth Sciences, Dartmouth College, Hanover, New Hampshire 03755, USA

ABSTRACT

Quantifying erosion rates and processes remains a central focus of studying the Earth's surface. Measurement of in situ–produced cosmogenic radionuclides (CRNs) enables a level of quantification that would otherwise be impossible or fraught with uncertainty and expense. Remarkable success stories punctuate the field over the last decade as CRN-based methodologies are pushed to new limits. Inherent to all is an assumption of steady-state rates and processes. This paper focuses on the use of cosmogenic ^{10}Be and ^{26}Al, extracted from quartz in bedrock, saprolite, and detrital material to quantify sediment production or erosion rates and processes. Previous results from two very different field areas are reviewed to highlight the potential for non-steady-state processes in shaping soil-mantled landscapes. With this potential in mind, a numerical model is presented, following a review of the CRN conceptual framework, to test the effects of non-steady-state erosion rates and processes on CRN concentrations. Results from this model focus on ^{10}Be concentrations accumulated under modeled variations in erosion rates with different ranges, frequencies, and styles of variability. In general, the higher the maximum erosion rate, the higher the impact on the CRN concentration and, therefore, the more likely that point measurements will capture the variable signal. Conversely, the higher the frequency of erosional variation, the less likely point measures are to accurately determine rates, but the closer the inferred rate is to the mean of the long-term erosion rate. Modeling results are applicable for point-specific erosion rates, but endorse the catchment-averaged approach for determining average rates. Potentially large uncertainties emphasize the need for careful sample selection, with adequate numbers of samples collected for quantifying the processes eroding the land. The two field examples show how analyzing enough samples can define a clear soil production function despite the potential for non-steady-state processes. The model presented here is ready for application to catchment-averaged processes, as well as modeling the role of muons in variable erosion rate scenarios.

Keywords: ^{10}Be and ^{26}Al, erosion, soil production, stochastic processes, landscape evolution.

[†]E-mail: arjun.heimsath@dartmouth.edu.

INTRODUCTION

In situ–produced cosmogenic radionuclides (CRNs) are used extensively to infer erosion rates acting upon a wide variety of landforms. Original application of this methodology focused on recently glaciated bedrock where the large amount of material removed during the Last Glacial Maximum provided a well-constrained case to determine nuclide production rates (Nishiizumi et al., 1986, 1989). The methodology is well developed (Lal, 1988, 1991) and well reviewed for geomorphic applications (Bierman, 1994; Bierman and Nichols, 2004; Cerling and Craig, 1994; Cockburn and Summerfield, 2004; Gosse and Phillips, 2001; Nishiizumi et al., 1993). Subsequent applications built upon the success of determining exposure ages of relatively unweathered rock to infer long-term, steady-state erosion rates from exposed bedrock surfaces (Bierman and Caffee, 2001; Bierman et al., 1999; Bierman and Turner, 1995; Small et al., 1997). More complicated use of CRN concentrations toward geomorphic applications involved determining river incision rates from exposure ages of strath terraces (Burbank et al., 1996; Pratt et al., 2002; Reusser et al., 2004; Weissel and Seidl, 1998), as well as from exposure ages inferred from concentrations in fluvial cobbles deposited on terraces (Anderson et al., 1996; Hancock et al., 1999; Perg et al., 2001; Repka et al., 1997). In situ concentrations of CRNs from bedrock beneath different depths of soil also led to the first determination of the rates of soil production (Heimsath et al., 1997) with subsequent and similar findings from the base of a retreating escarpment (Heimsath et al., 2000). A powerful methodology for determining average erosion rates for small watersheds was also developed, which relies on measuring CRN concentrations from stream sediments (Bierman and Steig, 1996; Brown et al., 1995b; Granger et al., 1996; Riebe et al., 2000). Each of these applications was relatively well constrained such that the exposure history of the sampled surfaces or small catchment areas could be reasonably inferred. Recent studies are pushing the limits of these constraints and are using burial age dating (Granger et al., 2001; Granger and Smith, 2000; Granger and Stock, 2004; Stock et al., 2005) as well as further applications of depth-profile dating (Perg et al., 2001; Schaller et al., 2004) to infer landscape-scale uplift and erosion rates. Similarly, application of the basin-averaged methodology is being applied across larger and larger catchments (Cockburn and Summerfield, 2004; Matmon et al., 2003a; Schaller et al., 2001), in more and more complicated erosional settings (Hewawasam et al., 2003; Wobus et al., 2005), and to develop landscape-scale sediment budgets (Clapp et al., 2000, 2002; Nichols et al., 2005). All in all, widespread use of CRNs enables a phenomenal amount of work on quantifying geomorphic rates and processes (Bierman and Nichols, 2004; Granger, 2002). Continued application of the CRN method of determining erosion rates and landscape evolution histories to more and more complicated geomorphic problems will depend on being able to more narrowly constrain the exposure history of the samples, a point emphasized by many of the above studies, but only tested by a few (Bierman and Steig, 1996; Lal, 1991; Small et al., 1997).

Here I briefly review the state of using CRN measurements (focusing specifically on in situ–produced ^{10}Be and ^{26}Al) to infer soil production and erosion rates. Using this context, I present a numerical model to test the effect of stochastic sediment production and transport processes on the accumulation of CRNs in eroding bedrock that can be either exposed or beneath a mobile soil mantle. This work was motivated by field observations from two very different field sites. First, a steep, soil-mantled hilly landscape in the Oregon Coast Range (Fig. 1) where geomorphic processes of erosion and soil production were identified to be nonuniform and potentially catastrophic (Heimsath et al., 2001b; Montgomery et al., 1998; Roering and Gerber, 2005; Schmidt, 1999). Measurements of the in situ–produced CRNs, ^{10}Be and ^{26}Al, from weathered bedrock beneath an actively eroding soil mantle led to determining an apparent soil production function, where soil production decreased exponentially with increasing soil thickness (Heimsath et al., 2001b). The second site, a gentle upland landscape eroding almost an order of magnitude more slowly (Fig. 2), was likely to have undergone dramatic changes in the dominant geomorphic processes due to Pleistocene climate changes (Heimsath et al., 2001a). Conclusions reached for both sites depended in part on assuming steady-state conditions for both the overlying soil mantle and the erosion rate (equivalent to the soil production rate) of the underlying bedrock.

Observations of how both the overlying soil thickness and the local production rate of soil, or erosion rate of exposed bedrock, may vary in a stochastic (or even predictable) ways across the landscape led to the development of the numerical model presented here to calculate CRN concentrations under non-steady-state conditions. Importantly, the model here differs from those presented by others (Bierman and Steig, 1996; Lal, 1991; Small et al., 1997) in that it integrates the governing differential equation for the CRN concentrations under changing conditions of erosion and overlying soil thickness (i.e., shielding from cosmic-ray penetration), using numerical techniques rather than solving the equations analytically. I present results specifically with the deviations from steady state for the Oregon and southeastern Australia cases in mind, using extensive in situ measurements of ^{10}Be and ^{26}Al for comparison with the modeling results. Furthermore, the model is freely available[1], and has broad application as a test for the potential use of CRN methodology for tackling the more complex geomorphic problems that are currently being pursued.

To develop this model, I first review the geomorphic conceptual framework and then explicitly derive the differential equations used in CRN applications. Brief field descriptions with a summary of results from the two field areas motivating this study are then presented, with the observations and data from the respective studies driving the modeling. Specific discussion into how soil production functions and erosion rates can vary across landscapes serves not only as a review of results from different field areas, but also as a launching point for future work. I

---

[1]Contact A.M.H. for free Matlab® code and instructions for model presented herein.

Figure 1. Photograph of a freshly clear-cut slope in the Oregon Coast Range near the field site used by Heimsath et al. (2001b). The clear-cut reveals exposed bedrock along the sharply convex-up ridge crests and the ridge-and-valley topography characteristic of the region. Light-colored patch at the uppermost extent of the valley to the far right is bedrock exposed by recent landsliding. Scale of the clear-cut at the ridgeline is ~200 m. Average slopes are ~45°.

Figure 2. Photograph of Frogs Hollow field area, looking north from the south side of the Bredbo River. Unchanneled valley to the left of the mid-photo ridge was sampled to yield average erosion rate shown by *C* on Figure 4; average and bedrock incision rates are of the Bredbo, flowing right to left in the midst of the trees in the foreground. Note outcropping tors especially prevalent upon the convex-up noses. The "tor profile" used for the data shown in Figure 6 is the highest visible tor on the ridge crest of the leftmost ridge. Results from the nuclide concentrations suggested "stripping" of the landscape at ca. 150 ka (Heimsath et al., 2001a). Relief of the unchanneled valley is ~17 m; the tor in the middle of the photograph is about three meters high and two and a half wide.

conclude, therefore, by placing the modeling results into a broader context by suggesting what the next steps are for application of such modeling efforts. Specifically, the model as developed here is evaluating how variable erosion rates affect CRN concentrations for point-specific samples without accounting for muogenic production. The next steps thus involve determining how variable and stochastic erosion rates affect CRN concentrations in sediments and also how the muogenic signature might be affected.

CONCEPTUAL FRAMEWORK AND MODEL

The Geomorphology

The focus here is on hilly and mountainous soil-mantled landscapes where the bedrock is actively converted to a continuous soil mantle (also referred to as regolith—the key is that the soil is the mobile layer). Importantly, saprolite, or weathered bedrock, is conceptualized as bedrock with the criterion that geomorphic processes do not physically mobilize it: It retains relict rock structure. When conditions of local steady state are assumed, the soil production rate equals the erosion rate, and the local soil thickness remains temporally constant (Heimsath et al., 1999). The bedrock-soil interface lowers spatially at the soil production rate, and the soil acts as a continuously moving layer removing sediment produced locally and transported from upslope (Heimsath et al., 1997) such that the lowering rate of the soil-bedrock interface is equivalent to the landscape lowering rate. Importantly, this rate can vary across the landscape, as discussed below, such that the landscape is not lowering at the same rate everywhere and is, therefore, out of dynamic equilibrium (Ahnert, 1967, 1987).

Cosmogenic nuclides are produced within the mineral grains present in the soil column as well as in a profile with depth in the underlying bedrock. Here we are specifically interested in the in situ–produced CRN concentrations in the bedrock at the soil-bedrock boundary, not in the soil. The sloping soil mantle is treated like an additional buffer against cosmic-ray penetration, and CRN production rates for a specific sample location are corrected to normalize against a flat, unburied surface (Dunne et al., 1999). Concentrations of both ^{10}Be and ^{26}Al measured from the bedrock or saprolite (chemically weathered, in-place bedrock) beneath the soil layer determine soil production rates (e.g., Heimsath et al., 1997, 1999; Small et al., 1999). It is especially important to recognize that under steady-state conditions the application of CRN measurements to determining soil production rates is identical to the procedure more commonly used to determine exposed bedrock erosion rates except for the explicit correction of CRN production rates for the overlying soil mantle.

Sediment produced from exposed bedrock as well as that eroded from the mobile soil layer is transported to channels and eventually removed from the landscape. Physical processes mix this sediment during transport such that a grab sample of sand from a sandbar in a channel is likely to contain sediment grains from eroding parts of the basin draining to that point. Following the well-constrained studies of Granger et al. (1996) and Bierman and Steig (1996), which showed that the CRN concentration of such a grab sample reflects a spatial average of the erosion rates acting in the basin, extensive application of this technique suggests it is robust across a wide range of geomorphic conditions (Clapp et al., 2000, 2002; Hewawasam et al., 2003; Matmon et al., 2003a, 2003b; Nichols et al., 2005; Schaller et al., 2001; Wobus et al., 2005). There are several critical assumptions, however, that were well articulated by Granger et al. (1996) and Bierman and Steig (1996). Perhaps most important are the assumptions that the collected sediments integrate the erosional processes operating across the landscape and that these processes are occurring close to steady state. The role of landslides in contributing pulses of sediment that are not spatially integrated as well as not being close to steady state is a potentially serious problem for this methodology, but testing this is beyond the scope of this paper and is being modeled with a different approach (Niemi et al., 2005). In the discussion below, I will address specifically how modeling results presented here may be used in an approach similar to Niemi et al. (2005).

The Cosmogenic Nuclide Approach

Here I review the derivation of the equations commonly used in geomorphic applications, to clarify the integration of the governing differential equation and its implications. This may be redundant to most experienced users, but the growing number of students of this technique warrants an explicit review. Similarly, while all users of CRNs apply the steady-state solution of Lal (1991), the derivation of the governing equations remains poorly understood despite the thorough review of Gosse and Phillips (2001). This derivation also builds the foundation for the numerical model presented below. Muogenic production of nuclides will be assumed to be small and will be left out of the derivation for the sake of simplicity and also ease of comparison with previous studies where muogenic production was not accounted for. It will follow that an extra term for nuclide production by muons (Gosse and Phillips, 2001; Granger and Smith, 2000; Stone et al., 1998b) is simple to add to the numerical model. This will be a critical addition for further applications of this model, as the impacts of variable erosion rates on the muogenic component of CRN concentrations is currently unknown. Following Lal (1991), for a flat bedrock target with no soil cover, the production rate of a radionuclide, $P(z, t)$ (atoms/g/yr), declines from the surface nuclide production rate, P_s, with depth, z (cm), such that

$$P(z,t) = P_s e^{-\rho z/\Lambda}, \quad (1)$$

where ρ is the density of the target material (g cm^{-3}), and Λ is the absorption mean free path for the nuclear interacting particles in the target (g cm^{-2}). An absorption coefficient (cm^{-1}) is defined for bedrock as $\mu = \rho/\Lambda$ to simplify all further equations. An additional coefficient is defined for the overlying soil mantle further dampening the production of nuclides as $\mu_s = \rho_s/\Lambda$, where ρ_s is the soil bulk density and is assumed constant based on field measurements in the well-mixed soils of the landscapes studied here. It is assumed that the nuclides are produced only by cosmic-ray nucleons and that the nuclide production rate is constant over time. The current debate over production rate uncertainties (Clark et al., 1995; Dunai, 2000; Nishiizumi et al., 1996; Stone et al., 1998a) and the likely contribution of muons to nuclide concentrations under moderate and high erosion rates (Brown et al., 1995a; Granger and Smith, 2000; Stone et al., 1998b) are not relevant to the derivation of this model because the application is to determine relative errors resulting from non-steady-state conditions of erosion.

All nuclides considered here decay radioactively at a rate proportional to the concentration, with a constant of proportionality, λ, which is specific to the nuclide. Thus, the concentration of nuclides in the rock horizon beneath a soil mantle with thickness h can be modeled as

$$\frac{dN(t)}{dt} = P_S \cdot e^{-\mu \cdot z(t) - \mu_S h} - N(t)\lambda. \tag{2}$$

For most applications of geomorphology, the concentration of nuclides is measured from the surface of the target material in an actively eroding environment. To evaluate equation 2 over the exposure history of the sample, the position with depth, $z(t)$, at time t of the sample must therefore be defined as a function of the erosion rate. The simplest case is that exposed rock has eroded at a steady rate over its exposure history (Lal, 1991; Nishiizumi et al., 1989). If the original position, generally assumed to be where there is no penetration of cosmic rays, of the sample under the ground is some depth, z_0, and the constant erosion rate is ε, then the sample position with time is

$$z(t) = z_0 - \varepsilon t. \tag{3}$$

Substituting equation 3 into equation 2 provides a governing differential equation for a bedrock surface, either below soil cover or not (in which case $h = 0$), that is eroding at a constant rate over time, and considering N as a function of time only,

$$\frac{dN(t)}{dt} = P_S e^{-\mu(z_0 - \varepsilon t) - \mu_S h} - \lambda N(t). \tag{4}$$

Grouping similar terms and simplifying for now by letting $h = 0$, yields

$$\frac{dN(t)}{dt} + \lambda N(t) = P_S e^{-\mu(z_0 - \varepsilon t)}. \tag{5}$$

Note that the left-hand side of this equation is equal to

$$\left(\frac{dN(t)}{dt}e^{\lambda t} + \lambda N(t)e^{\lambda t}\right) \Big/ e^{\lambda t},$$

which by virtue of the product rule is equal to

$$\frac{d}{dt}\left(N(t)e^{\lambda t}\right) \Big/ e^{\lambda t}.$$

Substituting this quantity into the above equation and multiplying both sides by $e^{\lambda t}$ yields

$$\frac{d}{dt}\left(N(t)e^{\lambda t}\right) = P_S e^{-\mu(z_0 - \varepsilon t)} e^{\lambda t}. \tag{6}$$

Grouping the exponentials on the right side of the equation into terms with and without the time variable,

$$\frac{d}{dt}\left(N(t)e^{\lambda t}\right) = P_S e^{-\mu z_0} e^{(\lambda + \mu \varepsilon)t}, \tag{7}$$

and integrating both sides with respect to time, t, yields, with N_0 as a constant,

$$N(t)e^{\lambda t} + N_0 = \int dt\, P_S e^{-\mu z_0} e^{(\lambda + \mu \varepsilon)t} = P_S e^{-\mu z_0} \int dt\, e^{(\lambda + \mu \varepsilon)t}. \tag{8}$$

Solving for $N(t)$ thus results in an expression that only requires solving a simple integral,

$$N(t) = -N_0 e^{-\lambda t} + P_S e^{-\mu z_0} e^{-\lambda t} \int dt\, e^{(\lambda + \mu \varepsilon)t}. \tag{9}$$

Under the steady-state assumption (i.e., that the erosion rate, ε, is constant) the integral can be solved analytically. If ε is a function of time, however, the integral must be solved numerically, as described below.

Assuming, for now, that the steady-state assumption holds, the concentration at a specific time, T, is determined by evaluating equation 9 at $t = T$ such that

$$N(T) = -N_0 e^{-\lambda T} + P_S e^{-\mu z_0} e^{-\lambda T} \int_0^T dt\, e^{(\lambda + \mu \varepsilon)t}. \tag{10}$$

At time T, a particle starting at depth z_0 will have risen to a depth $z = z_0 - \varepsilon T$. Substituting $z_0 = Z + \varepsilon T$ into equation 10 leads to

$$N(T) = -N_0 e^{-\lambda T} + P_S e^{-\mu(z + \varepsilon T)} e^{-\lambda T} \int_0^T dt\, e^{(\lambda + \mu \varepsilon)t}$$

$$= -N_0 e^{-\lambda T} + P_S e^{-\mu z} e^{-\mu \varepsilon T} e^{-\lambda T} \int_0^T dt\, e^{(\lambda + \mu \varepsilon)t}$$

$$= -N_0 e^{-\lambda T} + P_S e^{-\mu z} e^{-(\lambda + \mu \varepsilon)T} \int_0^T dt\, e^{(\lambda + \mu \varepsilon)t}. \tag{11}$$

The integral can now be solved analytically as follows:

$$N(T) = -N_0 e^{-\lambda T} + P_S e^{-\mu z} e^{-(\lambda + \mu \varepsilon)T} \left(\frac{1}{\lambda + \mu \varepsilon} e^{(\lambda + \mu \varepsilon)t} \Big|_0^T\right)$$

$$= -N_0 e^{-\lambda T} + P_S e^{-\mu z} \frac{e^{-(\lambda + \mu \varepsilon)T}}{\lambda + \mu \varepsilon} \left(e^{(\lambda + \mu \varepsilon)T} - 1\right)$$

$$= -N_0 e^{-\lambda T} + \frac{P_S e^{-\mu z}}{\lambda + \mu \varepsilon}\left(1 - e^{-(\lambda + \mu \varepsilon)T}\right). \tag{12}$$

Equation 12 provides a closed-form solution for concentration under the steady-state assumption (i.e., constant erosion, ε, over the sample exposure history). Finally, this reduces to the widely used solution as $T \gg (\lambda + \mu\varepsilon)^{-1}$ such that (e.g., Lal, 1991)

$$N = \frac{P_s e^{-\mu_s z}}{\lambda + \mu\varepsilon}, \quad (13)$$

which can be solved for the steady-state erosion rate by rearrangement:

$$\varepsilon = \frac{1}{\mu}\left(\frac{P_s e^{-\mu_s z}}{N} - \lambda\right). \quad (14)$$

Equation 14 is mostly used to interpret nuclide concentrations in the context of a surface eroding by processes assumed to be operating relatively constantly over time, such as grain-grain spallation of sandstone, thin exfoliation of exposed granite, and biogenic soil production from saprolite beneath an active soil mantle (in which case z is very close to zero, but the depth term for soil, h, will need to be accounted for in setting the nuclide production rate). This relationship is also used to interpret concentrations extracted from detrital sediments collected from channels to determine erosion rates averaged across the entire drainage basin. The model presented below does not, however, spatially integrate point-specific samples to determine basin-wide concentrations under variable erosion rate conditions. Recent modeling work suggests that the processes eroding the basin need not be in steady state as long as the basin is large enough to integrate signals from both the low-magnitude/high-frequency events (e.g., processes approaching steady state) and the low-frequency/high-magnitude events (e.g., landslides) (Niemi et al., 2005). While this model is not yet fully tested, the conclusions are intuitive and are discussed below.

The Numerical Model

I now extend the model equation 4, such that the effects of variable conditions in erosion and overlying soil depth can be evaluated. It will become clear that a temporally variable soil depth results in the same scenario as a variable erosion rate scenario, as the soil production rate, which depends on the soil depth, is equivalent to the local erosion rate. The model is then applied to hypothetical cases of temporally variable erosion rates, based on the field observations of Heimsath et al. (2001a, 2001b), as well as extended to potentially extreme scenarios to compare with the earlier models testing the effects of extreme events on nuclide concentrations (Bierman and Steig, 1996; Lal, 1991; Small et al., 1997).

Under conditions of non-steady-state erosion (i.e., the erosion rate, now denoted as $\varepsilon(t)$, is a function of time and can be modeled with different approximations), the initial differential equation for nuclide concentration can be written as

$$\frac{dN(t)}{dt} = P_s e^{-\mu(z_0 - \varepsilon(t))} - \lambda N(t). \quad (15)$$

Following the same derivation as done above leads to a similar expression as in equation 9:

$$N(t) = -N_0 e^{-\lambda t} + P_s e^{-\mu z_0} e^{-\lambda t} \int dt\ e^{\lambda t + \mu\varepsilon(t)}. \quad (16)$$

Note that, unlike the steady-state condition, the above integral does not have a closed-form solution for nontrivial erosion rates, $\varepsilon(t)$. This simply is because the erosion rate is a function of the integrating variable, t. To solve for the non-steady-state condition of equation 15, a numerical approximation of the differential equation is, therefore, used. Specifically, by summing both sides of the differential equation, the concentration at time T is approximated by the following:

$$N(T) = \sum_{t=0}^{t_0} P_s e^{-\mu z(t)} - \lambda N(t), \quad (17)$$

where $N(0) = N_0$ (the initial concentration, typically equal to zero), and the depth, $z(t)$, at time t_0 is

$$z(t_0) = \sum_{t=0}^{t_0} z - \varepsilon(t). \quad (18)$$

With this numerical solution, the nuclide concentration can be estimated for any arbitrary erosion function, $\varepsilon(t)$, that might approximate potential temporal variations in erosion observed in the field. It is important to note that as constructed here, the model is evaluating point-specific concentrations and inferred erosion rates. The obvious next step is to incorporate this point-specific model into a model that integrates the concentrations derived from all points upslope of any given catchment outlet. The approach could be similar to the one used by Niemi et al. (2005), except that every pixel in the catchment digital elevation model (DEM) would be assigned a CRN concentration based on its variable erosion rate history. Such a model would have to track the production and transport of sediment from all points and is, therefore, beyond the scope of this paper.

Equation 17 can also be modified easily to account for nuclide production by muons, as suggested above. In this case, an additional production term would be added to the right side and the modeling code adapted with the simple addition of the muogenic production term. For the purposes of this paper, this extension is not needed, which also makes results presented here easier to compare to other studies that have not accounted for muogenic production. Similarly, the dampening of nuclide production due to the overlying soil thickness, included in equation 1, is trivial in comparison to the oscillations in erosion rates modeled here and is therefore left out for simplicity. Specifically, if overlying soil thickness is varying stochastically or regularly, then the soil production rate will vary as a function of the change in thickness. Since soil production rates equal erosion rates, for the purposes of this paper it is sufficient to model variation in erosion rates. Specific forms for the variable erosion rate scenarios will be described following the field descriptions and summary of previous results.

FIELD AREAS AND PREVIOUS RESULTS

In this section, I briefly review the pertinent aspects of the field areas that motivated the model and present the soil production and erosion rates determined for both sites using both ^{10}Be and ^{26}Al concentrations from exposed and soil-mantled, weathered bedrock as well as stream sediments and bedrock. Both field sites differ in critical ways from those examined by similar studies across completely soil-mantled landscapes (e.g., Heimsath et al., 1997, 2000). Some of these differences will be reviewed here and a comparison between the sites will be drawn after application of the model, with specific discussion of the form and magnitude of the respective soil production functions.

Oregon Coast Range

The Oregon Coast Range is a well-studied region of the Pacific Northwest that exemplifies humid-temperate, soil-mantled landscapes where stochastic as well as steady-state erosional processes drive landscape development (e.g., Dietrich et al., 2003; Heimsath et al., 2001b; Montgomery et al., 1997, 1998, 2000; Roering and Gerber, 2005; Roering et al., 1999, 2001). The actively eroding landscape is characterized by ridge-and-valley topography (Fig. 1), and while there is some hint of long-term equilibrium in erosion rates, there is also clear indication of local disequilibrium across the landscape (Heimsath et al., 2001b; Reneau and Dietrich, 1991; Roering and Gerber, 2005). Bedrock is a relatively undeformed Eocene turbidite sandstone and siltstone called the Tyee Formation (Heller et al., 1985; Lovell, 1969; Snavely et al., 1964) and outcrops in unweathered units along sharply convex ridge crests. The soil-mantled ridges (a few meters wide at the crest and over one hundred meters down the crest to the channel) have high spatial variability of soil thicknesses such that there is little systematic variation between soil depth and topographic position (Heimsath et al., 2001b). Soil depths on slopes vary between zero and one to two meters in the upland areas with relatively unweathered bedrock often exposed in recently evacuated debris-flow or landslide scars in the hollows. Creeks and rivers draining the region are mixed alluvial-bedrock and are thought to be incising at roughly the same rate as the long-term average uplift rate of 100–300 m/m.y. documented by marine terrace records (Kelsey and Bockheim, 1994; Kelsey et al., 1994). Rainfall is high, averaging about two meters a year, and the dense forest, dominated by Douglas fir that can grow to be over forty meters tall, is actively cleared for timber harvesting.

Results relevant to this paper include a cosmogenic nuclide (both ^{26}Al and ^{10}Be) determined soil production function (Fig. 3) (Heimsath et al., 2001b). Apparent soil production is shown to be an inverse exponential function of overlying soil thickness, with a maximum inferred soil production rate of 268 m/m.y. occurring under soil depth that approaches zero. This soil production function is determined from saprolite samples (27 samples and eight duplicates) collected under soils thicker than ~20 cm, while erosion rate data from exposed bedrock (three samples and one duplicate) as well as from under soils thinner than 20 cm (three samples and two duplicates) average 160 m/m.y. Two samples and one duplicate of stream sediments determined a catchment-averaged erosion rate of 117 m/m.y. for the field area, which

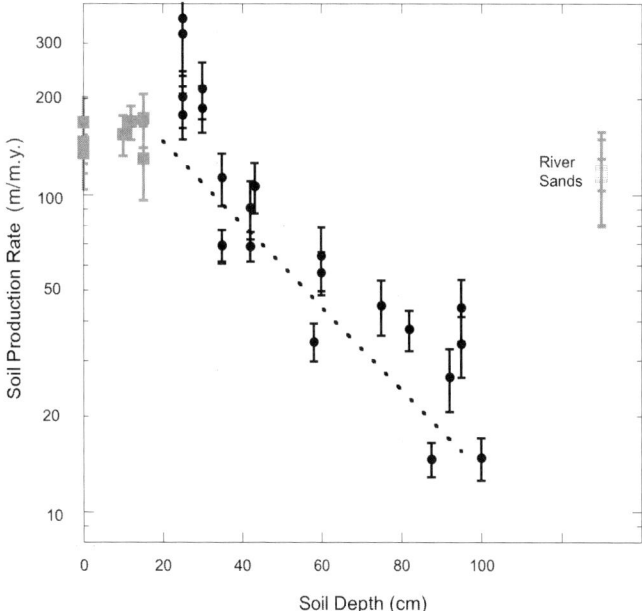

Figure 3. Apparent soil production rates (m/m.y.), calculated from the in situ–produced radionuclides ^{10}Be and ^{26}Al extracted from bedrock samples versus the observed normal soil depth, H, in cm. The soil production function, $\varepsilon(H) = (268 \pm 25)\, e^{-(0.03 \pm 0.02)H}$, is the variance-weighted least squares fit to the soil production rates under soil depths greater than 15 cm (plotted with black filled circles). Soil production rates for the shallower samples and the exposed bedrock samples (plotted with gray filled squares) were not used to determine this function because of lithologic differences, as discussed by Heimsath et al. (2001b). The average erosion rates for the catchment from nuclide concentrations from three stream sediment samples are plotted with open squares on the far right of the plot, as labeled. Rates are calculated from the concentrations of both nuclides with a few exceptions. Error bars are 1σ propagated from accelerator mass spectrometry (AMS), atomic absorption (AA), bulk density, absorption mean free path, and soil depth uncertainties. Modified from Heimsath et al. (2001b).

agreed well with other estimates of average erosion rates for the region (Reneau and Dietrich, 1991), but is much lower than the postfire rates determined from recently reanalyzed data (Roering and Gerber, 2005). Stochastic processes of tree throw and shallow landsliding can change dramatically local soil thicknesses over short timescales and, therefore, potentially affect the inferred rates of soil production, as well as the inferred long-term erosion rates (Heimsath et al., 2001b). These rates, as well as the apparent soil production function, enable modeling of landslide susceptibility for the region, which is especially important given the potentially fatal implications of debris flows on local homeowners. Similarly, long-term erosion rates from cosmogenic nuclides are significantly lower than short-term rates from sediment trap studies and suggest an increase in sediment removal from the landscape associated with timber harvesting. It is therefore both scientifically interesting and relevant to land management to assess the robustness of these data.

Southeastern Australian Highlands

Rolling soil-mantled hills dotted with outcropping tors are characteristic of the southeastern Australian highlands (Fig. 2). Land use results often in clearing of the dominant sclerophyll forests and subsequent gullying across the stony hills. Where undisturbed, shallow soils across the ridges are typically free from rilling and grade into thick, undissected colluvial fills in the hollows. Bedrock varies across the region between Ordovician metasediments and Devonian granites. Only the granites contain enough quartz for cosmogenic nuclide analyses, though abundant quartz veins in the metasediments can also be used. Gradients are gentle, averaging less than 20° in comparison with the nearly 45° slopes common in the Oregon Coast Range forests. Rainfall is a fraction of the Oregon Coast Range site and falls throughout the year to average between half to three-quarters of a meter a year. Temperatures average ~18–22 °C in the summer and 5–8 °C in the winter, with short periods of freezing temperatures and minor snowfall events. Soil production and erosion are primarily due to biogenic processes and creep, with some evidence of overland flow (Heimsath et al., 2001a).

Results relevant here are the soil production and erosion rates determined from cosmogenic nuclide (both ^{26}Al and ^{10}Be) concentrations (Fig. 4). Five samples of the granitic saprolite from under different soil depths led to the inference of an apparent soil production function for the Frogs Hollow site. Noticeably, the slope of the function is significantly steeper than the function determined in Oregon, as well as those reported for more completely soil-mantled landscapes in both California (Heimsath et al., 1997, 1999) and the coastal lowlands of southeastern Australia (Heimsath et al., 2000). This steep function was subsequently expanded with the collection of significantly more data from another highland site, which also led to the confirmation of the soil production function first reported from the lowland site (Fig. 5) (Heimsath et al., 2006). Importantly, the addition of new data from a greater range of soil depths and additional landscapes showed the robustness of the soil production function and its applicability across the soil-mantled landscapes of southeastern Australia. In any case, the initial quantification of soil production rates from the highland site enabled numeric modeling of landscape evolution and soil development for the region that was supported by field observations (Heimsath et al., 2001a). Soil production rates inferred from nuclide concentrations ranged from ~50 m/m.y. under 25 cm of soil to ~11 m/m.y. under 65 cm of soil. Average erosion rates for the landscape, determined from both ^{26}Al and ^{10}Be concentrations in sediments, were ~15 m/m.y. (denoted by C and Riv on Fig. 4), roughly an order of magnitude slower than the Oregon Coast Range. Incision rates for the river draining the landscape were determined from three samples (both ^{26}Al and ^{10}Be) from the fluvially sculpted and polished bedrock of the active channel bed and averaged 9 m/m.y., suggesting that the soil-mantled tributary catchment is eroding slightly more rapidly. The slightly higher rates from the tributary were explained by suggesting the landscape is responding to a pulse of incision, inferred from knickpoints in the channel's long profile, potentially driven by some period of increased erosion.

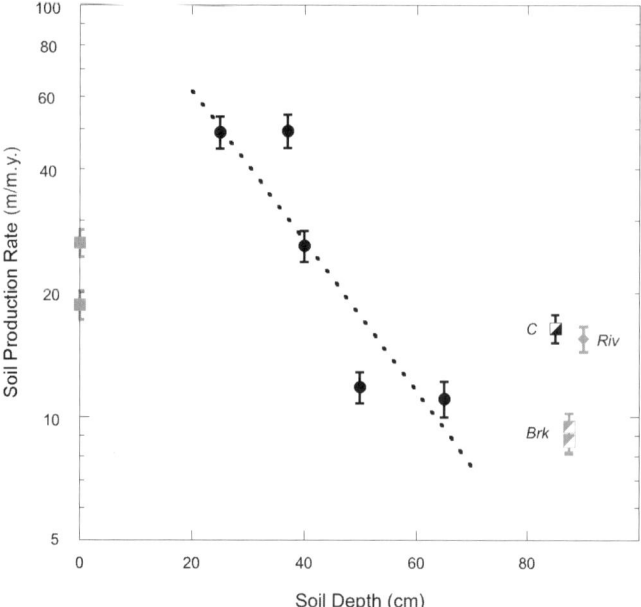

Figure 4. Apparent soil production rates versus observed normal soil depths, H, for Frogs Hollow. Filled circles show the soil production rates versus soil depth from both ^{26}Al and ^{10}Be concentrations. The variance-weighted best fit to the filled circles is $\varepsilon(H) = (143 \pm 20)\, e^{-(0.042 \pm 0.003)H}$, where soil production is in m/m.y. and soil depth is in cm. Gray filled squares plotted at zero depth are exposed bedrock samples at the ground surface, inferred to be emerging core stones. Error bars show 1σ error propagated from all sources of error (uncertainty in AA, AMS, bulk density, and soil depth measurements, and the attenuation length of the cosmic rays) except the uncertainty in nuclide production rates. Rates plotted on the far right of the plot, as labeled, show average erosion rates from (1) sediments from the Frogs Hollow catchment area (half-filled black square, labeled C), (2) sediments from the Bredbo River upstream of Frogs Hollow (gray diamond, labeled Riv), and (3) the bedrock incision rates of the Bredbo (four samples, gray half-filled squares, labeled Brk). Modified from Heimsath et al. (2001a).

What makes this highland site so interesting from a non-steady-state erosional perspective are the results from a profile of bedrock samples collected from the side of an outcropping granite tor. Our interest in long-term landscape development led to the development of a "Kuniometer," named for Kuni Nishiizumi, who came up with the idea, which is a profile sampled for nuclide concentrations from ground level to the highest point on a bedrock tor exposed in a soil-mantled landscape (Heimsath et al., 2000, 2001a). The idea is simple. Concentrations of nuclides from the profile sampled will depend on the relative rates of lowering between the soil-mantled landscape and the eroding bedrock that is outcropping. Steady-state lowering will lead to an increase in nuclide concentration with height above ground level, while some dramatic change in erosion that might have left the bedrock abruptly exposed will lead to a constant nuclide concentration

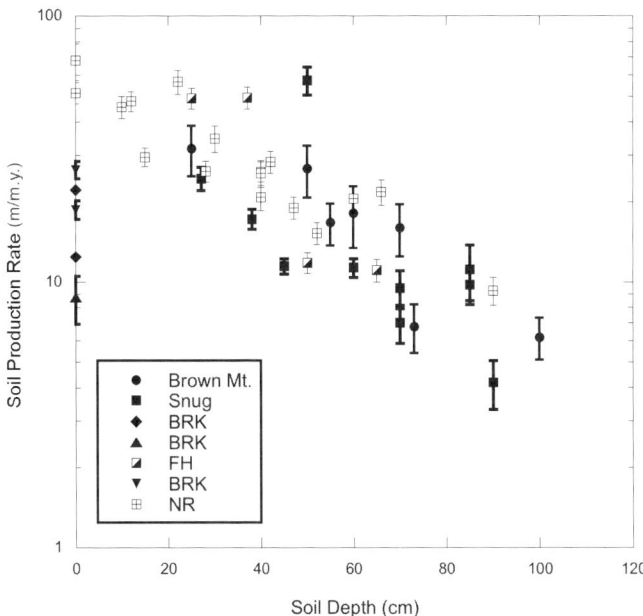
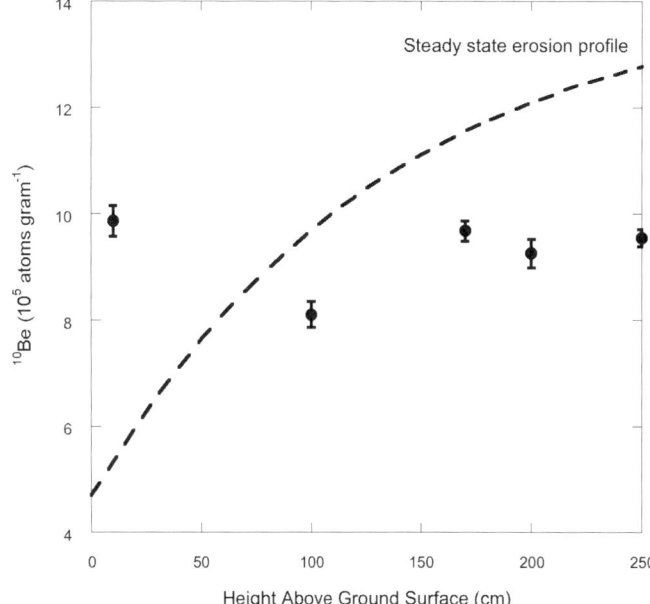

Figure 5. Apparent soil production rates plotted for all four southeastern Australia field sites discussed in Heimsath et al. (2006). Note the remarkable overlap between the two highland sites, Brown Mountain and Frogs Hollow (FH), which are ~50 km apart, but at roughly the same elevation and in the same physiographic region. Also note that when soil production rates from these highland sites are combined with the data from Heimsath et al. (2000) at the base of the escarpment (NR), as well as from the coastal lowlands (Snug), the soil production function appears to be quite robust. Erosion rates of bedrock (BRK) exposed at the soil surface are plotted for all sites with different symbols. Variance-weighted best fit to all the soil production data plotted here is $\varepsilon(H) = (53 \pm 2)\, e^{-(0.022 \pm 0.001)H}$. Error bars show the same sources of uncertainty as in Figures 3 and 4. Modified from Heimsath et al. (2006).

Figure 6. Observed concentrations of ^{10}Be (note, results from ^{26}Al yielded the same story) for tor profile sampled for exposed bedrock. Nuclide concentrations are plotted as closed black circles against height above the present ground surface, with measured nuclide concentrations normalized to sea level and the error bars showing 1σ propagated from AMS, AA, bulk density, and absorption mean free path uncertainties. The black dashed line plots the best-fit model prediction for the steady-state scenario from the upper Bega Valley site of Nunnock River of Heimsath et al. (2000), scaled to fit the Frogs Hollow data more closely and showing the marked deviation from observed concentrations. These observations of concentrations against height above ground are best explained by a model that posits complete emergence of the tor early in its exposure history, ca. 150 ka. Modified from Heimsath et al. (2001a).

with height. While the lowland site led to support for steady state (Heimsath et al., 2000), results from Frogs Hollow led to the suggestion of a dramatic stripping event in the late Pleistocene (Heimsath et al., 2001a). Specifically, the observed nuclide concentration profile suggests that a period of increased erosion ca. 150,000 yr ago led to the rapid exhumation of the tor by the removal of over two meters of easily erodible saprolite (Fig. 6). This evidence for dramatic changes in erosional regime leads to obvious concern for the potential effects on nuclide concentrations in the saprolite and, therefore, the inferred soil production rates.

RUNNING THE MODEL

Given these two dramatically different field examples (Oregon Coast Range and southeastern Australian highlands) for how erosion rates might vary over time, there are reasonable constraints for how to model the effects of potential non-steady-state scenarios on point samples used to determine soil production or erosion rates. Several forms of oscillating input erosion rates can be chosen from and are used to model (which is encoded in Matlab® and freely available upon request[2]) conditions of variable erosion. Additionally, there are two parameters that govern the resolution and run time of the model: the time step and the maximum time. The maximum time should be set such that the end of the model run reaches a steady-state nuclide concentration, or, in the case of an oscillating concentration, a steady-state oscillation with the imposed variable erosion rate. Conceptually, for steady-state erosion scenarios, the faster the erosion rate, the shorter the time to steady-state nuclide concentrations. The time step is the interval over which the model evaluates the numerical integration. Setting the time step governs, therefore, the efficiency of the numerical computations as well as the resolution of the integration. Naturally, the shorter the time step, the longer the computations take to complete, but the greater the resolution of the model output. For higher rates of erosion, the time step needs to be shortened to insure analytical accuracy. Setting both the time step and the maximum time involves editing the source code

---
[2]Contact A.M.H. for free Matlab® code and instructions for model presented herein.

for the model, and should be done prior to choosing the form of the model and the parameters governing the input for the model.

Setting the model parameters and the modeled erosion scenario is done by the command line in Matlab®. Each variable erosion rate functional form is based on field observations as well as a range of studies that have either speculated or documented the ways in which erosion rates and processes can change across a landscape. To check the mathematics of the numerical integration, the model also enables input of either zero erosion or steady-state erosion rates, functions that result in steadily decreasing steady-state nuclide concentrations with increasing erosion rate (Fig. 7) (Lal, 1991; Cerling and Craig, 1994). The variable erosion models are mathematically represented as the following: (1) step function, (2) rectified full wave, (3) square wave, (4) exponential, and (5) sawtooth. Examples of each of these functions are plotted in Figure 8 for erosion oscillations between 100 m/m.y. and 1000 m/m.y., showing two full cycles. Parameters governing the frequency and magnitude of each of these functions can be adjusted according to predicted, or hypothesized, variations in erosion rates.

Specifically, changing the amplitude of the function defines the maximum erosion rate for the specified model scenario (units of m/m.y.). Choosing the frequency for the oscillating functions divides the maximum time into the number of cycles as defined by the input. Finally, there is the option to choose a "pedestal" for the functions, which sets the minimum erosion rate for the cycles and defines that the oscillating erosion rates are nonzero, if such is the hypothesized scenario. These three parameters guide the model to test any potential scenario for variable erosion rates. For example, drawing on the potential variations for the Oregon Coast Range site described above, the recent work on the effects of fire (Roering and Gerber, 2005) on erosion rates can be evaluated in the context of potential effects on the nuclide concentrations used to determine soil production and average erosion rates. Roering and Gerber (2005) report that postfire erosion rates exceed long-term rates by a factor of six and that fire frequency is on the order of 100–200 yr from early to late Holocene, respectively. Modeling this scenario means positing an input function for the variable erosion rate, defining the maximum possible rate under the oscillating conditions, setting the frequency of fire "events" and setting the minimum, or background, erosion rate. For this example, an exponential scenario would be chosen following a conceptual model for impact of fire on sediment flux with time (Swanson, 1981), as reported in Roering and Gerber (2005). Maximum erosion rates would be set to range from 600 m/m.y. to as high as 1800 m/m.y. to capture the range of potential background erosion rates. Minimum erosion rates would be set to range from the long-term average of ~100 m/m.y. to 20 m/m.y., the minimum reported soil production rate from Heimsath et al. (2001b) (Fig. 3).

Motivation for the other variable erosion scenarios is also observation-based. The step function, first modeled by (Lal, 1991), then expanded by Small et al. (1997) to show the effect of exfoliation sheet size on inferred erosion rates, would, for an applicable example, be useful for the highland, southeastern Australia

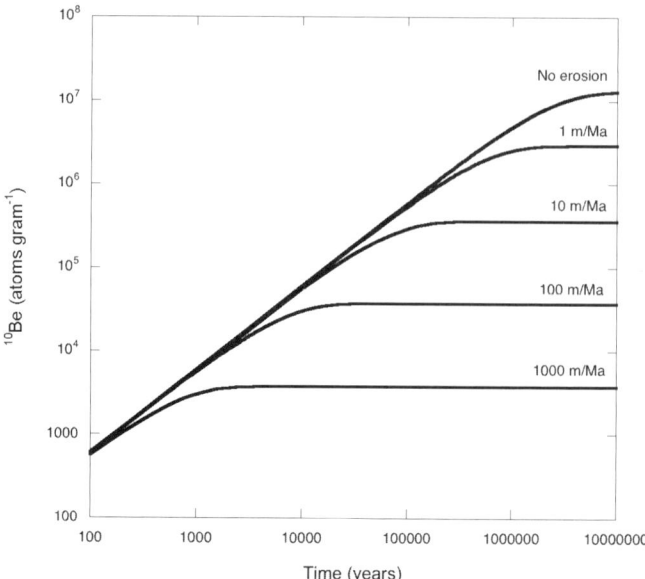

Figure 7. Modeled ^{10}Be accumulation in rock surface samples as a function of time and steady-state erosion rate. Concentration curves are labeled with the input erosion rate written above the resulting curve, varying from 0 to 1000 m/m.y. Note that the faster the erosion rate, the shorter the time needed for the nuclide concentration to reach steady state, and that this concentration is lower. Concentrations here and for the modeling results below are calculated for a sea-level ^{10}Be production rate of 6 atoms/g/yr, and would scale for production rates depending on sample location and shielding. Note that this figure is similar to ones commonly shown in textbooks describing the cosmogenic nuclide methodology and is implicit in the plots of Lal (1991), then shown plotted in this form by Cerling and Craig (1994).

(Frogs Hollow) site. This function would test specifically the effect of a potential Pleistocene "stripping" event as suggested by the tor data discussed above. I do not discuss results from modeling with this scenario as do both of the above studies, as well as the Bierman and Steig (1996), flesh out the important points quite well. The step function scenario might also be modeled by using either a square wave (as in Bierman and Steig, 1996), or a rectified full-wave function with a suitably long period and different rates of maximum erosion. Both the exponential and the sawtooth functions can also be used to model the potential for variable soil cover thickness, punctuated by erosional events effectively stripping the soil. The idea here being that these stripping events would strip the soil, increase the erosion rate to the maximum soil production rate, and the site would gradually return to a local steady state as soil thickness recovered. This return of soil thickness to some steady-state value could be either exponential or linear (sawtooth), which would drive the underlying erosion rate. Reasonable maximum rates of erosion for all the functions can be set either by some governing observation—e.g., the fire example—or by positing some potentially high rate and evaluating its effects on the nuclide concentration. Similarly, natural choices for setting the frequency of events would include fire or landslide recurrence intervals, or climatic change cycles.

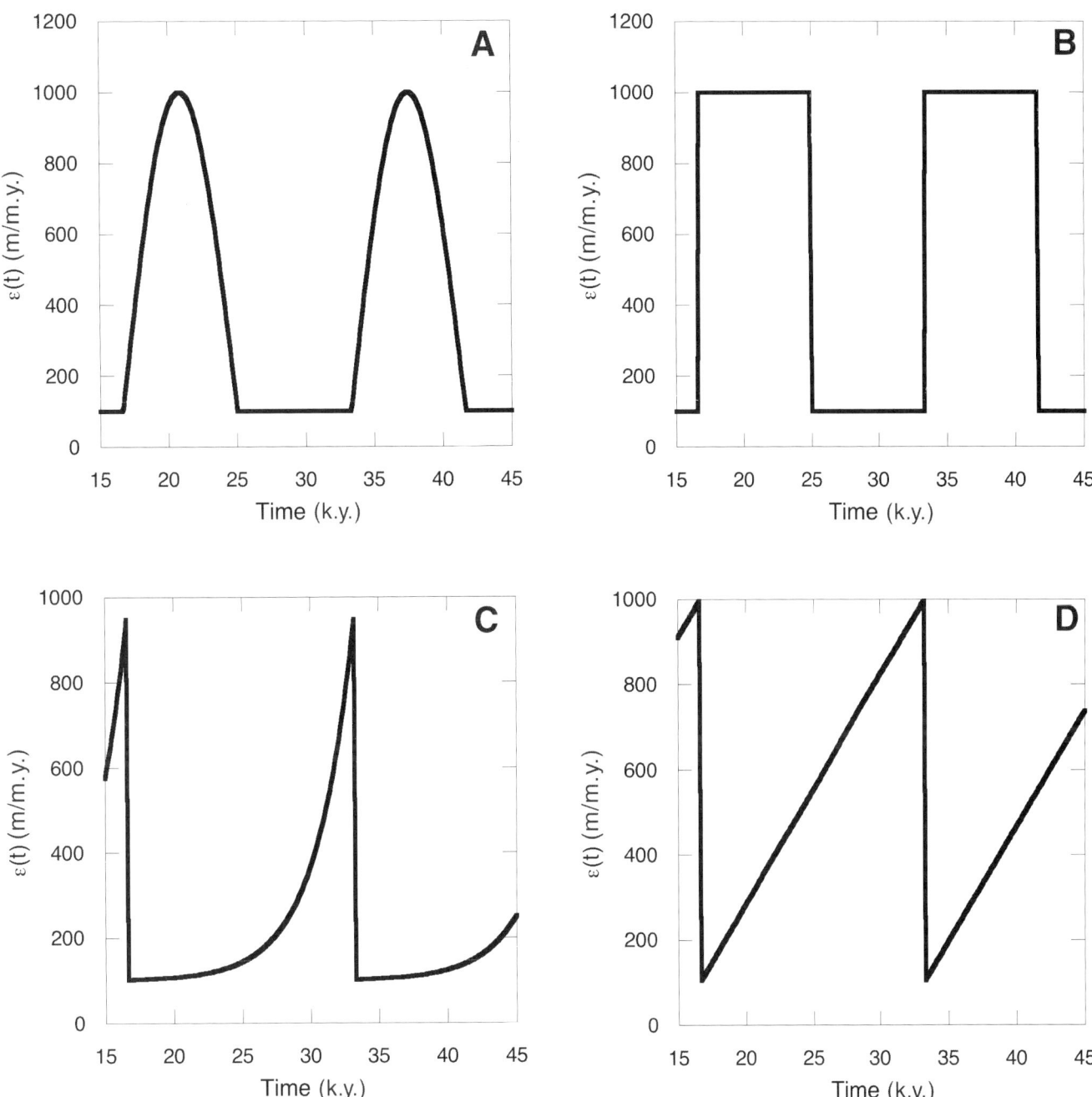

Figure 8. Input erosion rate functions plotted against time to show representative examples. Here the maximum erosion of the oscillations is set for 1000 m/m.y., while the "pedestal" is set at 100 m/m.y. Period is set for 15 k.y. in this example. A: Full-wave rectification. Erosion increases gradually to a peak, then drops back down to the background rate before rising again. B: Square wave. Similar to A, except that rise and fall of erosion is abrupt, best modeled with a square wave. Also, the time spent at the higher erosion rate is longer. C: Exponential. Erosion rate increases exponentially with time to a peak and then abruptly drops back to the base rate. D: Sawtooth. Erosion rate increases linearly with time to a peak and then drops abruptly back to the base rate. Both of the latter two functions can also be reversed such that the increase is abrupt and the return to base conditions is more gradual. For the purposes of the modeling examples here, the conclusions do not change.

MODELING RESULTS

Using the input erosion rates that oscillate as a function of time, as shown for example in Figure 8, the model computes the accumulation of the cosmogenic radionuclide, ^{10}Be, at a point. Perhaps the most relevant example of interest here, specifically for the Oregon Coast Range site, is when erosion rate can vary between 100 and 1000 m/m.y., as shown in Figure 8. The modeled ^{10}Be accumulations in samples subject to these oscillating erosion rates are shown in Figure 9, with Figures 9A and 9B showing the response to sudden changes in erosion, potentially due to a sudden change in climate or land use. Figure 9A shows how concentration varies under the full-wave rectification scenario. Nuclide concentration at a point increases and decreases gradually to and from a peak, which corresponds to the end of the time spent under the base erosion rate of 100 m/m.y. The lowest concentrations occur under the highest erosion rate. Figure 9B shows concentration varying under the square-wave oscillation, which shows a pattern similar to the full wave, except that the drop in concentration is more abrupt and the time spent at the lowest concentration set by the highest erosion rate is longer. The increase in nuclide concentration begins with the drop from 1000 m/m.y. to 100 m/m.y., while the rapid drop in concentration begins with the sudden increase in erosion rate.

Figures 9C and 9D show how concentrations vary with oscillating erosion rates that are likely to be similar to changing soil thickness at a given point. The oscillating erosion conditions can also be reversed, to show an immediate increase followed by an exponential or linear decrease in erosion. In Figure 9C, the response to the exponential change in erosion rate, the increase and decrease of concentration resembles a full-wave rectification, reflecting the exponential decrease in nuclide production with depth in the sample. Specifically, the peak in nuclide concentrations is reached just after the increase in erosion rate begins and subsequently decreases smoothly to the lowest concentration corresponding to the highest erosion rate. With the immediate decrease in erosion rate, the nuclide concentration increases steadily. Figure 9D shows a similar change in concentration occurring under a sawtooth variation in erosion, though with a more rapid increase and a lower peak concentration despite the same maximum input erosion rate.

Where this modeling exercise becomes interesting is using the modeled concentrations computed under the variable input erosion rates to infer an erosion rate. The idea here is to replicate the collection of a sample with the subsequent determination of an erosion rate from the measured nuclide concentration, comparing the inferred erosion rate with the "real" erosion rate, which, in this case, is the input erosion function. This is valid specifically for a point sample. To apply this exercise to the basin-averaged problem requires coupling this model for individual points with a model keeping track of how the sediment is generated and transported out of the basin (Niemi et al., 2005). In Figure 10, the variable erosion rates shown in Figure 8 are plotted again in the same order and at the same magnitude, and are now overlain by the inferred erosion, shown by dashed red curves, rates calculated from the modeled nuclide concentrations shown in Figure 9. There is a remarkable mirroring of the input erosion rates by the inferred rates, with slight offsets evident in all the predictions. In Figures 10A–10C, there is a factor of two overestimation of even the base rate of erosion, and Figure 10B shows a 10% overestimation of the peak erosion rates under the square-wave oscillations. Figure 10D shows that under a sawtooth oscillation the base rate is never inferred from the nuclide concentrations, while the peak rate seems to be well captured. For any given time for the full-wave scenario, Figure 10A, the inferred erosion rate is off by an average of 49%, with a standard deviation of 83%. The square-wave input function, Figure 10B, is less well captured at any given time, with an average deviation of 77% in the inferred erosion rate, with a standard deviation of 154%. With the exponential function, Figure 10C, the average error is 44%, with a standard deviation of 110%. Not surprisingly, given the closer fit of the two curves in Figure 10D, the inferred erosion rate is on average 28% off from the input rate, with a standard deviation of 108%.

While this result does not bode well for point-specific sample collection, it is worth thinking about how such erosion rate variations manifest themselves in the long-term signal and could potentially be captured by sampling catchment sediments. The implication of these imposed variations in erosion rates over the long term is that the average rate in input erosion for the full wave shown in Figure 10A is 384 m/m.y., while the average rate inferred from the nuclide concentrations over the same full cycle is 406 m/m.y. This represents a surprisingly small difference of 5.6% for the inferred erosion rate. For the square-wave scenario of Figure 10B, the long-term averages are 545 and 586 m/m.y., respectively for input and inferred rates, with the inferred overestimating the input by 7.6%. The long-term average for the input exponential variation, Figure 10C, is 242 m/m.y., while the long-term average for the inferred rate is 249 m/m.y., meaning an overestimation of only 5.2%. Despite the factor of four-plus overestimation of the low erosion rates due to the sawtooth variation of Figure 10D, the difference between the long-term averages is only 2.7%, with the inferred average rate of 579 m/m.y., compared to the input average rate of 551 m/m.y. These long-term average comparisons are most meaningful when considering the potential for widely variable erosion rates across a catchment where the inferred erosion rate is seeking to capture the spatial average for the basin. These comparisons are not as meaningful for determining whether a point-specific sample is likely to have captured an average rate.

The near mirroring of even the highest erosion rates for the oscillation between 100 and 1000 m/m.y. changes with a reduction in the magnitude of the amplitude of variation, or the maximum erosion rate that the sample is subjected to. When the same oscillating erosion functions are now set to range from 10 to 100 m/m.y., potentially a more reasonable scenario for the southeastern Australia site, the erosion rates inferred from the modeled nuclide concentration do not show the same dramatic fluctuations

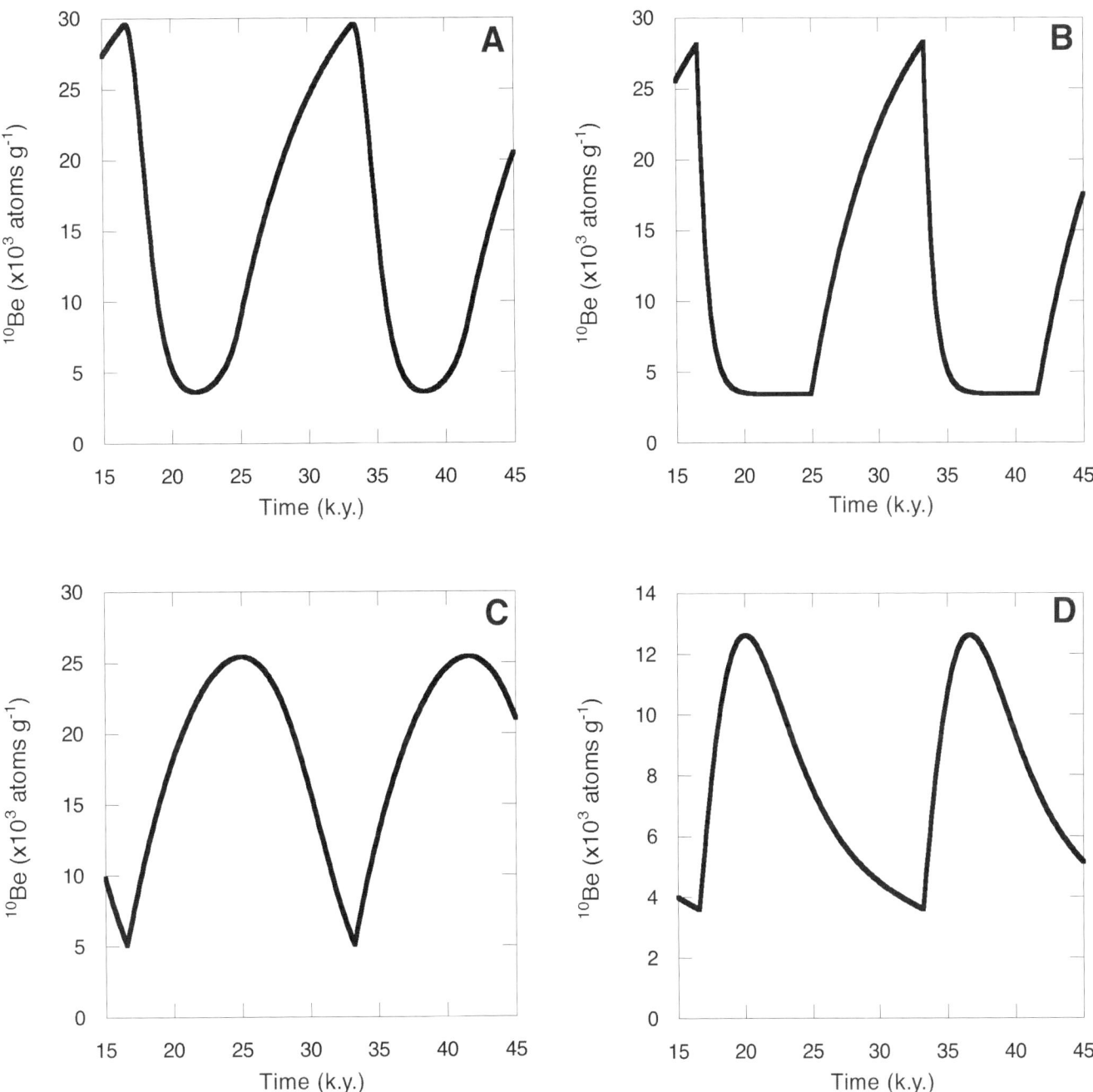

Figure 9. Modeled ^{10}Be accumulation in point samples (either exposed rock or at the soil-rock interface) subject to the oscillating erosion rates of Figure 8. A: Full-wave rectification. Concentration increases and decreases gradually to and from a peak. B: Square wave. Similar to A, except that the drop in concentration is more abrupt and the time spent at the base concentration set by the base rate is longer. C: Exponential. The increase and decrease of concentration resembles a full-wave rectification, reflecting the exponential decrease in nuclide production with depth in the sample. D: Sawtooth. Nuclide concentration increases more rapidly and decreases exponentially.

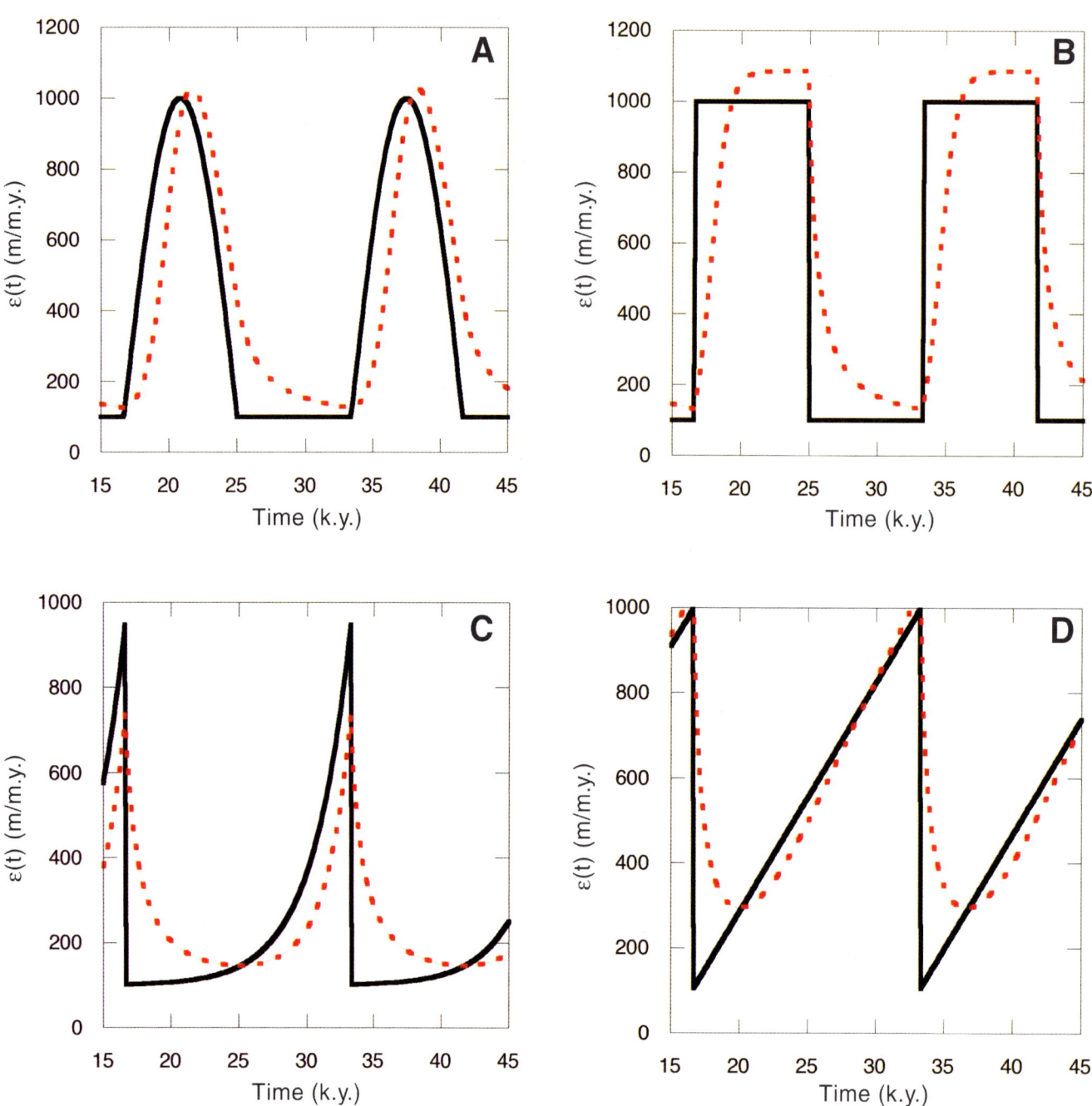

Figure 10. The input erosion rates modeled in Figure 8 are plotted again in the same order with the solid black curves, and are overlain by the inferred erosion rates, shown by dashed red curves, rates calculated from the modeled nuclide concentrations shown in Figure 9.

as the input rates. An additional way that the input erosion rate scenarios can change the modeled nuclide concentrations and, therefore, the inferred erosion rates, is with the frequency of oscillation. Figure 11 shows a reduction by an order of magnitude of the range of erosion rates and also a selection of some different modeled frequencies to illustrate what can now be expressed as generalizable results. The time window plotted reflects the longer time needed to achieve steady-state concentrations under the lower input erosion rates (see Fig. 7).

Figure 11 plots different frequencies for the input varying erosion rates, with Figure 11A showing the same frequency for the full-wave rectification as the input for Figure 10A. Figure 11B increases the frequency for the square wave by 30% to show three full cycles in 30 k.y., while Figure 11C decreases the frequency for the exponential function to one cycle every 20 k.y., and Figure 11D increases the frequency to one cycle every 2 k.y. for the sawtooth function, all for the same range in input erosion rates. The results on the inferred erosion rates, shown also with the red dashed lines in Figure 11, are illustrative and applicable for each of the input scenarios. Most immediately striking is that the inferred erosion rates do not cycle as strongly as they do when the input erosion rate increases to a higher rate. The second most apparent aspect is that the high-frequency change of the sawtooth input has very little effect on the inferred erosion rate at any time, meaning that the steady-state nuclide concentration is changing little under the input erosion rate variations. That this is reflected in the fact that the average for both the input and the inferred rates for Figure 11D is ~60 m/m.y. is surprising and has implications for the catchment-averaged sampling methodology, similar to the above scenario and as discussed below. The variation in rates shown by the square wave, Figure 11B, result in long-term averages that are also identical, with both the input and the inferred rates averaging 55 m/m.y. The long-term average inferred rate of 40 m/m.y. is only 5.2% higher than the input average of 38 m/m.y. for the full-wave rectification of Figure 11A, while the long-term average for the lower-frequency exponential input function is 26 m/m.y., compared to the 29 m/m.y. inferred from the resulting nuclide concentrations, an overestimate of 11%.

As expected from visual inspection of the plots shown by Figure 11, the inferred erosion rate at any given time is, however, not as likely to capture the variable input erosion as it was for the scenarios of Figure 10. Specifically, with the full-wave variation of Figure 11A, the average error is 147% with a standard deviation of 170%. With the square-wave input, Figure 11B, the inferred erosion rate is off on average by 204%, with a standard deviation of 249%. With the exponential scenario, Figure 11C, the average error for the single cycle is 86%, with a standard deviation of 104%. Perhaps most surprisingly, the average error for the high-frequency variation of the sawtooth shown in Figure 11D is 36%, with a standard deviation of 91%. Running similar scenarios for a wide range of frequencies and amplitudes leads to several pertinent conclusions drawn from the modeling results, which are summarized below.

CONCLUSIONS

This paper reviews and builds on results from two soil-mantled landscapes that are known to push the steady-state assumption in cosmogenic nuclide-based studies of erosion and soil production rates and processes. For different reasons, the Oregon Coast Range and the southeastern Australian highlands suggest the likelihood of variable erosion rates and processes. In Oregon, stochastic landsliding and periodic forest fires can significantly impact both the processes and the rates at which any given part of the landscape is eroding. On the highlands of Australia, dramatic variations in climate during the Pleistocene led to periods of increased erosion that may have left a legacy in the cosmogenic nuclide concentrations measured today. These examples of potentially variable erosion rates are not unique. Indeed, it seems likely that most landscapes are subject to similar exceptions to the steady-state assumption. With such a view in mind, this paper also reviews the details of the CRN approach to derive a model that can numerically simulate the accumulation of CRNs under non-steady-state conditions at a specific sample location. The model presented here enables testing the effects of these potential variations in erosion rate on the nuclide concentrations of samples and, therefore, on the inferred erosion rates obtained from the samples.

Results from the modeling exercises offer some valuable conclusions. First, the greater the maximum erosion rate, the more responsive the nuclide concentration is to the erosional variations. This means that for point-specific sampling, the exposure history of the sample is especially important to constrain. For the Oregon Coast Range, there is potential for each of the individual soil production rates to be off by as much as a factor of six, if, for example, a given sampling location had experienced a recent landslide that removed several tens of centimeters of saprolite and then was rapidly filled in with upslope-derived colluvium such that there was no surficial evidence of the event. Confidence in the apparent soil production function for the region is only gained by the agreement between several samples, in this case 21 from the weathered saprolite, that suggest a similar trend. Close examination of the data shows, however, that variations in soil production rate under similar soil thicknesses can be as great as a factor of three. In a landscape that is likely to be experiencing stochastic large-magnitude events, it is critical, therefore, to analyze enough samples to ensure that any local variations in rate do not obscure the long-term average rate. The southeastern Australian highlands example offers an example of how the initial soil production function, inferred from five point samples, was statistically different from the refined function, determined by twelve samples. Determining how many samples is enough to define a function such as this is difficult to predict, but modeling the potential local variations in rate can define a range of uncertainty to be expected.

A second conclusion, also relevant to an Oregon Coast Range–like landscape, is that the higher the frequency of the erosional variation, the more likely that the inferred erosion rate is closer to the long-term average. If, for example, every thousand to two thousand years, all the trees at a site burn thoroughly and

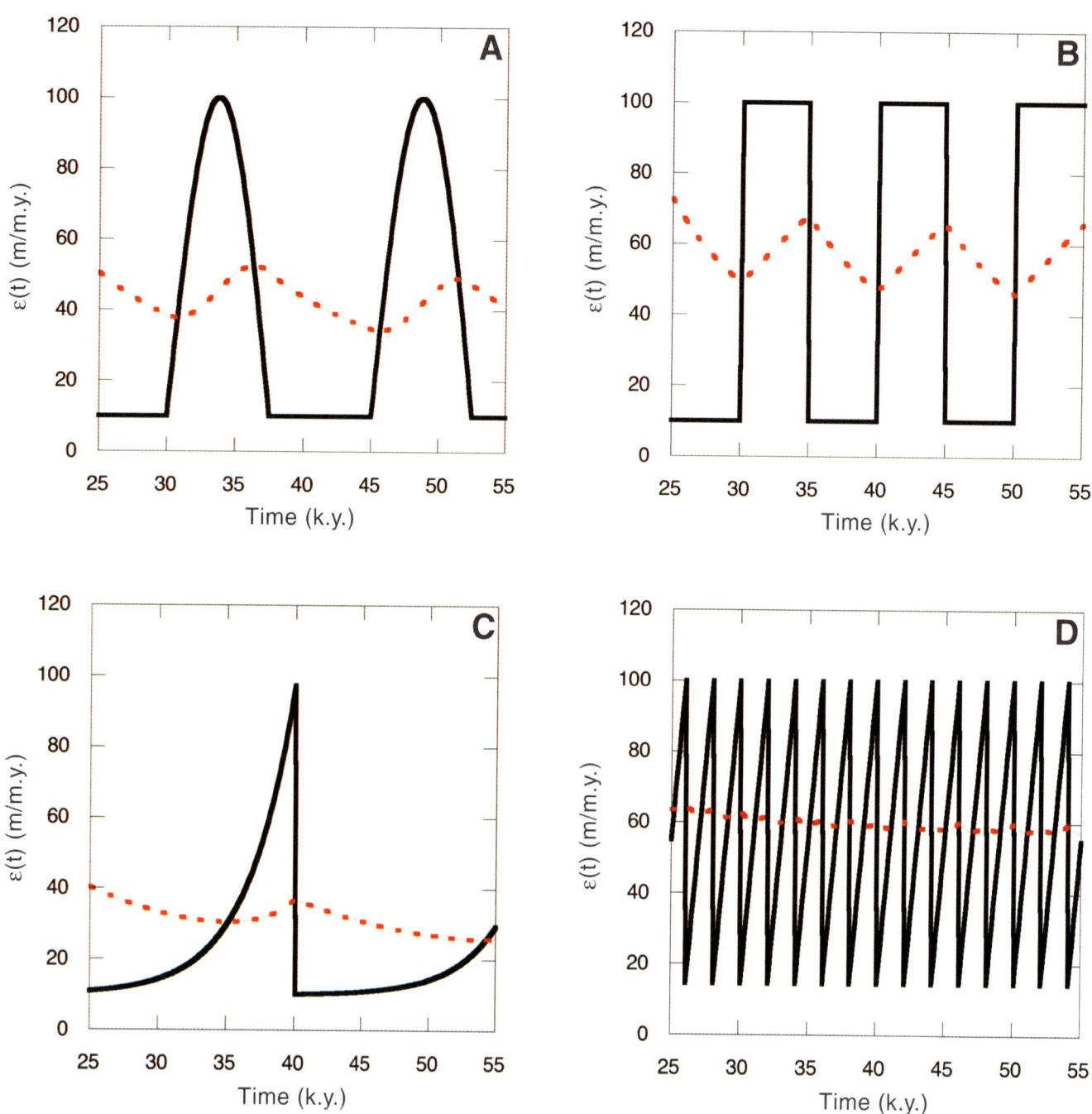

Figure 11. Similar to Figure 10, except that the range of erosion rate variation is from 10 to 100 m/m.y., and the frequency of cycles is variable for each scenario. Input erosion rates are shown by solid black lines, and erosion rates inferred from the modeled nuclide concentrations are shown by the dashed red curves. A: Full-wave rectification with the same 15 k.y. period as Figure 10A. B: Square wave with a period of 10 k.y. C: Exponential variation with a period of 20 k.y. D: Sawtooth variation with a period of 2 k.y.

erosion rates increase by a factor of six or more, it is likely that by the time the landscape has recovered (i.e., enough time has passed that the landscape appears to be in steady state) point-specific erosion or soil production rates will capture the average. This conclusion has important implications for using catchment-averaged samples to infer rates for an entire watershed, or drainage basin. Namely, it supports the methodology and does not draw a distinction between small and large catchments. An important next step is therefore to combine point-specific modeling with a catchment-averaged model and compare model outputs with actual measurements of both point-specific and average rates.

Reducing the magnitude of the variable erosion rate led to an even better agreement between the long-term average erosion rates, irrespective of input function. Again, the higher the frequency, the lower the impact of the input variable erosion functions on the output CRN concentrations and therefore on the inferred erosion rate, supporting the use of catchment-averaged samples. Specifically, this modeling result suggests that if a catchment is experiencing highly variable rates of erosion, due to dominance of different processes across the catchment, then using an integration of sediment derived from all points is more likely to capture the long-term average than any point sample is. Under the lower erosion rate conditions, however, the impact of the variable rate on the CRN concentration is not as great, such that inferred erosion rates are often off by a factor of two or more. Results from the examples shown here to explicitly address uncertainties raised from the field examples are robust and follow similar trends for either increasing or decreasing the magnitude and frequency of the input variation. Perhaps the saving grace of this is that more slowly eroding landscapes will preserve signatures of the anomalous events such that critical field sampling can avoid them. The amending of the southeastern Australia soil production function with the addition of more samples serves as a good example of how enough CRN measurements can help insure capturing the true form of the soil production function. An obvious implication of this is determining the cost-to-benefit ratio of a given CRN sampling scheme. It appears that using catchment-averaged samples to infer erosion rates is an accurate way, in both theory and practice, to quantify erosion rates. These average erosion rates continue to show their value across a broad range of geomorphic applications and will undoubtedly be used to further couple erosion with climatic, tectonic, and anthropogenic forcings. To more fully evaluate how catchment-averaged samples might reflect variable erosion rates, the model presented here can be applied to every point in a catchment and coupled with a sediment routing model. Despite the value and apparent accuracy of average rates determined from CRNs, determining the average rate of a catchment does not untangle the way that different processes interact on the slopes and in the channel of the catchment, such that there is still value in well-planned point sampling strategies. Quantifying the sediment production and transport processes that are contributing to these average erosion rates requires higher numbers of samples and will be more dramatically affected by potential local variations in rate. Specifically, for example, quantifying landslide and bedrock failure processes continues to be elusive and will require combining a point-specific with a catchment-averaged sampling scheme.

Finally, perhaps not surprisingly, geomorphology matters. That is, constraining the sampling site such that there is little likelihood of a high-magnitude/low-frequency event affecting the nuclide concentration is critical. While this is an obvious conclusion, and one that many have articulated previously, the wider and wider application of CRN studies to quantifying surface processes warrants further emphasis of this point. This is particularly relevant to geomorphic settings that are not as well constrained as a soil-mantled, convex-up hillslope or a landslide-free first- or second-order catchment. One potential approach is to collect large numbers of samples, but that quickly becomes cost and time prohibitive. Another approach is to tackle the more complicated landscapes by examining the catchment-averaged rates, but this approach leads to empiricism rather than determinism. For example, catchments that are not uniformly soil-mantled are eroding by a range of processes. Determining the catchment-averaged rates for such diverse drainages have provided empirical support for a wide number of recent studies, but have not been able to distinguish the process-specific contributions of the sediment load. Similarly, the range of timescales over which the catchment-averaged rates are applicable has yet to be shown definitively. It is likely that here the application of multiple analytical tools—combining CRN studies with short-lived isotope studies and low-temperature thermochronometry, for example—will make the most headway. Perhaps the most difficult geomorphology to constrain is the dating of features, whether by burial-age dating in caves or thick deposits, or by profile dating on geomorphic surfaces, simply because the landscapes contributing the sediments being dated are gone or are changed significantly. Irrespective of all the uncertainties and potential pitfalls, it remains amazing how well the application of CRNs to untangling the way Earth's surface is eroding is doing.

ACKNOWLEDGMENTS

Special thanks to Belinda Barnes for initial help with the mathematical formulation of the model and Hany Farid for invaluable help writing the Matlab® code. Kabir Heimsath, Josh Roering, Damien Kelleher, and Geoff Hunt assisted with the fieldwork across the two main study areas. Follow-up work in southeastern Australia has benefited greatly from collaboration with Ron Amundson. Ideas explored and results summarized here reflect work done in close collaboration with Bill Dietrich, John Chappell, Kuni Nishiizumi, and Bob Finkel. Thanks to Lionel Siame, Didier Bourlès, and Erik Brown, the GSA Volume Editors for their efforts to compile this volume, and to Milan Pavich and an anonymous reviewer for their valuable comments on a previous version of this manuscript. All cosmogenic results reported here were partially performed under the auspices of the U.S. Department of Energy by Lawrence Livermore National Laboratory under contract W-7405-Eng-48. Funding for this work was from National Science Foundation grant EAR-0239655.

REFERENCES CITED

Ahnert, F., 1967, The role of the equilibrium concept in the interpretation of landforms of fluvial erosion and deposition, in Macar, P., ed., L'evolution des versants: Liege, University of Liege, p. 23–41.

Ahnert, F., 1987, Approaches to dynamic equilibrium in theoretical simulations of slope development: Earth Surface Processes and Landforms, v. 12, p. 3–15.

Anderson, R.S., Repka, J.L., and Dick, G.S., 1996, Explicit treatment of inheritance in dating depositional surfaces using in situ Be-10 and Al-26: Geology, v. 24, p. 47–51, doi: 10.1130/0091-7613(1996)024<0047:ETOIID>2.3.CO;2.

Bierman, P., and Steig, E.J., 1996, Estimating rates of denudation using cosmogenic isotope abundances in sediment: Earth Surface Processes and Landforms, v. 21, p. 125–139, doi: 10.1002/(SICI)1096-9837(199602)21:2<125::AID-ESP511>3.0.CO;2-8.

Bierman, P.R., 1994, Using in situ produced cosmogenic isotopes to estimate rates of landscape evolution: A review from the geomorphic perspective: Journal of Geophysical Research, B, Solid Earth and Planets, v. 99, no. 7, p. 13,885–13,896, doi: 10.1029/94JB00459.

Bierman, P.R., and Caffee, M., 2001, Slow rates of rock surface erosion and sediment production across the Namib Desert and escarpment, southern Africa: American Journal of Science, v. 301, no. April/May, p. 326–358.

Bierman, P.R., and Nichols, K.K., 2004, Rock to sediment—Slope to sea with Be-10—Rates of landscape change: Annual Review of Earth and Planetary Sciences, v. 32, p. 215–255, doi: 10.1146/annurev.earth.32.101802.120539.

Bierman, P.R., and Turner, J., 1995, ^{10}Be and ^{26}Al evidence for exceptionally low rates of Australian bedrock erosion and the likely existence of pre-Pleistocene landscapes: Quaternary Research, v. 44, no. 3, p. 378–382.

Bierman, P.R., Marsella, K.A., Patterson, C., Thompson Davis, P., and Caffee, M., 1999, Mid-Pleistocene cosmogenic minimum-age limits for pre-Wisconsin glacial surfaces in southwestern Minnesota and southern Baffin Island: A multiple nuclide approach: Geomorphology, v. 27, no. 1-2, p. 25–39, doi: 10.1016/S0169-555X(98)00088-9.

Brown, E.T., Bourlès, D.L., Colin, F., Raisbeck, G.M., Yiou, F., and Desgarceaux, S., 1995a, Evidence for muon-induced production of ^{10}Be in near-surface rocks from the Congo: Geophysical Research Letters, v. 22, no. 6, p. 703–706, doi: 10.1029/95GL00167.

Brown, E.T., Stallard, R.F., Larsen, M.C., Raisbeck, G.M., and Yiou, F., 1995b, Denudation rates determined from the accumulation of in situ-produced ^{10}Be in the Luquillo Experimental Forest, Puerto Rico: Earth and Planetary Science Letters, v. 129, p. 193–202, doi: 10.1016/0012-821X(94)00249-X.

Burbank, D.W., Leland, J., Fielding, E., Anderson, R.S., Brozovic, N., Reid, M.R., and Duncan, C., 1996, Bedrock incision, rock uplift and threshold hillslopes in the northwestern Himalayas: Nature, v. 379, no. 6565, p. 505–510, doi: 10.1038/379505a0.

Cerling, T.E., and Craig, H., 1994, Geomorphology and in-situ cosmogenic isotopes: Annual Review of Earth and Planetary Sciences, v. 22, p. 273–317, doi: 10.1146/annurev.ea.22.050194.001421.

Clapp, E.M., Bierman, P.R., Schick, A.P., Lekach, J., Enzel, Y., and Caffee, M., 2000, Sediment yield exceeds sediment production in arid region drainage basins: Geology, v. 28, p. 995–998, doi: 10.1130/0091-7613(2000)028<0995:SYESPI>2.3.CO;2.

Clapp, E.M., Bierman, P.R., and Caffee, M., 2002, Using Be-10 and Al-26 to determine sediment generation rates and identify sediment source areas in an arid region drainage basin: Geomorphology, v. 45, no. 1-2, p. 89–104, doi: 10.1016/S0169-555X(01)00191-X.

Clark, D.H., Bierman, P.R., and Larsen, P., 1995, Improving in situ cosmogenic chronometers: Quaternary Research, v. 44, p. 367–377, doi: 10.1006/qres.1995.1081.

Cockburn, H.A.P., and Summerfield, M.A., 2004, Geomorphological applications of cosmogenic isotope analysis: Progress in Physical Geography, v. 28, no. 1, p. 1–42, doi: 10.1191/0309133304pp395oa.

Dietrich, W.E., Bellugi, D., Heimsath, A.M., Roering, J.J., Sklar, L., and Stock, J.D., 2003, Geomorphic transport laws for predicting landscape form and dynamics, in Wilcock, P., and Iverson, R., eds., Prediction in geomorphology: Washington, D.C., American Geophysical Union, p. 103–132.

Dunai, T.J., 2000, Scaling factors for production rates of in situ produced cosmogenic nuclides: A critical reevaluation: Earth and Planetary Science Letters, v. 176, p. 157–169, doi: 10.1016/S0012-821X(99)00310-6.

Dunne, J., Elmore, D., and Muzikar, P., 1999, Scaling factors for the rates of production of cosmogenic nuclides for geometric shielding and attenuation at depth on sloped surfaces: Geomorphology, v. 27, no. 1-2, p. 3–11, doi: 10.1016/S0169-555X(98)00086-5.

Gosse, J.C., and Phillips, F.M., 2001, Terrestrial in situ cosmogenic nuclides: Theory and application: Quaternary Science Reviews, v. 20, p. 1475–1560, doi: 10.1016/S0277-3791(00)00171-2.

Granger, D.E., 2002, Spatially averaged erosion rates from cosmogenic nuclides in sediments: Ten years later: Geochimica et Cosmochimica Acta, v. 66, no. 15A, p. A288–A288.

Granger, D.E., and Smith, A.L., 2000, Dating buried sediments using radioactive decay and muogenic production of ^{26}Al and ^{10}Be: Nuclear Instruments and Methods in Physics Research, Section B, Beam Interactions with Materials and Atoms, v. 172, no. 1-4, p. 824–828.

Granger, D.E., and Stock, G.M., 2004, Using cave deposits as geologic tiltmeters: Application to postglacial rebound of the Sierra Nevada, California: Geophysical Research Letters, v. 31, no. 22, doi: 10.1029/2004GL021403.

Granger, D.E., Kirchner, J.W., and Finkel, R., 1996, Spatially averaged long-term erosion rates measured from in situ-produced cosmogenic nuclides in alluvial sediment: Journal of Geology, v. 104, no. 3, p. 249–257.

Granger, D.E., Fabel, D., and Palmer, A.N., 2001, Pliocene-Pleistocene incision of the Green River, Kentucky, determined from radioactive decay of cosmogenic ^{26}Al and ^{10}Be in Mammoth Cave sediments: Geological Society of America Bulletin, v. 113, p. 825–836, doi: 10.1130/0016-7606(2001)113<0825:PPIOTG>2.0.CO;2.

Hancock, G.S., Anderson, R.S., Chadwick, O.A., and Finkel, R.C., 1999, Dating terraces with ^{10}Be and ^{26}Al profiles: Application to the Wind River, Wyoming: Geomorphology, v. 27, no. 1-2, p. 41–60, doi: 10.1016/S0169-555X(98)00089-0.

Heimsath, A.M., Dietrich, W.E., Nishiizumi, K., and Finkel, R.C., 1997, The soil production function and landscape equilibrium: Nature, v. 388, p. 358–361, doi: 10.1038/41056.

Heimsath, A.M., Dietrich, W.E., Nishiizumi, K., and Finkel, R.C., 1999, Cosmogenic nuclides, topography, and the spatial variation of soil depth: Geomorphology, v. 27, no. 1-2, p. 151–172, doi: 10.1016/S0169-555X(98)00095-6.

Heimsath, A.M., Chappell, J., Dietrich, W.E., Nishiizumi, K., and Finkel, R.C., 2000, Soil production on a retreating escarpment in southeastern Australia: Geology, v. 28, p. 787–790, doi: 10.1130/0091-7613(2000)028<0787:SPOARE>2.3.CO;2.

Heimsath, A.M., Chappell, J., Dietrich, W.E., Nishiizumi, K., and Finkel, R.C., 2001a, Late Quaternary erosion in southeastern Australia: A field example using cosmogenic nuclides: Quaternary International, v. 83-85, p. 169–185, doi: 10.1016/S1040-6182(01)00038-6.

Heimsath, A.M., Dietrich, W.E., Nishiizumi, K., and Finkel, R.C., 2001b, Stochastic processes of soil production and transport: Erosion rates, topographic variation, and cosmogenic nuclides in the Oregon Coast Range: Earth Surface Processes and Landforms, v. 26, p. 531–552, doi: 10.1002/esp.209.

Heimsath, A.M., Chappell, J., Finkel, R.C., Fifield, L.K., and Alimanovic, A., 2006, Escarpment erosion and landscape evolution in southeastern Australia, in Willett, S.D., et al., eds., Tectonics, Climate, and Landscape Evolution: Geological Society of America Special Paper 398, p. 173–190, doi: 10.1130/2006.2398(10).

Heller, P.L., Peterman, Z.E., O'Neil, J.R., and Shafiqullah, M., 1985, Isotopic provenance of sandstones from the Eocene Tyee Formation, Oregon Coast Range: Geological Society of America Bulletin, v. 96, p. 770–780, doi: 10.1130/0016-7606(1985)96<770:IPOSFT>2.0.CO;2.

Hewawasam, T., von Blanckenburg, F., Schaller, M., and Kubik, P., 2003, Increase of human over natural erosion rates in tropical highlands constrained by cosmogenic nuclides: Geology, v. 31, p. 597–600, doi: 10.1130/0091-7613(2003)031<0597:IOHONE>2.0.CO;2.

Kelsey, H.M., and Bockheim, J.G., 1994, Coastal landscape evolution as a function of eustasy and surface uplift rates, Cascadia margin, southern Oregon: Geological Society of America Bulletin, v. 106, p. 840–854, doi: 10.1130/0016-7606(1994)106<0840:CLEAAF>2.3.CO;2.

Kelsey, H.M., Engebretson, D.C., Mitchell, C.E., and Ticknor, R.L., 1994, Topographic form of the Coast Ranges of the Cascadia margin in relation to coastal uplift rates and plate subduction: Journal of Geophysical Research, v. 99, p. 12,245–12,255, doi: 10.1029/93JB03236.

Lal, D., 1988, In situ-produced cosmogenic isotopes in terrestrial rocks: Annual Review of Earth and Planetary Sciences, v. 16, p. 355–388.

Lal, D., 1991, Cosmic ray labeling of erosion surfaces: In situ nuclide production rates and erosion models: Earth and Planetary Science Letters, v. 104, p. 424–439, doi: 10.1016/0012-821X(91)90220-C.

Lovell, J.P.B., 1969, Tyee Formation: Undeformed turbidites and their lateral equivalents: Mineralogy and paleogeography: Geological Society of America Bulletin, v. 80, p. 9–22.

Matmon, A., Bierman, P.R., Larsen, J., Southworth, S., Pavich, M., and Caffee, M., 2003a, Temporally and spatially uniform rates of erosion in the southern Appalachian Great Smoky Mountains: Geology, v. 31, p. 155–158.

Matmon, A., Bierman, P.R., Larsen, J., Southworth, S., Pavich, M., Finkel, R., and Caffee, M., 2003b, Erosion of an ancient mountain range, the Great Smoky Mountains, North Carolina and Tennessee: American Journal of Science, v. 303, no. 9, p. 817–855.

Montgomery, D.R., Dietrich, W.E., Torres, R., Anderson, S.P., Heffner, J.T., and Loague, K., 1997, Hydrologic response of a steep, unchanneled valley to natural and applied rainfall: Water Resources Research, v. 33, no. 1, p. 91–109, doi: 10.1029/96WR02985.

Montgomery, D.R., Sullivan, K., and Greenberg, H.M., 1998, Regional test of a model for shallow landsliding: Hydrological Processes, v. 12, no. 6, p. 943–955, doi: 10.1002/(SICI)1099-1085(199805)12:6<943::AID-HYP664>3.0.CO;2-Z.

Montgomery, D.R., Schmidt, K.M., Greenberg, H.M., and Dietrich, W.E., 2000, Forest clearing and regional landsliding: Geology, v. 28, p. 311–314, doi: 10.1130/0091-7613(2000)028<0311:FCARL>2.3.CO;2.

Nichols, K.K., Bierman, P.R., Caffee, M., Finkel, R.C., and Larsen, J., 2005, Cosmogenically enabled sediment budgeting: Geology, v. 33, p. 133–136, doi: 10.1130/G21006.1.

Niemi, N.A., Oskin, M., Burbank, D.W., Heimsath, A.M., and Gabet, E.J., 2005, Effects of bedrock landsliding on cosmogenically determined erosion rates: Earth and Planetary Science Letters, v. 237, p. 480–498, doi: 10.1016/j.epsl.2005.07.009.

Nishiizumi, K., Lal, D., Klein, J., Middleton, R., and Arnold, J.R., 1986, Production of ^{10}Be and ^{26}Al by cosmic rays in terrestrial quartz in situ and implications for erosion rates: Nature, v. 319, no. 6049, p. 134–136, doi: 10.1038/319134a0.

Nishiizumi, K., Winterer, E.L., Kohl, C.P., Klein, J., Middleton, R., Lal, D., and Arnold, J.R., 1989, Cosmic ray production rates of ^{10}Be and ^{26}Al in quartz from glacially polished rocks: Journal of Geophysical Research, v. 94, no. B12, p. 17,907–17,915.

Nishiizumi, K., Kohl, C.P., Arnold, J.R., Dorn, R., Klein, J., Fink, D., Middleton, R., and Lal, D., 1993, Role of in situ cosmogenic nuclides ^{10}Be and ^{26}Al in the study of diverse geomorphic processes: Earth Surface Processes and Landforms, v. 18, p. 407–425.

Nishiizumi, K., Finkel, R.C., Klein, J., and Kohl, C.P., 1996, Cosmogenic production of ^{7}Be and ^{10}Be in water targets: Journal of Geophysical Research, v. 101, no. B10, p. 22,225–22,232, doi: 10.1029/96JB02270.

Perg, L.A., Anderson, R.S., and Finkel, R.C., 2001, Use of a new ^{10}Be and ^{26}Al inventory method to date marine terraces, Santa Cruz, California, USA: Geology, v. 29, p. 879–882, doi: 10.1130/0091-7613(2001)029<0879:UOANBA>2.0.CO;2.

Pratt, B., Burbank, D.W., Heimsath, A., and Ojha, T., 2002, Impulsive alluviation during early Holocene strengthened monsoons, central Nepal Himalaya: Geology, v. 30, p. 911–914, doi: 10.1130/0091-7613(2002)030<0911:IADEHS>2.0.CO;2.

Reneau, S.L., and Dietrich, W.E., 1991, Erosion rates in the southern Oregon Coast Range: Evidence for an equilibrium between hillslope erosion and sediment yield: Earth Surface Processes and Landforms, v. 16, no. 4, p. 307–322.

Repka, J.L., Anderson, R.S., and Finkel, R.C., 1997, Cosmogenic dating of fluvial terraces, Fremont River, Utah: Earth and Planetary Science Letters, v. 152, no. 1-4, p. 59–73, doi: 10.1016/S0012-821X(97)00149-0.

Reusser, L.J., Bierman, P.R., Pavich, M.J., Zen, E.A., Larsen, J., and Finkel, R., 2004, Rapid late Pleistocene incision of Atlantic passive-margin river gorges: Science, v. 305, no. 5683, p. 499–502, doi: 10.1126/science.1097780.

Riebe, C.S., Kirchner, J.W., Granger, D.E., and Finkel, R.C., 2000, Erosional equilibrium and disequilibrium in the Sierra Nevada, inferred from cosmogenic Al-26 and Be-10 in alluvial sediment: Geology, v. 28, p. 803–806, doi: 10.1130/0091-7613(2000)028<0803:EEADIT>2.3.CO;2.

Roering, J.J., and Gerber, M., 2005, Fire and the evolution of steep, soil-mantled landscapes: Geology, v. 33, p. 349–352, doi: 10.1130/G21260.1.

Roering, J.J., Kirchner, J.W., and Dietrich, W.E., 1999, Evidence for non-linear, diffusive sediment transport on hillslopes and implications for landscape morphology: Water Resources Research, v. 35, no. 3, p. 853–870, doi: 10.1029/1998WR900090.

Roering, J.J., Kirchner, J.W., and Dietrich, W.E., 2001, Hillslope evolution by nonlinear, slope-dependent transport: Steady state morphology and equilibrium adjustment timescales: Journal of Geophysical Research, B, Solid Earth and Planets, v. 106, no. B8, p. 16,499–16,513, doi: 10.1029/2001JB000323.

Schaller, M., von Blanckenburg, F., Hovius, N., and Kubik, P.W., 2001, Large-scale erosion rates from in situ-produced cosmogenic nuclides in European river sediments: Earth and Planetary Science Letters, v. 188, no. 3-4, p. 441–458, doi: 10.1016/S0012-821X(01)00320-X.

Schaller, M., von Blanckenburg, F., Hovius, N., Veldkamp, A., van den Berg, M.W., and Kubik, P.W., 2004, Paleoerosion rates from cosmogenic Be-10 in a 1.3 Ma terrace sequence: Response of the river Meuse to changes in climate and rock uplift: Journal of Geology, v. 112, no. 2, p. 127–144, doi: 10.1086/381654.

Schmidt, K., 1999, Root strength, colluvial soil depth, and colluvial transport on landslide-prone hillslopes: Seattle, Washington, University of Washington, 257 p.

Small, E.E., Anderson, R.S., Repka, J.L., and Finkel, R., 1997, Erosion rates of alpine bedrock summit surfaces deduced from in situ ^{10}Be and ^{26}Al: Earth and Planetary Science Letters, v. 150, no. 3-4, p. 413–425, doi: 10.1016/S0012-821X(97)00092-7.

Small, E.E., Anderson, R.S., and Hancock, G.S., 1999, Estimates of the rate of regolith production using ^{10}Be and ^{26}Al from an alpine hillslope: Geomorphology, v. 27, no. 1-2, p. 131–150, doi: 10.1016/S0169-555X(98)00094-4.

Snavely, P.D., Wagner, H.C., and MacLeod, N.S., 1964, Rhythmic-bedded eugeosynclinal deposits of the Tyee Formation: Oregon Coast Range: Kansas Geological Survey Bulletin, v. 169, p. 461–480.

Stock, G.M., Anderson, R.S., and Finkel, R.C., 2005, Rates of erosion and topographic evolution of the Sierra Nevada, California, inferred from cosmogenic ^{26}Al and ^{10}Be concentrations: Earth Surface Processes and Landforms, v. 30, p. 985–1006.

Stone, J.O., Ballantyne, C.K., and Fifield, L.K., 1998a, Exposure dating and validation of periglacial weathering limits, northwest Scotland: Geology, v. 26, p. 587–590, doi: 10.1130/0091-7613(1998)026<0587:EDAVOP>2.3.CO;2.

Stone, J.O., Evans, J.M., Fifield, L.K., Allan, G.L., and Cresswell, R.G., 1998b, Cosmogenic chlorine-36 production in calcite by muons: Geochimica et Cosmochimica Acta, v. 62, no. 3, p. 433–454, doi: 10.1016/S0016-7037(97)00369-4.

Swanson, F.J., 1981, Fire and geomorphic processes, in Mooney, H.A., et al., eds., Proceedings, Fire Regimes and Ecosystems Conference: Honolulu, Hawaii, U.S. Department of Agriculture, Forest Service, p. 401–420.

Weissel, J.K., and Seidl, M.A., 1998, Inland propagation of erosional escarpments and river profile evolution across the southeast Australian passive continental margin, in Tinkler, K.J., and Wohl, E.E., eds., Rivers over rock: Fluvial processes in bedrock channels: Washington, D.C., American Geophysical Union, Geophysical Monograph 107, p. 189–206.

Wobus, C.W., Heimsath, A.M., Whipple, K.X., and Hodges, K.V., 2005, Active surface thrust faulting in the central Nepalese Himalaya: Nature, v. 434, p. 1008–1011, doi: 10.1038/nature03499.

MANUSCRIPT ACCEPTED BY THE SOCIETY 11 APRIL 2006

Geological Society of America
Special Paper 415
2006

Exposure dating (^{10}Be, ^{26}Al) of natural terrain landslides in Hong Kong, China

Roderick J. Sewell[†]
*Hong Kong Geological Survey, Geotechnical Engineering Office, Civil Engineering and Development Department,
101 Princess Margaret Road, Kowloon, Hong Kong SAR, China*

Timothy T. Barrows[‡]
*Department of Nuclear Physics, Research School of Physical Sciences and Engineering,
Australian National University, ACT, 0200, Canberra, Australia*

S. Diarmad G. Campbell[§]
*Hong Kong Geological Survey, Geotechnical Engineering Office, Civil Engineering and Development Department,
101 Princess Margaret Road, Kowloon, Hong Kong SAR, China*

L. Keith Fifield[#]
*Department of Nuclear Physics, Research School of Physical Sciences and Engineering,
Australian National University, ACT, 0200, Canberra, Australia*

ABSTRACT

We successfully apply exposure dating using cosmogenic nuclides to natural terrain landslides in Hong Kong. Forty-five samples from eight landslide sites were exposure dated using ^{10}Be, and a subset of six samples was also dated using ^{26}Al. The sites comprised four large, deep-seated landslides featuring well-preserved rock scarps and associated debris lobes; two sites of rock and boulder fall; and two sites where scarps only are preserved. All of the deep-seated landslides gave ages within the last 50,000 yr, and the largest landslide gave an age of ~32,000 yr. The youngest (~2000 yr) and oldest (~57,000 yr) landslide events dated came from the two sites of rock and boulder fall. Exposure ages from the deep-seated landslide scarps generally gave the most internally consistent ages for the landslides. However, only in rare cases did the landslide scarp ages overlap with those of boulders in the associated debris. Generally, boulders in the debris appeared to contain significant inheritance of cosmogenic nuclides from previous exposure and so yielded ages greater than those from the scarps. Surface exposure ages of ~285,000 yr from boulders in the debris of

[†]E-mail: jsewell@cedd.gov.hk.
[‡]E-mail: tim.barrows@anu.edu.au.
[§]Present address: British Geological Survey, Murchison House, West Mains Road, Edinburgh, UK; e-mail: sdgc@bgs.ac.uk
[#]E-mail: keith.fifield@anu.edu.au.

Sewell, R.J., Barrows, T.T., Campbell, S.D.G., and Fifield, L.K., 2006, Exposure dating (^{10}Be, ^{26}Al) of natural terrain landslides in Hong Kong, China, *in* Siame, L.L., Bourlès, D.L., and Brown, E.T., eds., In Situ–Produced Cosmogenic Nuclides and Quantification of Geological Processes: Geological Society of America Special Paper 415, p. 131–146, doi: 10.1130/2006.2415(08). For permission to copy, contact editing@geosociety.org. © 2006 Geological Society of America. All rights reserved.

two deep-seated landslides provide minimum ages considered to represent the original rock surfaces. This study has shown that it is possible to measure exposure ages of surfaces associated with large landslides from 70,000 yr down to a few thousand years old, despite low cosmogenic isotope production rates in Hong Kong due to low latitude and low altitude.

Keywords: Al-26, Be-10, cosmogenic elements, exposure age, Hong Kong, landslides.

INTRODUCTION

In situ–produced cosmogenic nuclides (^{10}Be, ^{26}Al) are now widely used for dating geomorphological events that have occurred within the last few million years. The most common application is for exposure dating of glacial surfaces (e.g., Bierman et al., 1999; Barrows et al., 2001, 2002). Cosmogenic nuclides have also been used to date (1) fault movements (e.g., Zreda and Noller, 1998), (2) volcanic landforms (e.g., Licciardi et al., 1999), (3) meteorite impacts (e.g., Phillips et al., 1991), and (4) periglacial deposits (e.g., Barrows et al., 2004).

Dating ancient landslides has historically presented a significant challenge for traditional dating methods. The relatively short life-span of trees and shrubs, usually limited to a few decades, means that estimates of landslide ages based on the assessment of vegetation cover and revegetation rates may only be reliable for very young (<300 yr) landslides. Dating buried organic material using radiocarbon is possible in only some circumstances. Exposure dating, however, has the distinct advantage over other dating methods of directly dating rock surfaces, so no assumptions need be made, as are required with radiocarbon dating as to whether the organic material was contemporaneous with the landslide event. Prior to this study, there have been relatively few attempts to exposure date ancient landslides (e.g., Bierman et al., 1995; Nichols et al., 2000). Ballantyne et al. (1998) used exposure dating to investigate the timing of postglacial landslides in Scotland. In Austria, cosmogenic nuclide data from a landslide were used in an attempt to calibrate the ^{10}Be production rate at Köfels landslide (Kubik et al., 1998). Successful dating of the landslide using radiocarbon allowed a production rate to be calculated that is quite similar to that derived elsewhere (Stone, 2000).

Approximately 50% of Hong Kong's land area (1103 km^2, the area of the Hong Kong Special Administrative Region, China) comprises steep natural terrain[1] sloping at angles greater than 30°. With most of the population being concentrated in the flat-lying urban areas (215 km^2), and with the shortage of land available for housing and economic activities, new developments increasingly encroach on the surrounding natural terrain, leading to intense pressures on the environment. Approximately 300 natural terrain landslides occur in Hong Kong each year (King, 1999), and between the years 1982 and 1999, ~800 natural terrain landslides were reported (King, 1999) as affecting urban developments. Thus, landsliding is an important mechanism controlling landscape evolution in Hong Kong and has significant economic implications. Despite the difficulty in dating ancient landslides, larger relict features in Hong Kong have generally been considered to be several 10s to 100s, rather than 1000s, of years old (Evans, 1998; Murray, 2001; Ng et al., 2003). These notional ages have fallen within the approximate design life of engineering structures in Hong Kong, and therefore have influenced the development of landslide hazard risk management strategies (Ng et al., 2003). However, more accurate estimates of landslide age are now required if improved assessment and reliable and cost-effective mitigation of natural terrain hazards is to be achieved.

One of the Hong Kong SAR Government's policy objectives is to advance the understanding of the causes and mechanisms of natural terrain hazards, so as to improve the reliability of landslide preventive or remedial works. Assessment and mitigation of natural terrain hazards depend on the interpretation of historical landslide data and consideration of the possible return periods of natural terrain failures. The estimation of return periods, based on historical landslide data, assumes some constancy in the environmental controls of the landslides. Given the extreme changes in sea level and known climatic fluctuations that have occurred during the last 10–15,000 yr (Thomas, 1994; Fyfe et al., 2000), this assumption requires confirmation. Therefore, knowledge of relict natural terrain landslide chronologies is essential not only for providing more reliable constraints on landscape evolution and estimation of return periods, but also for understanding the causes and mechanisms of landslides.

In this paper, we apply exposure dating to large, deep-seated landslides, and substantial rockfall events in Hong Kong. This is part of a larger dating study of natural terrain landslides in Hong Kong (Sewell and Campbell, 2005), which includes the establishment of relative chronologies based on aerial photograph interpretation, and dating of landslide deposits using optically stimulated luminescence (OSL) and radiocarbon (^{14}C-AMS). We demonstrate that exposure dating using ^{10}Be and ^{26}Al can be applied successfully to landslides in Hong Kong. Furthermore, the sensitivity of the analytical techniques is sufficiently high to measure exposure ages of surfaces only a few thousand years old, despite the seemingly unfavorable circumstances in Hong Kong of low latitude and low altitude.

---

[1]Natural terrain follows the definition given in Ng et al. (2003), namely, "terrain that has not been modified substantially by human activity but includes areas where grazing, hill fires, and deforestation may have occurred."

SAMPLING STRATEGY

Exposure dating is based on the accumulation of cosmogenic nuclides in a rock surface first exposed to cosmic radiation. Most cosmogenic nuclides are created in the uppermost 3 m of rock. To date landslides, inheritance from any previous exposure must be avoided. Accordingly, the landslides must be of sufficient magnitude to generate freshly exposed surfaces. The surface to be dated also needs to have remained stable, i.e., in the same geometry, during its exposure history, and to have experienced negligible weathering.

There are two options for applying exposure dating to landslides: dating scarp faces and dating blocks within debris lobes. Potentially, exposure ages both from the scarp and from blocks within the debris could give the age of detachment of the landslide from its source area. However, several factors can bias exposure ages in these settings. For scarp faces, exposure ages can overestimate the age of the landslide if a shallow (<3 m deep) surface is sampled. Similarly, exposure ages for blocks sourced from the original land surface will also appear too old. Sampling scarp faces, where there is reliable evidence that the material is in situ, has the advantage that the surface has been continually exposed in the same geometry. However, blocks within, or on top of, a debris lobe may have moved after deposition, or even been exhumed from beneath the original surface of the debris. In both scenarios, exposure ages could underestimate the age of deposition of the debris and hence of landsliding. Similarly, if material has weathered from the sampled rock surface, that age of deposition would again be an underestimate.

In this study, 45 samples were dated using ^{10}Be. A subset of six samples was also dated using ^{26}Al to investigate reproducibility of results and to identify the possibility of multiple exposure histories on blocks in the debris lobes. The availability of excellent low-level (3900 ft) and high-level (10,000 ft) aerial photographs of Hong Kong, taken in 1963 and 1964 when the general vegetation cover was much less dense than at the present day (due mainly to the activities of the largely rural population), greatly assisted (1) the identification of relict natural terrain landslides, (2) the establishment of relative chronologies among clusters of relict landslides, and (3) the development of sampling strategies for exposure dating of relict landslides.

Where possible, samples were collected both from scarp faces and from the associated debris lobes. We were careful to select rocks that did not show any unusual weathering, to minimize the possibility of inheritance. The selected deep-seated landslide sites generally exhibited >4 m high rock scarps. In cases where these could be associated with obvious debris lobes, the debris often contained very large (>5 m diameter) boulders, providing ideal sites for exposure dating and strong evidence that the lobes were the products of single large events. Multiple samples were collected from various elevations across the scarp surface, and from the flat upper surfaces of large boulders within the debris, to investigate the effects of inheritance.

Generally, the landslide features selected for exposure dating were located within clusters of overlapping natural terrain landslides. The results of direct dating, therefore, provided an opportunity to assess the relative chronology of the overlapping landslides previously established by aerial photograph interpretation. Accordingly, the detailed characteristics of the sample sites on the following figures include the location and extent of surrounding landslide features. Site data for the samples collected for the study are summarized in Table 1.

SITES

Eight natural terrain landslide sites from across Hong Kong were selected for study (Fig. 1). These comprised four large, deep-seated landslides featuring well-preserved scarps, mainly in rock, and associated debris lobes (Sham Wat Debris Lobe, Sunset Peak West, Tsing Yi, and Ap Lei Chau); two sites of rock and boulder fall (Lion Rock and Fei Ngo Shan); and two deep-seated landslides with scarps preserved, but no debris that could be related to the scars with confidence (Lai Cho Road and Mid-Levels). The sites were selected for a variety of reasons: (1) They were the subject of recent detailed investigations of instability (e.g., Lai Cho Road), (2) they were previously investigated sites associated with larger integrated studies (e.g., Sham Wat Debris Lobe, Mid-Levels, Ap Lei Chau), and (3) they were close to existing developments (e.g., Lion Rock, Fei Ngo Shan, Tsing Yi, Sunset Peak West).

Deep-Seated Scarp and Debris Lobe Sites

Sham Wat Debris Lobe

The largest landslide feature we studied is located at Sham Wat in western Lantau Island, and is known as the "Sham Wat Debris Lobe" (Fig. 2). The debris lobe was previously the subject of a detailed investigation (King, 1998), including a preliminary dating study (King, 2001). On the basis of standard ("average large aliquot") optically stimulated luminescence dating methods, the apparent age of the deposit was reported as ca. 35 ka. However, Murray (2001) considered this age to be a gross overestimate because using the dose distribution method, based on the assumption of incomplete bleaching, the deposit returned ages as young as 1.4 ± 0.2 ka, with a likely age range between 1 and 23 ka.

There are two main parts to the Sham Wat Debris Lobe, the main large northern debris lobe, with an estimated volume of 850,000 m³, and a subordinate southern lobe with an estimated volume of 150,000 m³. The main lobe has a well-defined source area marked by a prominent 10 m high rock escarpment, composed of porphyritic rhyolite. Numerous large boulders protrude from the surface of the debris lobe and these, together with the main scarp region, provided suitable sites for exposure dating. Ten samples (Table 1) were collected from the prominent rock escarpment forming the source area for the main lobe (SWD-01 to SWD-05), as well as large boulders (SWD-06 to SWD-10) within the associated debris (Fig. 2). Samples across the scarp

TABLE 1. SITE DATA

Sample	Latitude (°N)	Longitude (°E)	Altitude (m)	Setting	Scaling factor (nucleons)	Scaling factor (muons)	Horizon correction	Thickness (cm)	Thickness (correction)
Deep-seated Scarps and Debris Lobes									
Sham Wat Debris Lobe									
SWD-01	22°15′33.53	113°52′56.39	358	Scar	0.930	0.840	0.894	4.0	0.968
SWD-02	22°15′33.53	113°52′56.39	358	Scar	0.930	0.840	0.894	4.7	0.963
SWD-03	22°15′33.30	113°52′57.16	350	Scar	0.920	0.840	0.952	4.1	0.967
SWD-04	22°15′33.30	113°52′57.16	350	Scar	0.920	0.840	0.910	4.0	0.968
SWD-05	22°15′33.30	113°52′57.68	340	Scar	0.920	0.840	0.979	3.5	0.972
SWD-06	22°15′41.76	113°52′57.56	75	Boulder	0.750	0.740	0.984	2.0	0.984
SWD-07	22°15′43.29	113°52′57.84	69	Boulder	0.750	0.730	0.985	2.25	0.982
SWD-08	22°15′47.68	113°53′00.27	38	Boulder	0.730	0.720	0.986	3.5	0.972
SWD-09	22°15′44.76	113°53′01.53	45	Boulder	0.730	0.730	0.986	2.5	0.980
SWD-10	22°15′41.33	113°52′54.52	90	Boulder	0.760	0.740	0.985	1.0	0.992
Sunset Peak West									
TCR-01	22°15′13.97	113°56′43.72	480	Scar	1.020	0.890	0.852	4.0	0.968
TCR-02	22°15′14.00	113°56′43.83	486	Scar	1.020	0.900	0.852	6.0	0.954
TCR-03	22°15′14.00	113°56′43.76	482	Scar	1.020	0.900	0.950	3.5	0.972
TCR-04	22°15′13.84	113°56′43.83	490	Scar	1.030	0.900	0.851	2.8	0.977
TCR-05	22°15′17.71	113°56′25.28	215	Boulder	0.830	0.790	0.987	6.0	0.954
TCR-06	22°15′17.71	113°56′25.28	215	Boulder	0.830	0.790	0.986	2.2	0.982
TCR-07	22°15′17.20	113°56′26.15	220	Boulder	0.840	0.790	0.987	4.0	0.968
TCR-08	22°15′16.45	113°56′26.46	225	Boulder	0.840	0.790	0.985	1.75	0.986
TCR-09	22°15′16.45	113°56′26.46	225	Boulder	0.840	0.790	0.985	4.2	0.967
TCR-10	22°15′16.28	113°56′25.94	223	Boulder	0.840	0.790	0.985	2.0	0.984
Tsing Yi									
TY-01	22°20′17.95	114°06′11.59	138	Boulder	0.786	0.760	0.947	4.5	0.964
TY-02	22°20′17.37	114°06′10.68	151	Boulder	0.790	0.760	0.931	2.2	0.982
TY-03	22°20′15.97	114°06′09.49	173	Boulder	0.810	0.770	0.922	2.0	0.984
TY-04	22°20′15.42	114°06′08.37	190	Scar	0.820	0.780	0.905	4.75	0.963
TY-05	22°20′14.89	114°06′05.09	255	Scar	0.860	0.800	0.720	1.5	0.988
TY-07	22°20′14.83	114°06′07.81	216	Scar	0.830	0.790	0.942	5.0	0.961
Ap Lei Chau									
ALC-01	22°14′17.14	114°09′27.22	161	Scar	0.800	0.770	0.966	2.5	0.980
ALC-02	22°14′17.21	114°09′26.27	173	Scar	0.810	0.770	0.924	2.75	0.978
ALC-03	22°14′16.33	114°09′27.22	155	Scar	0.800	0.770	0.879	2.5	0.980
ALC-04	22°14′18.22	114°09′28.97	137	Boulder	0.790	0.760	0.968	5.0	0.961
Rock and Boulder Falls									
Lion Rock									
LR-01	22°20′55.04	114°11′17.65	176	Boulder	0.810	0.770	0.973	1.5	0.988
LR-02	22°20′57.64	114°11′14.57	208	Boulder	0.830	0.790	0.947	1.5	0.988
LR-03	22°20′55.98	114°11′13.84	190	Boulder	0.820	0.780	0.964	4.5	0.964
Fei Ngo Shan									
FNS-01	22°20′15.69	114°13′02.13	162	Boulder	0.800	0.770	0.968	3.0	0.976
FNS-02	22°20′14.87	114°13′03.36	158	Boulder	0.800	0.770	0.975	2.2	0.982
FNS-03	22°20′15.30	114°13′03.53	162	Boulder	0.800	0.770	0.975	1.75	0.986
Deap-Seated Scarps									
Lai Cho Road									
LCR-01	22°21′05.03	114°07′45.89	109	Scar	0.770	0.750	0.847	3.5	0.972
LCR-02	22°21′05.00	114°07′45.82	110	Scar	0.770	0.750	0.703	5.7	0.956
LCR-03	22°21′05.19	114°07′46.10	111	Scar	0.770	0.750	0.854	2.0	0.984
LCR-04	22°21′05.26	114°07′46.06	109	Scar	0.770	0.750	0.789	6.5	0.950
Mid-Levels									
ML-01	22°16′38.69	114°08′35.48	420	Scar	0.970	0.870	0.815	2.5	0.980
ML-02	22°16′38.78	114°08′35.44	400	Scar	0.960	0.860	0.815	3.0	0.976
ML-03	22°16′38.65	114°08′35.37	420	Scar	0.970	0.870	0.855	3.7	0.970
ML-04	22°16′38.82	114°08′35.20	400	Scar	0.960	0.860	0.758	4.3	0.966
ML-05	22°16′38.82	114°08′35.20	400	Scar	0.960	0.860	0.758	3.0	0.976

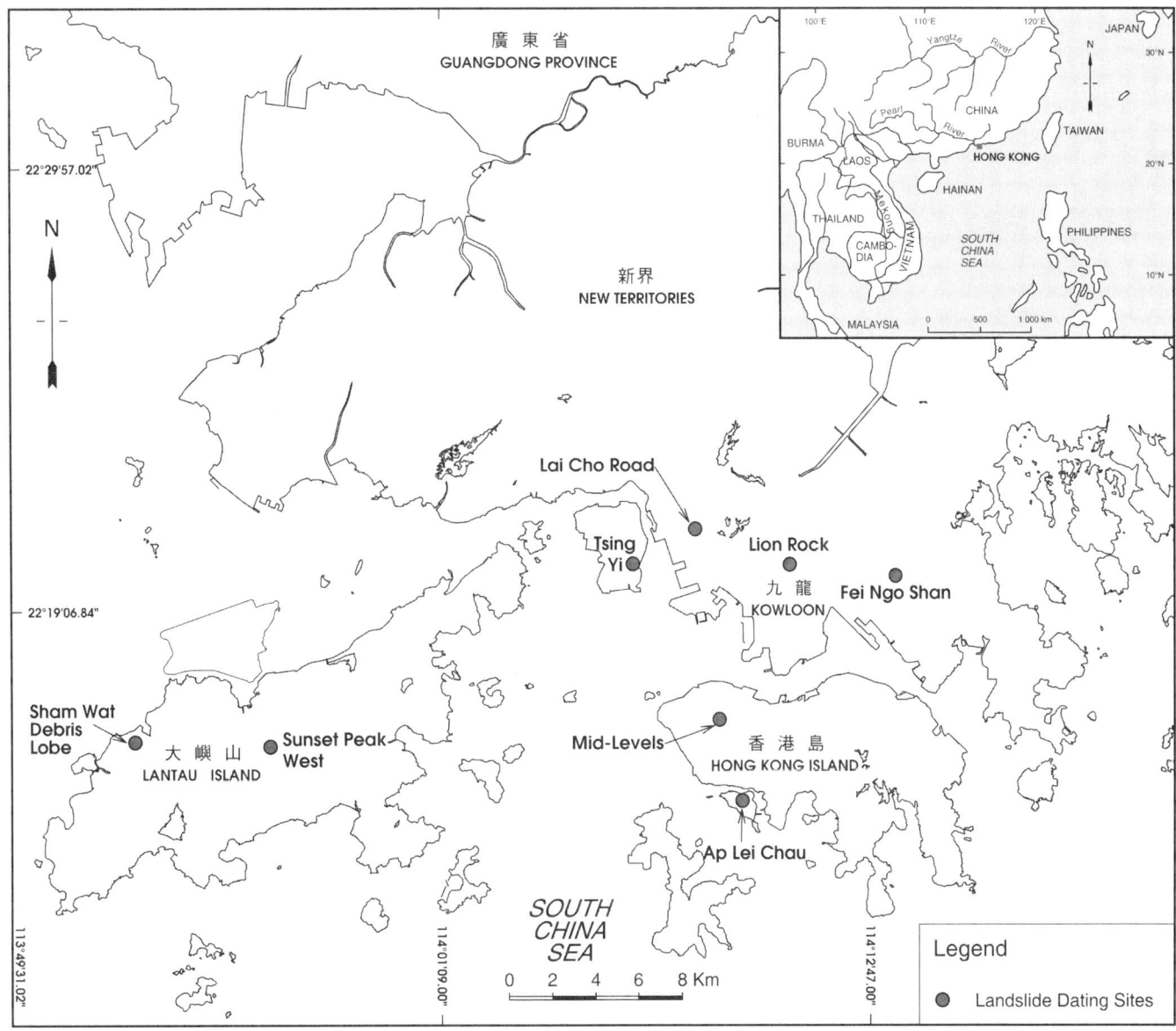

Figure 1. Location map of landslide sites selected for the dating study in the Hong Kong Special Administrative Region.

surface came from steep (62°–83°) well-jointed faces, more or less at similar elevations (340–358 mPD).[2] Large boulders sampled were generally well distributed across the surface of the debris lobe, had a minimum volume of 15 m³, and displayed similar weathering characteristics. Sample elevations in the debris lobe varied from 38 to 90 mPD.

Sunset Peak West

A large, deep-seated, natural terrain landslide, with a well-preserved main scarp and debris lobe, was studied on the western flanks of Sunset Peak, central Lantau Island (Fig. 3). The main rock type is coarse to fine ash rhyolitic tuff. Ten samples (Table 1) for exposure dating were collected from the scarp and debris lobe regions of the landslide (Fig. 3). Three samples (TCR-01, TCR-02, and TCR-04) came from elevations of 480 mPD, 486 mPD, and 490 mPD, respectively, on the moderately inclined (53°) scarp surface. One sample (TCR-03, 482 mPD) came from an adjacent, inclined (40°) rock surface on the upper western flank of the scarp, although this was considered not to have contributed to the landslide, and the sample was collected mainly for control purposes. Six samples (TCR-05 to TCR-10) were collected from large (20 m³) boulders within the debris lobe. The boulders were generally well distributed, with similar internal features (e.g., flow banding) and lithologies to the rocks in the scarp area. Sample

---

[2]Principal datum (PD) in Hong Kong is 1.23 m below mean sea level.

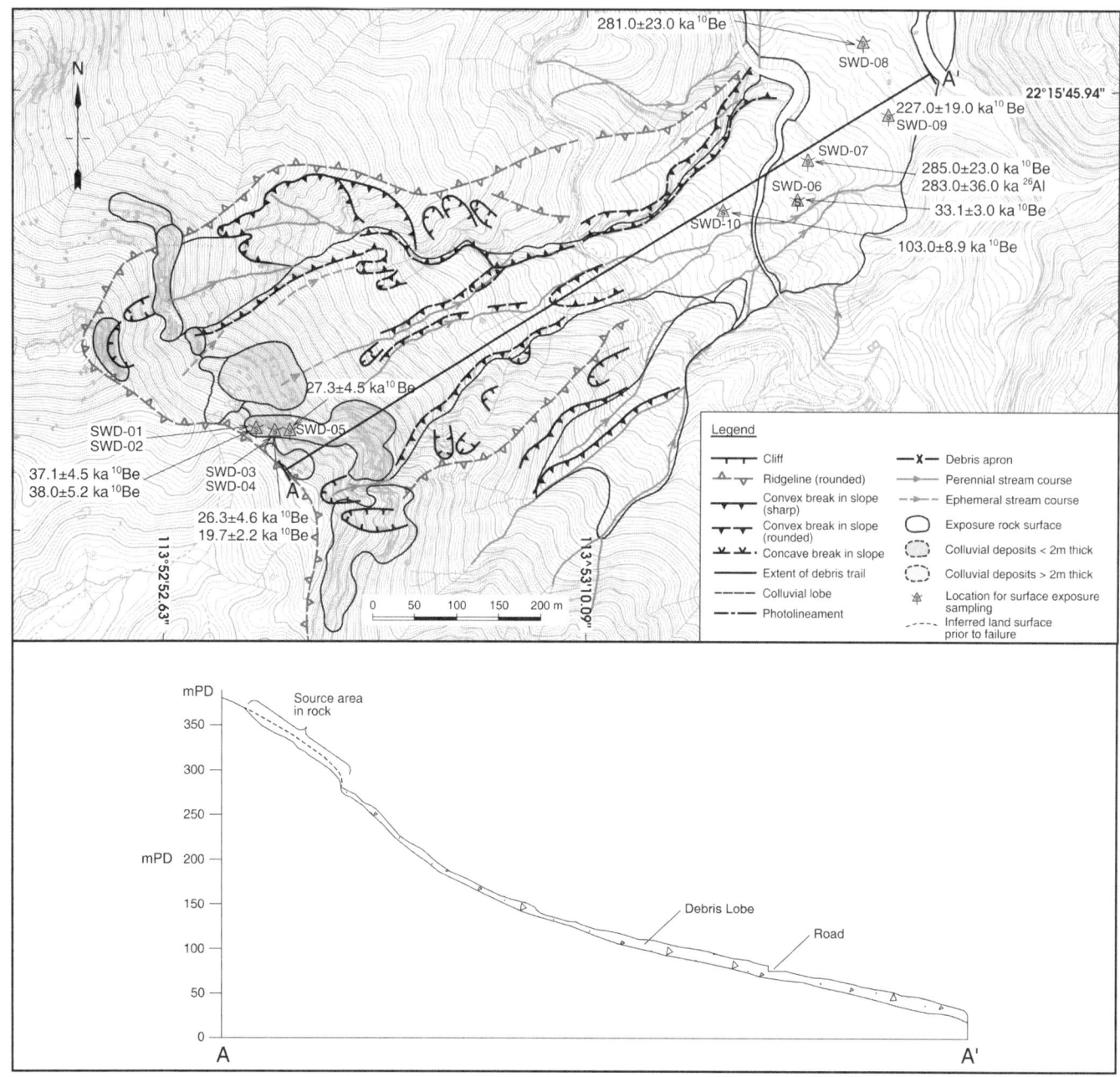

Figure 2. Geomorphological features, sample locations, and dating results for the Sham Wat Debris Lobe.

elevations in the debris lobe varied between 215 and 225 mPD. Samples TCR-08 and TCR-09 came from different positions on the same boulder to test for possible inheritance.

Tsing Yi

Two large landslide features were studied on the southeast side of Tsing Yi (Fig. 1). These relict features comprise two steep (>46°) cliff areas (sheeting joint escarpment nos. 1 [SJ1] and 2 [SJ2]) composed of medium-grained granite, each characterized by distinctive sheeting joints, and with a large accumulation of boulder-bearing debris downslope (Fig. 4).

Six samples (Table 1) were analyzed from the site. Two samples (TY-01 and TY-02) came from very large (50 m³) boulders from the debris lobe no. 4 (DL4) debris at elevations of 138 mPD and 151 mPD, while a third boulder (TY-03) was sampled from the colluvium no. 1 (CL1) debris adjacent to the SJ1 cliff. Two samples from the cliff area (TY-04 and TY-07) came from SJ1 rock surfaces at elevations of 190 mPD and 216 mPD,

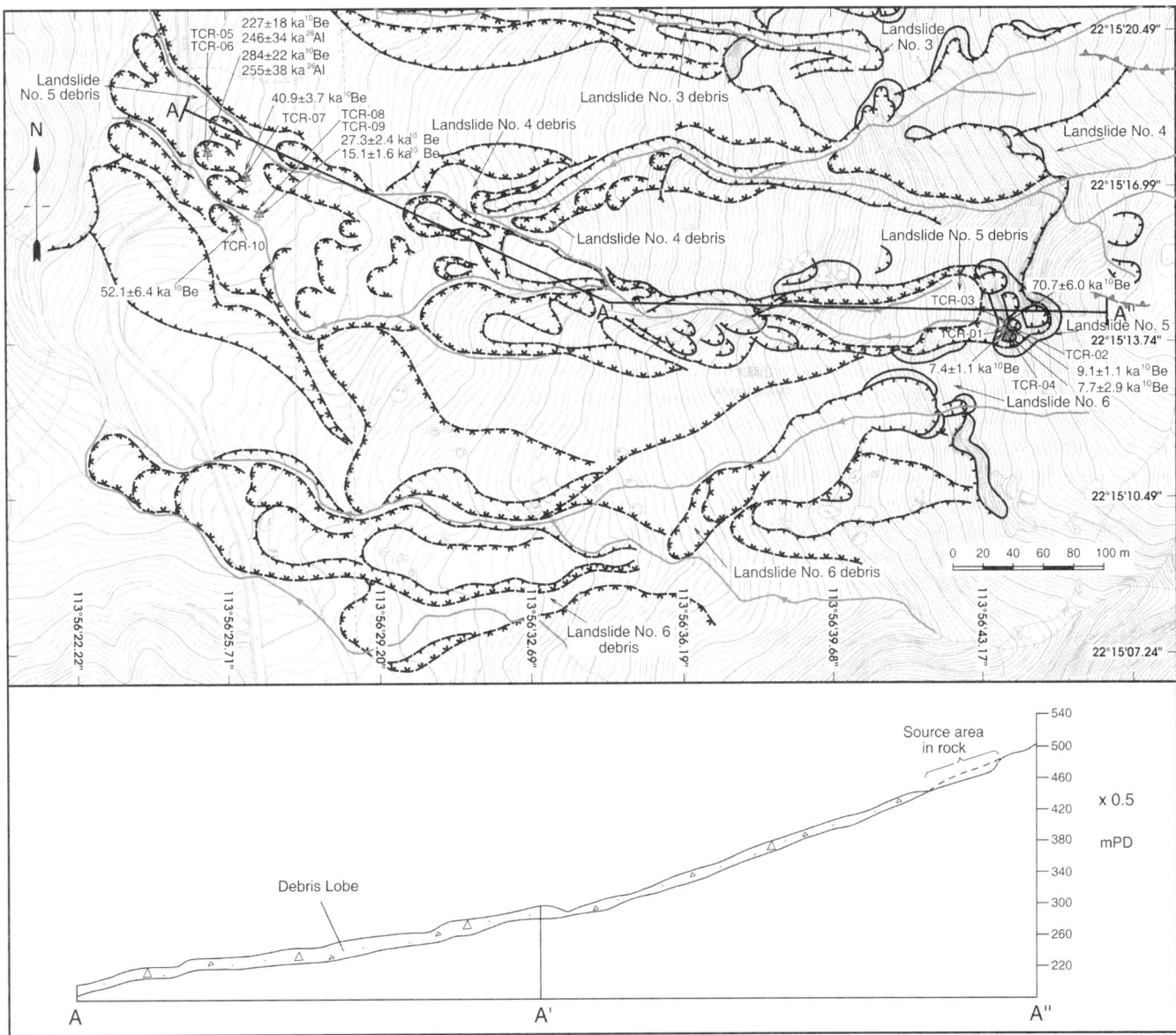

Figure 3. Geomorphological features, sample locations, and dating results for the Sunset Peak West landslide site. Legend as for Figure 2.

respectively, and one sample (TY-05) came from a rock surface on the cliff area SJ2 at an elevation of 255 mPD.

Ap Lei Chau

A large landslide (relict landslide no. 1 [RLS1]) underlain by eutaxitic fine ash vitric rhyolitic tuff, with a clearly defined main scarp and debris lobe, is located on Ap Lei Chau, to the southwest of Hong Kong Island (Fig. 5). A total of four samples (Table 1) were collected. These comprised two samples (ALC-01 and ALC-02) from the main scarp, one sample (ALC-03) from an unrelated adjacent scarp for control purposes, and one sample (ALC-04) from a boulder on the debris lobe of the main scarp. Sample ALC-01 came from a sheeting joint exposed at an eleva-tion of 161 mPD and inclined at 30° on the edge of the main scarp. The sampled rock surface was strongly weathered, and elsewhere partly covered by a relatively thin (0.2 m) layer of surface debris, and so might be anticipated to yield an age by exposure dating that underestimated to some extent the age of formation of the surface. Sample ALC-02 came from a sheeting joint, inclined at 36°, along the crest of the main scarp, at an elevation of 173 mPD. Compared with ALC-01, the sampled rock surface was distinctly less weathered. Sample ALC-03 came from near the base of a prominent rock cliff at an elevation of 155 mPD, ~40 m southeast of the main scarp. The boulder sampled on the debris lobe of the main scarp (ALC-04) is ~4 m^3 and lies at an elevation of 137 mPD.

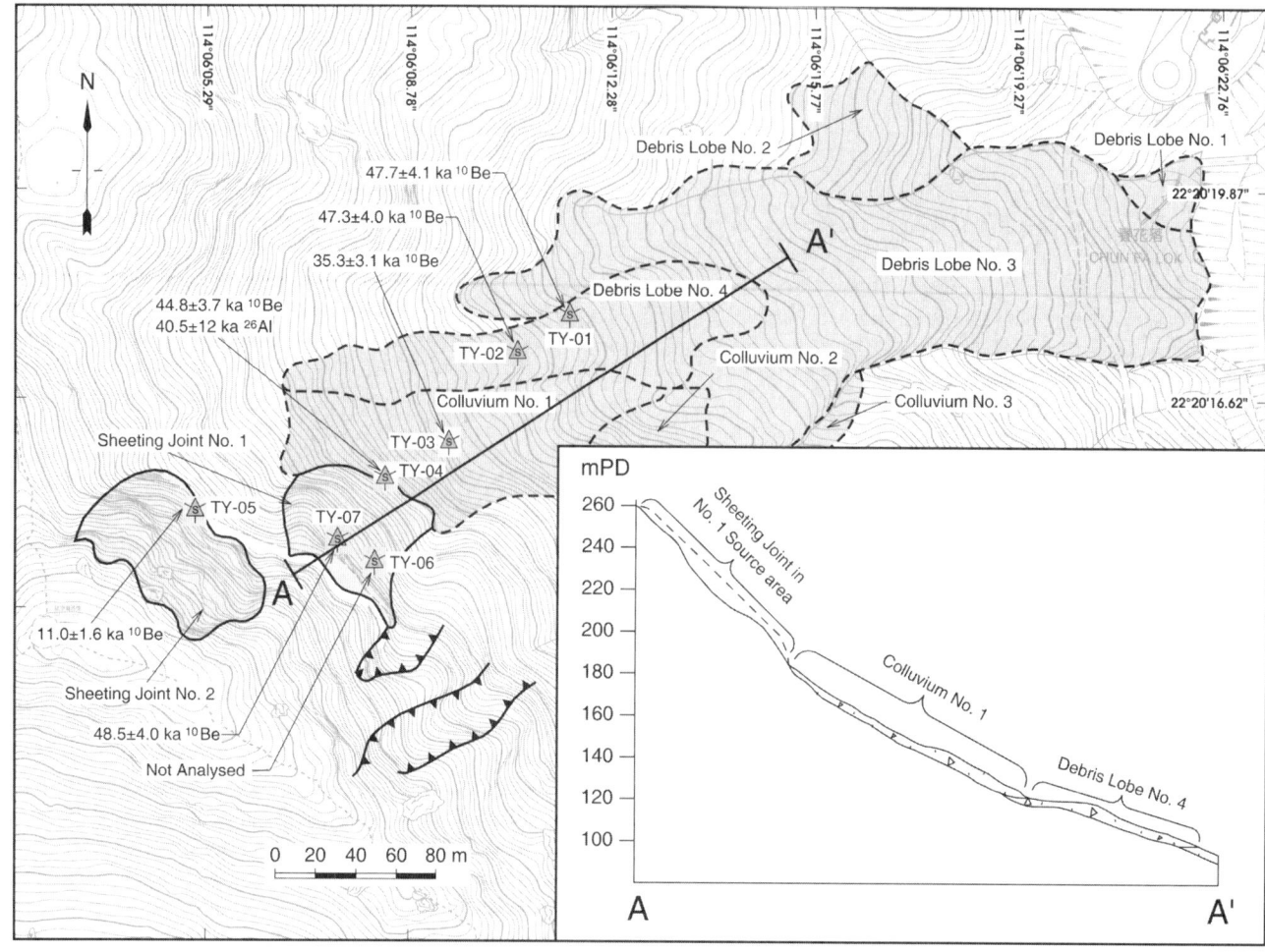

Figure 4. Geomorphological features, sample locations, and dating results for the Tsing Yi landslide site. Legend as for Figure 2.

Rock and Boulder Fall Sites

Lion Rock

Three extremely large (1000 m³) granite boulders (Table 1), ~300 m apart, were sampled below Lion Rock, in northern Kowloon (Fig. 1). Two of the boulders (LR-02 and LR-03) lie within, or on top of, the same colluvial unit (Fig. 6), whereas the third boulder (LR-01) appears to lie on the boundary between two distinctive colluvial units. The colluvial deposits are situated below very steep (>50°) natural terrain forming the summit of Lion Rock and are likely to represent debris fan aprons. The boulders may reflect rockfalls from Lion Rock itself. However, neither the relative ages of emplacement of the boulders nor their exact sources could be confidently determined, although the lithology of the boulders is the same as that forming the upper cliffs of Lion Rock.

Samples were collected at elevations between 176 and 208 mPD from the uppermost central portions of the boulders, and from rock faces inclined between 14° and 32°. A small portion of the upper face of LR-01 is covered by colluvium No. 3 (Fig. 6), which also appears to wrap around the boulder to the east, suggesting that LR-01 predated deposition of colluvium No. 3.

Fei Ngo Shan

In eastern Kowloon, three very large (5–10 m diameter) boulders (Table 1) within a single large colluvial unit below the southwest flanks of Fei Ngo Shan (Fig. 7) were sampled. The boulders are composed of block bearing coarse ash crystal tuff, similar to that forming the upper cliffs of Fei Ngo Shan, although the exact source of detachment could not be located. The three boulders were sampled at approximately the same elevation (158–162 mPD). Samples were taken from the uppermost central portions of the boulders, from rock faces inclined between 8° and 30°.

Deep-Seated Scarp Sites

Lai Cho Road

Samples were collected from a large composite natural terrain escarpment above Lai Cho Road, in western Kowloon

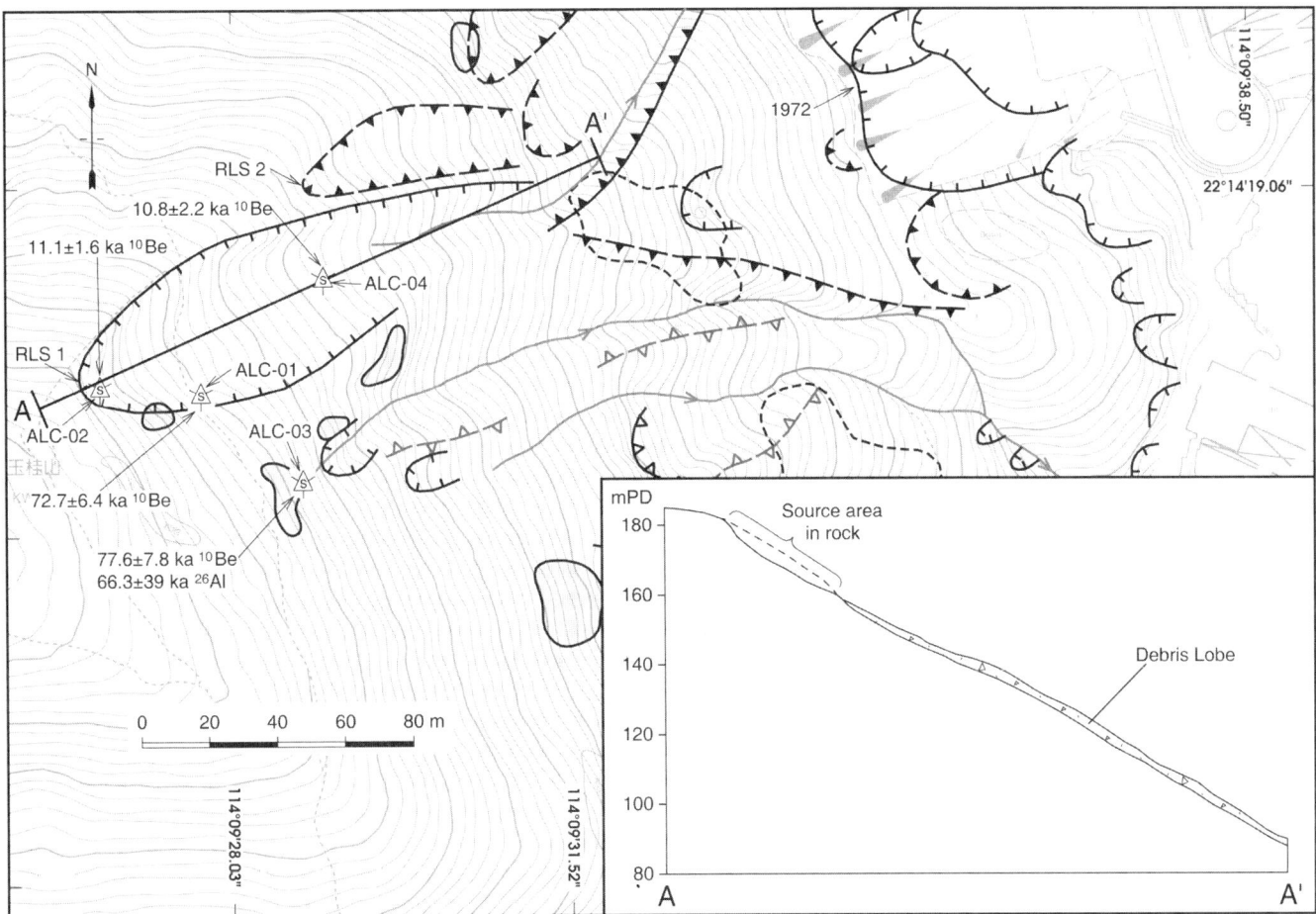

Figure 5. Geomorphological features, sample locations, and dating results for the Ap Lei Chau landslide site. Legend as for Figure 2.

(Fig. 1), featuring two clearly defined steep scarps (A1 and A2 in Fig. 8) with sharp lateral boundaries. The escarpment is underlain by coarse-grained granite and is located above a recently formed major cut-slope. Neither of the two scarps is associated with recognizable debris lobes, and man-made excavations have probably removed these. Three samples (LCR-01, LCR-03, and LCR-04, Table 1) were taken from the northern and southern flanks of the smaller (A2) scarp, close to the intersection with the A1 scarp, whereas one sample (LCR-02) was taken from the A1 scarp, 10 m southwest of this intersection (Fig. 8). Samples were collected from 3 m deep rock scarp surfaces inclined at 36° to 45°, and at elevations varying from 109 to 111 mPD. Based on aerial photograph interpretation, the A2 scarp truncates the A1 scarp and therefore is younger.

Mid-Levels

The Mid-Levels landslide comprises a prominent natural terrain landslide feature with a single well-defined 5 m deep main scarp, located immediately northwest of Victoria Peak on Hong Kong Island (Fig. 1). The scarp lies at an elevation of ~400 mPD (Fig. 9). Although the ground immediately below the Mid-Levels scarp is moderately sloping, no directly associated landslide debris was identified downslope from the scar.

Five samples were collected for exposure dating from the main scarp area (Fig. 9). All of the samples were collected from relatively steep (60°–80°) faces, mostly at 3 m or greater below the inferred original rock surface. One sample (ML-04) was collected within less than 3 m of the inferred prefailure land surface to test for possible inheritance.

ANALYTICAL METHODS

Sample preparation was carried out at the Australian National University and followed established protocols (Barrows et al., 2001). All exposure ages are calculated conventionally assuming no significant erosion since initial exposure. We used a ^{10}Be production rate of 5.1 and a ^{26}Al production rate of 31.1 atoms/g/yr (Stone, 2000). The isotope ratios of ^{10}Be/Be and ^{26}Al/Al were measured by accelerator mass spectrometry on the 14UD accelerator at the Australian National University (Fifield et al., 1994). Production rate calculations and ages for ^{10}Be and ^{26}Al respectively are presented in Tables 2 and 3.

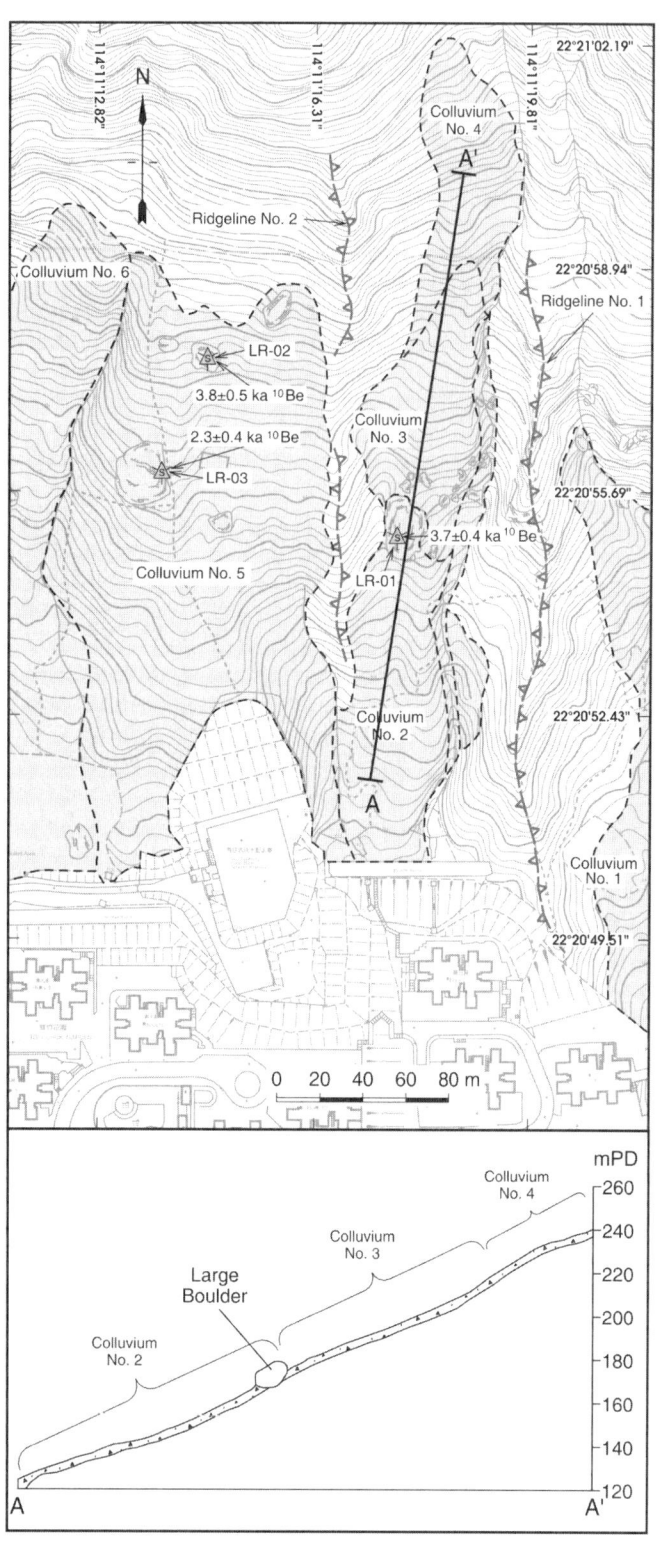

Figure 6. Geomorphological features, sample locations, and dating results for the Lion Rock landslide site. Legend as for Figure 2.

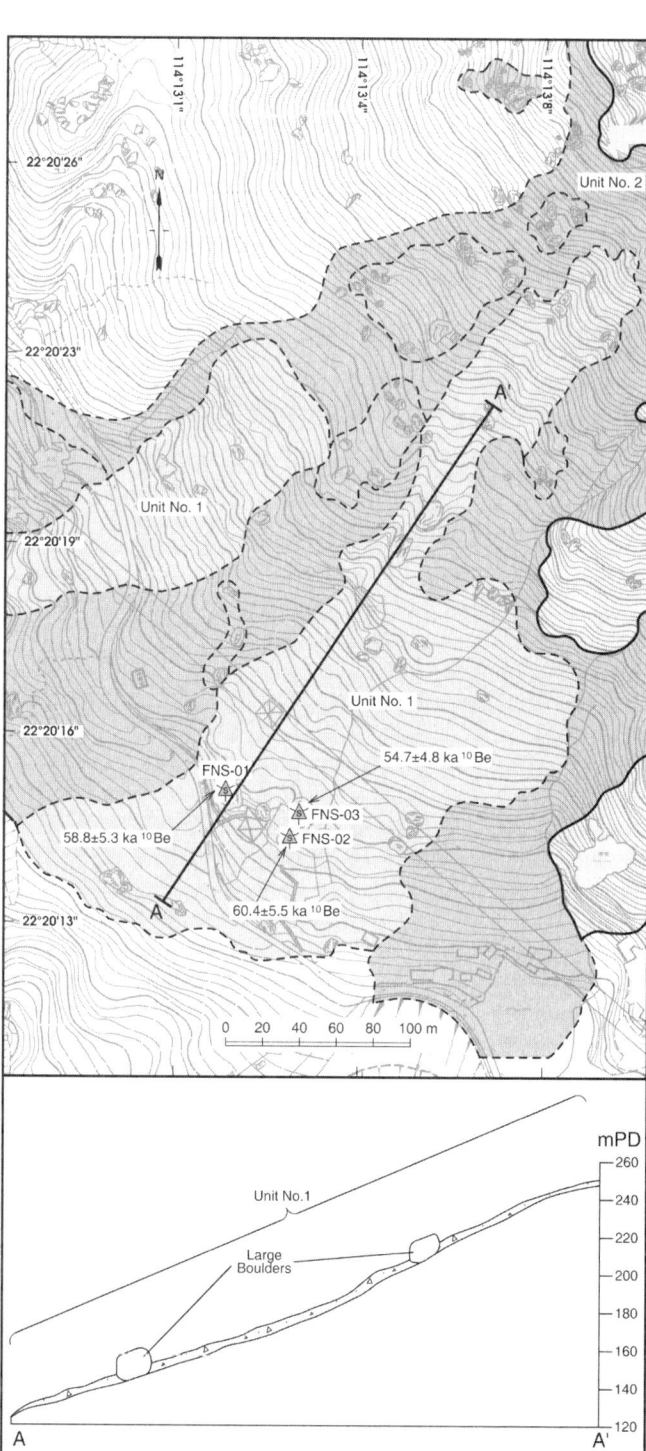

Figure 7. Geomorphological features, sample locations, and dating results for the Fei Ngo Shan landslide site. Legend as for Figure 2.

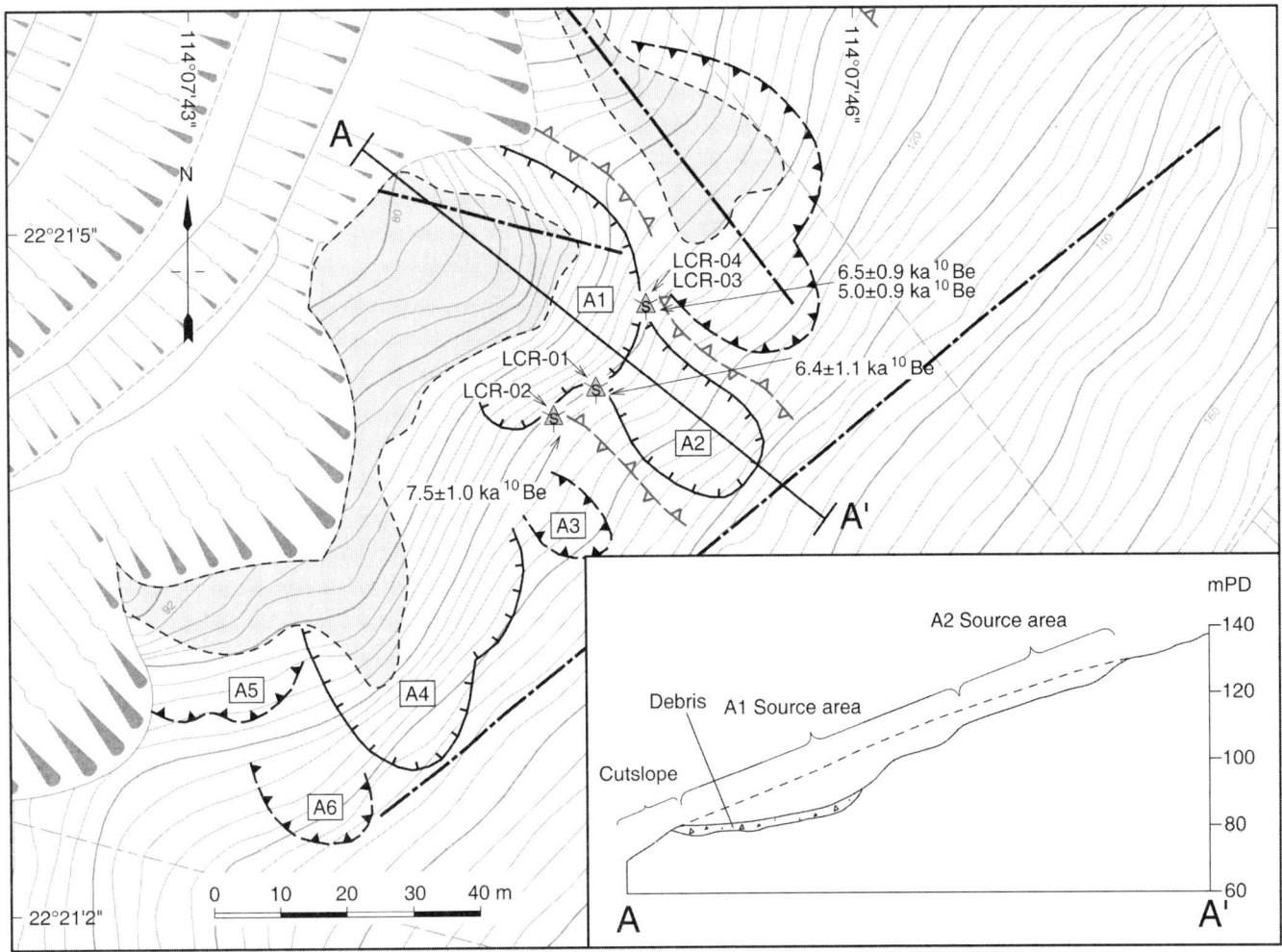

Figure 8. Geomorphological features, sample locations, and dating results for the Lai Cho Road landslide site. Legend as for Figure 2.

RESULTS

Deep-Seated Scarp and Debris Lobe Sites

Sham Wat Debris Lobe

The five ^{10}Be exposure ages (Table 2) from the scarp/source area of the main lobe (Fig. 2) range from 19.7 ± 2.2 ka to 38.0 ± 5.2 ka. The five exposure ages from blocks on the debris lobe are much more variable, ranging from 33.1 ± 3.0 ka to 285 ± 23 ka. A single exposure age using ^{26}Al (283 ± 36 ka) (Table 3) was also obtained on one sample (SWD-07). This is in good agreement with the ^{10}Be age (285 ± 23 ka), indicating no significant burial history. The oldest exposure age (285 ± 23 ka) is interpreted as providing a minimum apparent age for the original ground surface. The youngest exposure age falls within the range of the four exposure ages from the scarp. A normalized X^2 test on these five exposure ages is consistent with a single population with dispersion in the data due to random errors ($X^2 = 1.34$). The weighted mean for this population is 32.3 ± 5.4 ka and provides the best estimate for the age of the landslide.

Sunset Peak West

Three exposure ages (Table 2) from the scarp/source area of the landslide (Fig. 3) range in age from 9.1 ± 1.1 ka to 7.4 ± 1.1 ka. The youngest sample (7.4 ± 1.1 ka) was collected from the base of the main scarp, well below (>3 m) the reconstructed original rock surface. This is unlikely to preserve an inherited cosmogenic signature from prior to the landslide. Another sample (TCR-04) from the eastern flank has a similar age (7.7 ± 2.9). The oldest age (9.1 ± 1.1 ka) comes from a sample (TCR-02) on the upper part of western flank of the scarp, closer to the former rock surface, and is therefore more likely to include some inheritance. A sample from the rock surface adjacent to the main scarp (TCR-03) provides a minimum age of the original land surface prior to slope failure (70.7 ± 6.0 ka). The three other samples from the scarp/source area are likely to come from a single population ($X^2 = 0.63$), with a mean age of 8.2 ± 0.9 ka.

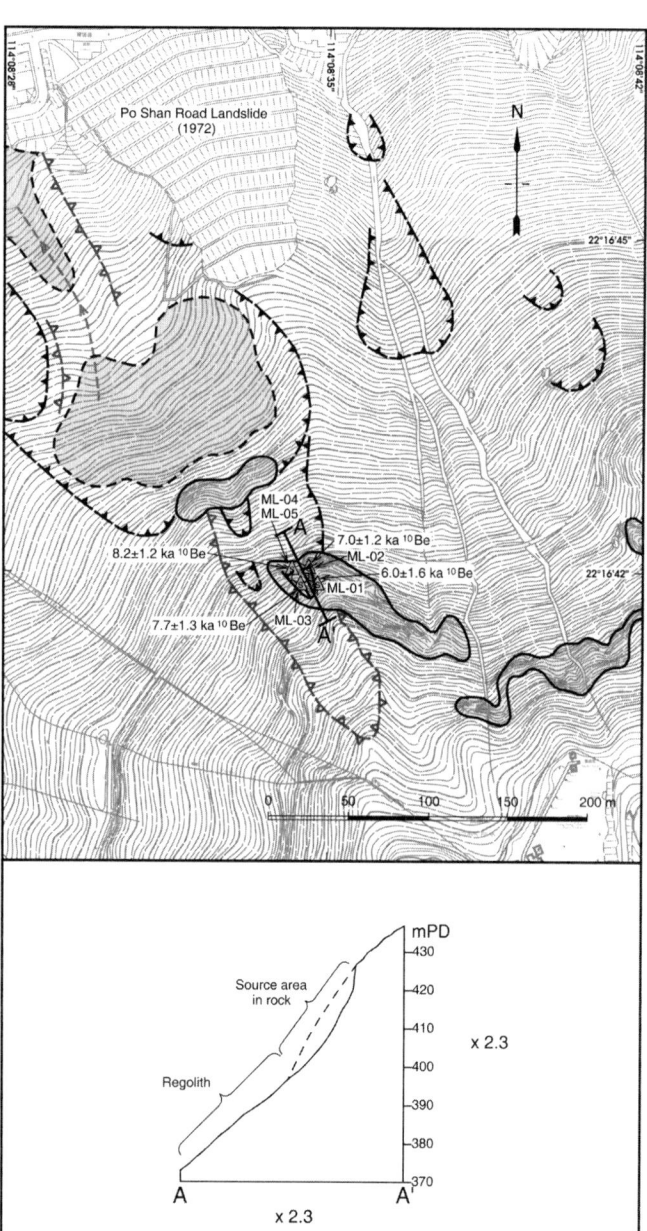

Figure 9. Geomorphological features, sample locations, and dating results for the Mid-Levels landslide site. Legend as for Figure 2.

TABLE 2. ^{10}Be EXPOSURE AGES

Sample	Lab code	^{10}Be ($\times 10^3$ g^{-1})	Production rate (g^{-1}yr^{-1})	Age (ka)
Deep-seated Scarps and Debris Lobes				
Sham Wat Debris Lobe				
SWD-01	ANU-M234-19	148 ± 15	4.0	37.1 ± 4.5
SWD-02	ANU-M226B-07	151 ± 18	4.0	38.0 ± 5.2
SWD-03	ANU-M234-10	111 ± 18	4.2	26.3 ± 4.6
SWD-04	ANU-M224B-30	80 ± 7	4.1	19.7 ± 2.2
SWD-05	ANU-M234-09	119 ± 18	4.4	27.3 ± 4.5
SWD-06	ANU-M234-05	120 ± 7	3.6	33.1 ± 3.0
SWD-07	ANU-M234-06	973 ± 36	3.6	285 ± 23
SWD-08	ANU-M224-26	926 ± 41	3.5	281 ± 23
SWD-09	ANU-M224-27	762 ± 35	3.5	227 ± 19
SWD-10	ANU-M226B-08	376 ± 19	3.7	103 ± 8.9
Sunset Peak West				
TCR-01	ANU-M224-25	31 ± 4	4.2	7.4 ± 1.1
TCR-02	ANU-M226B-05	38 ± 4	4.1	9.1 ± 1.1
TCR-03	ANU-M224-24	328 ± 16	4.7	70.7 ± 6.0
TCR-04	ANU-M233-09	33 ± 12	4.3	7.7 ± 2.9
TCR-05	ANU-M234-15	845 ± 29	3.9	227 ± 18
TCR-06	ANU-M233-10	1072 ± 37	4.0	284 ± 22
TCR-07	ANU-M224-30	163 ± 9	4.0	40.9 ± 3.7
TCR-08	ANU-M234-07	111 ± 6	4.1	27.3 ± 2.4
TCR-09	ANU-M226B-06	60 ± 5	4.0	15.1 ± 1.6
TCR-10	ANU-M234-08	210 ± 21	4.1	52.1 ± 6.4
Tsing Yi				
TY-01	ANU-M233-02	170 ± 9	3.6	47.7 ± 4.1
TY-02	ANU-M225-05	170 ± 8	3.6	47.3 ± 4.0
TY-03	ANU-M233-03	129 ± 7	3.7	35.3 ± 3.1
TY-04	ANU-M233-20	159 ± 7	3.6	44.8 ± 3.7
TY-05	ANU-M225-04	34 ± 4	3.1	11.0 ± 1.6
TY-07	ANU-M233-04	181 ± 8	3.8	48.5 ± 4.0
Ap Lei Chau				
ALC-01	ANU-M234-13	271 ± 15	3.8	72.7 ± 6.4
ALC-02	ANU-M225-12	41 ± 5	3.7	11.1 ± 1.6
ALC-03	ANU-M234-14	263 ± 19	3.5	77.6 ± 7.8
ALC-04	ANU-M225-09	40 ± 7	3.7	10.8 ± 2.2
Rock and Boulder Falls				
Lion Rock				
LR-01	ANU-M225-08	15 ± 2	3.9	3.7 ± 0.4
LR-02	ANU-M225-07	15 ± 2	3.9	3.8 ± 0.5
LR-03	ANU-M233-19	9 ± 2	3.9	2.3 ± 0.4
Fei Ngo Shan				
FNS-01	ANU-M233-16	220 ± 13	3.8	58.8 ± 5.3
FNS-02	ANU-M225-03	229 ± 14	3.8	60.4 ± 5.5
FNS-03	ANU-M233-17	208 ± 11	3.9	54.7 ± 4.8
Deep-Seated Scarps				
Lai Cho Road				
LCR-01	ANU-M224-30	20 ± 3	3.2	6.4 ± 1.1
LCR-02	ANU-M233-07	19 ± 2	2.6	7.5 ± 1.0
LCR-03	ANU-M224B-27	21 ± 3	3.2	6.5 ± 0.9
LCR-04	ANU-M233-08	15 ± 2	2.9	5.0 ± 0.9
Mid-Levels				
ML-01	ANU-M234-02	23 ± 6	3.9	6.0 ± 1.6
ML-02	ANU-M224B-28	27 ± 4	3.8	7.0 ± 1.2
ML-03	ANU-M234-03	31 ± 5	4.0	7.7 ± 1.3
ML-04	ANU-M224-29	134	3.5	<38
ML-05	ANU-M234-04	29 ± 4	3.6	8.2 ± 1.2

TABLE 3. ^{26}Al EXPOSURE AGES

Sample	Lab code	Al (ppm)	^{26}Al ($\times 10^3$ g^{-1})	Production Rate (g^{-1}yr^{-1})	Age (ka)
TCR-05	ANU-M237-21	211 ± 11	5211 ± 624	23.9	246 ± 34
TCR-06	ANU-M237-19	176 ± 9	5545 ± 733	24.6	255 ± 38
TY-04	ANU-M237-16	93 ± 5	867 ± 249	21.8	40.5 ± 12
SWD-07	ANU-M237-17	522 ± 26	5485 ± 590	22.2	283 ± 36
ALC-03	ANU-M237-20	155 ± 8	1352 ± 783	21.1	66.3 ± 39

Four large boulders, protruding from the debris lobe of the landslide have exposure ages (Table 2) ranging from 15.1 ± 1.6 ka (TCR-09) to 284 ± 22 ka (TCR-06) (Fig. 3). The two oldest ages (227 ± 18 ka and 284 ± 22 ka) came from samples taken from opposite ends of the same block (TCR-05 and TCR-06 respectively). The independent exposure ages using ^{26}Al (246 ± 34 ka and 255 ± 38 ka) (Table 3) for these samples are in very good agreement with the ^{10}Be exposure ages and indicate no significant burial. The two youngest ages from the debris lobe (27.3 ± 2.4 and 15.1 ± 1.6 ka) also come from two samples (TCR-08 and TCR-09 respectively) from the same block. The two remaining blocks have intermediate exposure ages (40.9 ± 3.7 [TCR-07], 52.1 ± 6.4 ka [TCR-10]). The difference between the paired ages from the same blocks, the range of ages between the four blocks, and the much younger ages from the scarp face, together indicate significant inheritance of cosmogenic isotopes in these blocks from previous exposure. The shallow nature of this landslide means that much of the material was probably sourced from close to the original rock face. The oldest exposure age (284 ± 22 ka) provides a minimum age for this original surface. The mean age from the source area (8.2 ± 0.9 ka) provides the best estimate for the age for the landslide.

Tsing Yi

The two boulders (TY-01 and TY-02) within the DL4 debris lobe have almost identical exposures ages (Table 2) of 47.7 ± 4.1 ka and 47.3 ± 4.0 ka, respectively (Fig. 4). A sample from the third boulder (TY-03), within the adjacent CL1 debris, has a younger exposure age of 35.3 ± 3.1 ka. The two samples (TY-04 and TY-07) from the SJ1 cliff have exposure ages of 44.8 ± 3.7 ka and 48.5 ± 4.0 ka, respectively (Table 2). A duplicate of sample TY-04 was also analyzed for ^{26}Al (Table 3). This gave an exposure age of 40.5 ± 12 ka, agreeing with the ^{10}Be exposure age. A sample (TY-05) from the SJ2 cliff has an exposure age of 11.0 ± 1.6 ka (Table 2).

The exposure ages for the Tsing Yi site are remarkably consistent and indicate that most of the debris in the DL4 debris lobe was derived from the SJ1 escarpment (Fig. 4). The five exposure ages from SJ1 and DL4 are very likely to be from a single population ($X^2 = 0.18$), with a mean age of 47.0 ± 1.6 ka. A single outlier, represented by a boulder age of 35.3 ± 3.1 ka (TY-03), suggests that this boulder may have fallen more recently from the cliff. Alternatively, this boulder may have rolled during its exposure history. The younger exposure age for SJ2 (11.0 ± 1.6 ka) indicates a more recent landslide/rockfall event from this source area.

Ap Lei Chau

The sample from the highest face (ALC-02) gave the youngest exposure age (11.1 ± 1.6 ka) (Fig. 5; Table 2). This age closely matches the exposure age of the sample from the boulder (ALC-04, 10.8 ± 2.2 ka) (Table 2), indicating that it is the likely source area for that debris. Together, these two ages are very likely to record the same event ($X^2 = 0.01$) with a mean age of 11.0 ± 1.3. The sample on the eastern edge of the main scarp (ALC-01) and the sample from the adjacent scarp feature (ALC-03) gave similar exposure ages of 72.7 ± 6.4 ka and 77.6 ± 7.8 ka (Table 2). The ALC-03 sample also gave an exposure age of 66.3 ± 39 ka using ^{26}Al (Table 3), which overlaps the ^{10}Be exposure age. There is no evidence of debris below the ALC-03 scarp feature, which is very degraded and poorly defined on aerial photographs.

Rock and Boulder Fall Sites

Lion Rock

The ages indicate that the boulders were all detached from their source areas within the past 4000 yr. Two samples (LR-02 and LR-03) on the same colluvial unit gave exposure ages of 3.8 ± 0.5 ka and 2.3 ± 0.4 ka, respectively (Table 2). The third sample (LR-01) within an adjacent unit gave an age of 3.7 ± 0.4 ka (Table 2). The exposure ages of LR-01 and LR-02 overlap, and although the boulders appear to have detached from slightly different source areas on Lion Rock, they appear to have done so at approximately the same time. This is likely to represent the best estimate for the timing of a major boulder fall event. Boulder LR-03 (2.3 ± 0.4 ka) probably represents a separate, more recent, rockfall.

Fei Ngo Shan

All three boulders gave virtually identical exposure ages, within 1σ error (Table 2). Sample FNS-01 gave an age of 58.8 ± 5.3 ka, whereas samples FNS-02 and FNS-03 have exposure ages of 60.4 ± 5.5 ka and 54.7 ± 4.8 ka, respectively. The three exposure ages are likely to constitute a single population ($X^2 = 0.34$), with a mean age of 57.7 ± 3.0 ka. Due to the complexity and general degradation of the upper flanks of Fei Ngo Shan, the source of the boulders is uncertain. However, the identical lithology of the boulders and absolute age data suggest that the large boulders probably fell together, or within a short period of each other, from the same source area.

Deep-Seated Scarp Sites

Lai Cho Road

The three exposure ages from the smaller A2 scarp range from 6.5 ± 0.9 ka to 5.0 ± 0.9 ka (Table 2). It is likely that these three ages constitute a single population ($X^2 = 0.83$) with a mean age of 5.9 ± 0.8 ka. The exposure age from the larger scarp is 7.5 ± 1.0 ka, suggesting that this feature is slightly older than the smaller scarp. However, two ages from the smaller landslide scarp overlap the age from the larger scarp. This raises the possibility that both scarps were formed during a single event ($X^2 = 1.2$). The weighted mean for these four ages is 6.3 ± 1.0 ka, which gives the best age estimate for a single landslide.

Mid-Levels

Aside from one sample (ML-04) that did not run well, the remaining four exposure ages (Table 2) group closely between 6.0 ± 1.6 ka (ML-01) and 8.2 ± 1.2 ka (ML-05). These four ages probably constitute a single population ($X^2 = 0.47$), with a mean age of 7.4 ± 1.0 ka. Therefore, this is our best estimate for the age of this landslide.

DISCUSSION

The exposure ages reported here provide the first direct estimates of the timing of landscape evolution by landsliding in Hong Kong. They reveal a history of deep-seated landsliding dating back at least 50,000 yr, which may be the upper limit for recognition of such features in the Hong Kong weathering environment. The largest landslide event occurred ca. 32,000 yr ago and entailed deposition of some 1,000,000 m^3 of debris almost to sea level. Rock and boulder falls involving "mega-scale" boulders have also been active in the evolution of the Hong Kong landscape. The exposure ages suggest that such events have occurred sporadically between ca. 57,000 and ca. 2000 yr ago. There is potential evidence in debris from two of the deep-seated landslides studied of ancient landscape surfaces in Hong Kong that are at least 280,000 yr old. The exposure ages have provided important validation of the relative chronology identified from aerial photograph interpretation, particularly in areas where clusters of relict natural terrain landslides are observed. Overall, this study has demonstrated the successful application of cosmogenic nuclide analysis using ^{10}Be and ^{26}Al to the dating of landslides in Hong Kong from a few tens of thousands of years down to only a few thousand years, despite low production rates because of low latitude and low altitude.

Deep-Seated Scarps and Debris Lobes

For the scarp and debris lobe sites, we generally found poor agreement between exposure ages in the debris and the scarp areas. The wide range of ages from the debris lobes at the Sham Wat Debris Lobe and Sunset Peak West sites indicates that much of the surface of these lobes is composed of material that was derived from levels that were close to the ground surface before the landslide. Intriguingly, at both sites there is an increase in boulder age with distance from the scarp, suggesting that the main mode of failure was by large-scale debris avalanche. This corroborates the findings of King (1998) at the Sham Wat site, who proposed that the landslide incorporated considerable fine material from the weathered profile to form a slurry flow that emplaced boulders of up to 20 m^3, far from their source. Multiple samples were required from the scarp to identify the depth of the former land surface, whereas multiple samples were required from the debris lobes to identify blocks with inheritance. At both sites, the most consistent, and presumably most reliable, ages for the landslide came from the scarp face, with the unweathered scarps providing the best ages to constrain the landslide.

At the Sham Wat Debris Lobe, the samples along the scarp face were collected at slightly different elevations, but all were from more than 2.5 m below the crest of the scarp. Sample SWD-04 gave a relatively young age of 19.7 ± 2.2 ka, and we strongly suspect that this rock surface spalled sometime after the landslide. The age of ~32,000 yr for the Sham Wat Debris Lobe is comparable to the 35,000 yr age reported by King (2001), based on standard optically stimulated luminescence dating methods. The surface exposure ages strongly suggest that the landslide occurred as a single large event rather than as a result of multiple smaller events. None of the surface exposure ages reported here and elsewhere (Sewell and Campbell, 2005) suggest a significant transport event for the Sham Wat Debris Lobe within the last 10,000 yr, as proposed by Murray (2001) on the basis of the dose distribution luminescence dating method.

At the Sunset Peak West landslide site, the depth of the former land surface was investigated by collecting three samples from the scarp surface at slightly different depths in relation to the crest of the scarp. A sample from a rock surface adjacent to the main scarp was also analyzed to test whether this surface could have formed at the same time as the main scarp. The results from the scarp surface for Sunset Peak West showed a slight increase in age from the deepest to the shallowest, indicating that the latter samples are likely to include some inheritance. Despite the presence of some inheritance, the mean exposure age (8.2 ± 0.9 ka) for the three samples from the scarp/source area provides the best estimate for the age for the landslide. The exposure age of the adjacent rock surface indicates either that it was very shallow before the landslide or that it represents a much older surface. It is intriguing that the oldest exposure age (284 ± 22 ka) in debris at the Sunset Peak West landslide is similar to that of debris in the Sham Wat Debris Lobe. Thus, there is potential evidence from two of the sites for ancient landscape surfaces in Hong Kong of at least 280,000 yr old (i.e., surfaces of this age will have been weathered and come into equilibrium with the "erosion" rate).

Owing to the substantial degradation of the scarp/source area at the Tsing Yi landslide site, it was uncertain which scarp surfaces contributed to the various debris lobes, even after meticulous aerial photograph study. Therefore, multiple samples were collected from the debris lobes, as well as from the overlooking sheeted rock surfaces, in an attempt to identify (1) the timing of the landslide event (or events) and (2) the possible source areas of the debris using the surface exposure ages. The detailed interpretation of aerial photographs, which permitted accurate delineation of the debris lobes, provided invaluable assistance in the selection of the samples sites. The exposure ages reveal unequivocally that most of the debris was derived from the nearest sheeting joint escarpment (SJ1) during a single event ca. 47,000 yr ago, and that a more recent landslide event occurred from the adjacent SJ2 escarpment at ca. 11,000 yr ago. The Tsing Yi site is the oldest deep-seated landslide feature successfully dated in this study with a definable scarp area and debris lobe. An age of 50,000 yr may be close to the upper limit for recognition of such features in Hong Kong by aerial photograph interpretation.

Despite a clearly defined main scarp and debris lobe identified from aerial photographs, the Ap Lei Chau landslide yielded only two suitable rock surfaces for exposure dating in the scarp area, and only one suitable boulder surface in the debris. Therefore, an additional sample was collected from an adjacent, very degraded scarp to test (1) the relative age determined from aerial photograph interpretation and (2) whether this scarp recorded a much older landslide event. Our results have yielded excellent agreement between the scarp and boulder exposure ages, and indicate that the Ap Lei Chau landslide scar has been well preserved over the past 11,000 yr. Not surprisingly, the adjacent, highly degraded scarp feature turned out to be considerably older.

Rock and Boulder Falls

The strategy at the sites of rockfall (Fei Ngo Shan and Lion Rock) was to sample large fallen boulders to determine (1) whether they were emplaced at the same time and (2) the relative age of surrounding colluvium. The application of cosmogenic nuclide exposure dating to these very large, generally tabular boulders was remarkably successful and was assisted by the availability of unobstructed, nearly horizontal surfaces suitable for sampling on the tops of the boulders. The geomorphology at both sites indicated that the boulder falls were not accompanied by emplacement of major colluvial lobes, and that the boulders either rolled, bounced, or slid into position, or combinations of all three mechanisms.

Exposure ages at the Fei Ngo Shan site reveal that the boulders most likely fell together, or within a short period of each other, suggesting that an exceptional trigger event (i.e., a seismic event) may have been responsible. The results from the Fei Ngo Shan site demonstrate the usefulness of dating clusters of large fallen boulders, even where there is little indication of the point of detachment. The well-correlated ages also suggest that the boulders were sourced from a relatively deep-seated failure, where inheritance was negligible.

The exposure ages at the Lion Rock site reveal that large isolated boulders have detached from the cliffs forming Lion Rock in the relatively recent past. Two of the boulders "within" one colluvial unit appear to have fallen at slightly different times (ca. 3700 and ca. 2300 yr ago). The exposure ages also indicate that the "host" colluvium is likely to predate both boulder fall events. A third large boulder "in" an adjacent colluvial unit is likely to have fallen at the same time as the 3700 yr old boulder, suggesting that an exceptional trigger event (i.e., seismic event) may also have been responsible. The surface exposure ages from the three large boulders at the Lion Rock site are the youngest recorded in this dating study. They indicate that a rockfall event (or events) occurred between ca. 2000 and 4000 yr ago resulting in the emplacement of very large boulders on the slopes below Lion Rock.

Deep-Seated Scarps

The results from the Lai Cho Road and Mid-Level landslide sites demonstrate that apparent landslide scarp features can be reliably dated using cosmogenic isotopes, even where associated debris lobes are not preserved, and even though the scarps are relatively young (e.g., <10,000 yr). At both sites, the exposure ages obtained from various positions on the scarp surfaces are generally in good agreement.

At the Lai Cho Road site, exposure dating of an adjacent scarp has confirmed the relative age of an older and much larger landslide feature identified from aerial photograph interpretation.

CONCLUSIONS

This study has demonstrated that exposure dating using ^{10}Be and ^{26}Al can be successfully applied to dating of exposed rock surfaces associated with gravitational deformations such as large, deep-seated landslides, and rock and boulder falls. Exposure dating of the landslide scarp areas has produced the most reliable results.

In Hong Kong, well-defined ancient landslide features identified from aerial photograph interpretation and with scars greater than 20 m wide appear on the basis of exposure dating to be generally 1000s rather than 100s of years old. The maximum age of recognition by aerial photograph interpretation of these features in Hong Kong's weathering environment may be ~50,000 yr, but more data are required to substantiate this. However, there is some potential exposure dating evidence of ancient landscape surfaces in Hong Kong at least as old as 280,000 yr. This indicates, perhaps surprisingly given Hong Kong's present climate, that in some moderately steeply sloping parts of Hong Kong, the landscape has remained remarkably stable for long periods of time. However, it has also been subject to occasional large-scale failures. Whether such failures have been generated by extreme rainfall events, as are most present landslides in Hong Kong, and/or by other agencies, such as seismic events, as has already been alluded to in the case of the large rockfall events studied, remains to be established.

The findings in this study have direct relevance to natural terrain hazard studies carried out in Hong Kong in relation to areas of future development. Precise dating can provide reliable constraints on return periods for natural terrain landslides. The potential impact this might have on design of mitigation measures in the urban environment, however, requires considerably more investigation.

ACKNOWLEDGMENTS

This paper is published with the permission of the Director of Civil Engineering and Development and the Head of the Geotechnical Engineering Office, Hong Kong Special Administrative Region, China. We thank A. Dias of Maunsell Geotechnical Services for assistance with sampling and V. Levchenko for assistance in the laboratory.

REFERENCES CITED

Ballantyne, C.K., Stone, J.O., and Fifield, L.K., 1998, Cosmogenic Cl-36 dating of post-glacial landsliding at The Storr, Isle of Skye, Scotland: The Holocene, v. 8, p. 347–351, doi: 10.1191/095968398666797200.

Barrows, T.T., Stone, J.O., Fifield, L.K., and Cresswell, R.G., 2001, Late Pleistocene glaciation of the Kosciuszko massif, Snowy Mountains, Australia: Quaternary Research, v. 55, p. 179–189, doi: 10.1006/qres.2001.2216.

Barrows, T.T., Stone, J.O., Fifield, L.K., and Cresswell, R.G., 2002, The timing of the Last Glacial Maximum in Australia: Quaternary Science Reviews, v. 21, p. 159–173, doi: 10.1016/S0277-3791(01)00109-3.

Barrows, T.T., Stone, J.O., and Fifield, L.K., 2004, Exposure ages for Pleistocene periglacial deposits in Australia: Quaternary Science Reviews, v. 23, p. 697–708, doi: 10.1016/j.quascirev.2003.10.011.

Bierman, P.R., Gillespie, A.R., and Caffee, M.W., 1995, Cosmogenic ages for earthquake recurrence intervals and debris flow fan deposition, Owens Valley, California: Science, v. 270, p. 447–450.

Bierman, P.R., Marsella, K.A., Patterson, C., Thompson Davis, P., and Caffee, M., 1999, Mid-Pleistocene cosmogenic minimum-age limits for pre-Wisconsinan glacial surfaces in southwestern Minnesota and southern Baffin Island: A multiple nuclide approach: Geomorphology, v. 27, p. 25–39, doi: 10.1016/S0169-555X(98)00088-9.

Evans, N.C., 1998, The natural terrain landslide study, *in* Li, K.S., et al., eds., Proceedings, Annual Seminar on Slope Engineering in Hong Kong: Rotterdam, Balkema, p. 137–144.

Fifield, L.K., Allan, G.L., Stone, J.O., and Ophel, T.R., 1994, The ANU AMS system and research program: Nuclear Instruments and Methods in Physics Research, v. 92, p. 85–88, doi: 10.1016/0168-583X(94)95982-X.

Fyfe, J.A., Shaw, R., Campbell, S.D.G., Lai, K.W., and Kirk, P.A., 2000, The Quaternary geology of Hong Kong: Geotechnical Engineering Office, Civil Engineering Department, Hong Kong SAR Government (ISBN 962-02-0298-8), 209 p.

King, J.P., 1998, Natural terrain landslide study: The Sham Wat Debris Lobe, Interim Report: Geotechnical Engineering Office, Civil Engineering Department, Hong Kong SAR Government, Technical Note TN 6/98, 37 p.

King, J.P., 1999, Natural terrain landslide study—The natural terrain landslide inventory: Geotechnical Engineering Office, Civil Engineering Department, Hong Kong SAR Government, GEO Report No. 74, 127 p.

King, J.P., 2001, Luminescence dating of colluvium and landslide deposits in Hong Kong: Geotechnical Engineering Office, Civil Engineering Department, Hong Kong SAR Government, GEO Report No. 134, 65 p.

Kubik, P.W., Ivy-Ochs, S., Masarik, J., Frank, M., and Schlucter, C., 1998, ^{10}Be and ^{26}Al production rates deduced from an instantaneous event within the dendro-calibration curve, the landslide of Köfels, Otz Valley, Austria: Earth and Planetary Science Letters, v. 161, p. 231–241, doi: 10.1016/S0012-821X(98)00153-8.

Licciardi, J.M., Kurz, M.D., Clark, P.U., and Brook, E.J., 1999, Calibration of cosmogenic ^{3}He production rates from Holocene lava flows in Oregon, USA, and effects of the Earth's magnetic field: Earth and Planetary Science Letters, v. 172, no. 3-4, p. 261–271, doi: 10.1016/S0012-821X(99)00204-6.

Murray, A., 2001, Luminescence dating of a debris lobe from Sham Wat, Hong Kong, *in* King, J.P., ed., Section 1: Luminescence dating of colluvium and landslide deposits in Hong Kong: Geotechnical Engineering Office, Civil Engineering Department, Hong Kong SAR Government, GEO Report No. 134, p. 29–49.

Ng, K.C., Parry, S., King, J.P., Franks, C.A.M., and Shaw, R., 2003, Guidelines for natural terrain hazard studies: Geotechnical Engineering Office, Civil Engineering Department, Hong Kong SAR Government, GEO Report No. 138, 138 p.

Nichols, K.K., Bierman, P.R., and Caffee, M., 2000, The Blackhawk keeps its secrets: Landslide dating using in situ 10-Be: Geological Society of America Abstracts with Programs, v. 32, no. 7, p. A-400.

Phillips, F.M., Zreda, M.G., Smith, S.S., Elmore, D., Kubrik, P.W., Dorn, R.I., and Roddy, D.J., 1991, Age and geomorphic history of Meteor Crater, Arizona, from cosmogenic ^{36}Cl and ^{14}C in rock varnish: Geochimica et Cosmochimica Acta, v. 55, no. 9, p. 2695–2698, doi: 10.1016/0016-7037(91)90387-K.

Sewell, R.J., and Campbell, S.D.G., 2005, Report on the dating of natural terrain landslides in Hong Kong: Geotechnical Engineering Office, Civil Engineering and Development Department, Hong Kong SAR Government, GEO Report No. 170, 151 p.

Stone, J.O., 2000, Air pressure and cosmogenic isotope production: Journal of Geophysical Research, v. 105, p. 23,753–23,759, doi: 10.1029/2000JB900181.

Thomas, M.F., 1994, Geomorphology in the tropics: A study of weathering and denudation in low latitudes: Chichester, John Wiley and Sons, p. 193–224.

Zreda, M.G., and Noller, F.M., 1998, Ages of prehistoric earthquakes revealed by cosmogenic chlorine-36 in a bedrock fault scarp at Hebgen Lake: Science, v. 282, no. 5391, p. 1097–1099, doi: 10.1126/science.282.5391.1097.

Manuscript Accepted by the Society 11 April 2006